Non-Military Warfare

This volume analyzes the phenomenon of non-military warfare in theory and practice, including its relation to military warfare, and how states can understand and counter this activity.

War has traditionally been understood as soldiers with weapons fighting on a battlefield. Today, however, it is difficult to say who is a combatant, what weapons are, or even where the battlefield ends. Industrial-scale information operations and hacking attacks can be both an integral part of and a substitute for military operations. This book, which brings together scholars from different fields, goes beyond conceptual debates and provides a focused analysis of the advent of non-military warfare and its impact on strategy. The empirical chapters investigate non-military warfare primarily as conducted by the United States, Russia, and China in the areas of politics, cyber, diasporas, technology, and law. The theoretical chapters seek to answer broader questions about how we should think about victory, defeat, and participation in non-military warfare, and when it can complement and take the place of military operations. As modern warfare is moving into non-military domains, understanding this shift is vital to state survival in the 21st century.

This book will be of much interest to students of strategic studies, defense studies, foreign policy, and International Relations, as well as professional practitioners.

Oscar Jonsson is the director of phronesis analysis, a senior non-resident associate fellow at NATO Defense College, and an associate senior lecturer at the Swedish Defence University. He is the author of three books and holds a PhD from the Department of War Studies, King's College London.

Ilmari Käihkö is university researcher at the University of Helsinki, Finland, an associate professor of war studies in the Department of War Studies, Swedish Defence University, and a veteran of the Finnish Defence Forces.

Routledge Advances in Defence Studies

Series editor: Timothy Clack
University of Oxford, UK

Routledge Advances in Defence Studies is a multi-disciplinary series examining innovations, disruptions, counter-culture histories, and unconventional approaches to understanding contemporary forms, challenges, logics, frameworks, and technologies of national defence. This is the first series explicitly dedicated to examining the impact of radical change on national security and the construction of theoretical and imagined disruptions to existing structures, practices, and behaviours in the defence community of practice. The purpose of this series is to establish a first-class intellectual home for conceptually challenging and empirically authoritative studies that offer insight, clarity, and sustained focus.

The Russia-Ukraine War
Towards Resilient Fighting Power
Viktoriya Fedorchak

Defence Planning for Small and Middle Powers
Rethinking Force Development in an Age of Disruption
Edited by Tim Sweijs, Saskia van Genugten and Frans Osinga

Operational Art
Theory and Practice
Milan Vego

European Total Defence
Past, Present and Future
Edited by Gjermund Forfang Rongved

Non-Military Warfare
A War of Our Time
Edited by Oscar Jonsson and Ilmari Käihkö

For more information about this series, please visit:https://www.routledge.com/Routledge-Advances-in-Defence-Studies/book-series/RAIDS

Non-Military Warfare

A War of Our Time

Edited by Oscar Jonsson and Ilmari Käihkö

Routledge
Taylor & Francis Group
LONDON AND NEW YORK

First published 2026
by Routledge
4 Park Square, Milton Park, Abingdon, Oxon OX14 4RN

and by Routledge
605 Third Avenue, New York, NY 10158

Routledge is an imprint of the Taylor & Francis Group, an informa business

British Library Cataloguing-in-Publication Data
A catalogue record for this book is available from the British Library

ISBN: 9781041019206 (hbk)
ISBN: 9781041019251 (pbk)
ISBN: 9781003616993 (ebk)

DOI: 10.4324/9781003616993

Typeset in Times New Roman
by Apex CoVantage LLC

Contents

Acknowledgments

The editors would like to thank the Swedish Civil Contingencies Agency for supporting the project that culminated in this anthology.

Contributors

Jan Angstrom is a professor of war studies at the Swedish Defence University. He earned his PhD from the Department of War Studies, King's College London. He has published widely on the use of force in modern war for the past two decades.

Justin Key Canfil is an assistant professor of international relations at Carnegie Mellon University. His research concerns the politics of science and technology. He held fellowships at the Harvard Kennedy School, Princeton University, the Council on Foreign Relations, the Wilson Center, and the Columbia-Harvard China and the World Program. He holds a PhD in political science from Columbia University.

Sanshiro Hosaka is a research fellow at the International Centre for Defence and Security. His interests include strategic narrative, political technology, Soviet/Russian active measures, and intelligence history. He has been widely published in academic journals and think-tank reports. In 2023, his monograph on Russian intelligence history and tradecraft won the 32nd "Yamamoto Shichihei" humanities and social science publication award.

Benjamin Jensen, PhD, is the Frank E. Petersen Chair for Emerging Technology at the Marine Corps University, School of Advanced Warfighting, and a senior fellow in the Futures Lab at the Center for Strategic and International Studies. His most recent book is *Information in War: Military Innovation, Battle Networks, and the Future of Artificial Intelligence* (Georgetown University Press, 2022).

Oscar Jonsson (ed.) is the director of phronesis analysis, a senior non-resident associate fellow at NATO Defense College, and an associate senior lecturer at the Swedish Defence University. He has spent over a dozen years researching Russian military and non-military warfare. He is the author of three books and holds a PhD from the Department of War Studies, King's College London.

Ilmari Käihkö (ed.) is a university researcher at the University of Helsinki, Finland, an associate professor of war studies in the Department of War Studies, Swedish Defence University, and a veteran of the Finnish Defence Forces. His research focuses on strategy, or the creation, control, and use of force, as well as the ethnographic methods underpinning such study.

Elsa B. Kania is a PhD candidate in Harvard University's Department of Government. Her research concentrates on China's military strategy, defense innovation, and emerging capabilities. She is also an adjunct senior fellow with the Center for a New American Security's Technology and National Security Program. She serves as an officer in the U.S. Navy Reserve. Her views are her own.

Nina Kollars is an associate professor in the Cyber and Innovation Policy Institute of the U.S. Naval War College and the director/co-founder of the Maritime Hacking Village. She is a former advisor to the USD (R&E) Heidi Shyu. She is a certified executive bourbon steward and fan of cigars on occasion.

Jack McDonald is a senior lecturer in war studies in the Department of War Studies, King's College London. He is Director of the Centre for Science and Security Studies, a global leader in studying security challenges related to emerging technology. He is the author of three books on law and ethics of war and emerging technology.

Michael Poznansky is an associate professor in the Strategic and Operational Research Department and a core faculty member in the Cyber and Innovation Policy Institute at the U.S. Naval War College. He is the author of *In the Shadow of International Law: Secrecy and Regime Change in the Postwar World* (Oxford University Press, 2020).

Audrye Wong is an assistant professor of political science and international relations at the University of Southern California and Jeane Kirkpatrick Fellow at the American Enterprise Institute. Her research examines China's use of economic statecraft, propaganda and disinformation, and foreign influence operations. She received her PhD in security studies from Princeton University's School of Public and International Affairs.

Abbreviations

APT	Advanced Persistent Threat
CCP	Chinese Communist Party
C4ISR	Command, Control, Communications, Computers, Intelligence, Surveillance, Reconnaissance
CIA	Central Intelligence Agency, U.S.
CISA	Cybersecurity and Infrastructure Security Agency
CSF	Cybersecurity firms
DHS	Department of Homeland Security, U.S.
DOD	Department of Defense, U.S.
DOI	Operational Information Division, Russian Federation
DPH	Direct Participation in Hostilities
DoJ	Department of Justice
DPR	Donetsk People's Republic
FBI	Federal Bureau of Investigation
FROC	Federation of Returned Overseas Chinese
FSB	Federal Security Service, Russian Federation
GRU	Main Intelligence Directorate of the General Staff, Russian Federation
ICB	international crisis behavior
ICT	Information and Communication Technology
LPR	Lugansk People's Republic
MID	Militarized Interstate Dispute
MIT	Massachusetts Institute of Technology
NATO	North Atlantic Treaty Organization
NSC	National Security Council, U.S.
OPZhZ	Opposition Platform – For Life, Ukraine
PLA	People's Liberation Army, China
RMA	Revolution in Military Affairs
SBU	Security Service of Ukraine
SVR	Foreign Intelligence Service, Russian Federation

Tables and Figures

Tables

Figures

1 Introduction

A War of Our Time

Ilmari Käihkö and Oscar Jonsson

The question of how to understand the war of our time is perennial for both scholars and policymakers. As the Prussian general and military theorist Carl von Clausewitz argued, understanding the kind of war one engages in is nothing less than "the first, the grandest, and the most decisive act of judgment."[1] Nonetheless, analyzing war is a Sisyphean task because war is constantly changing in line with society, politics, technology, and economics. Correctly gauging how war has changed is a critical precondition for success in war and for national defense. These are the goals to which this anthology aims to contribute.

The arguably most famous way to understand war comes from Clausewitz himself. According to Clausewitz, war "is an act of violence intended to compel our opponent to fulfill our will."[2] The nature of war was thus violence, the "literal meaning" of war is fighting.[3] While Clausewitz considered the violent nature of war constant, he saw the character of war – how it is fought – as ever-changing. This distinction became the foundation for subsequent understandings of war.

During the past decades, a great chorus of observers has claimed that war has transformed. From the 1990s up until 2010s the debate was dominated by those who saw that globalization, the emergence of non-state actors and technology had profoundly changed the character of war.[4] The impact of these processes has not abated, but has, if anything, strengthened as Western policymakers began to shift their attention back to revanchist great powers, especially China and Russia.

The general trend after the Cold War was a steady decrease of large-scale overt use of organized violence, especially between great powers.[5] Restrained by adversaries' deterrence capabilities, mutual dependencies caused by trade, and international norms about non-use of force, states increasingly competed through non-military means rather than military ones. Consequently, the utility of non-military means was elevated, often to surpass that of traditional military means.

This was particularly the case in understanding of Russian actions against Ukraine between 2014 and 2022. The Russian invasion and annexation of the Crimean Peninsula and support of insurgency in Ukraine's eastern Donbas region in the spring of 2014 was interpreted by some as successful use of non-military means against its smaller neighbor. The large-scale Russian invasion of Ukraine in February 2022 may at first appear to reverse the trend and return focus to conventional warfare. Even though the bulk of the fighting in Ukraine is done by

DOI: 10.4324/9781003616993-1

military means, the Russian theory of victory was underpinned by a strategy of non-military means to break Ukraine's will and to disorganize and disable Ukrainians' will and ability to resist. Moreover, aside from direct military assistance, the Western response to Russian aggression has come in the economic, cyber, energy, and informational domains.

As victory on the battlefield remained elusive for both Ukraine and Russia, it was increasingly hoped that the use of non-military means would succeed in what traditional military means could not. The war, therefore, only reinforces the importance of understanding the modern balance between military and non-military means as this anthology proposes to do. Even further, as the largest war in Europe after World War II, Ukraine offers an important case study of the utility of both military and non-military means in war. While both Russia and those supporting Ukraine sought to contain the war, Russia has answered Western sanctions, economic warfare, and military support to Ukraine with non-military means, including subversion, sabotage, energy blackmail, and instrumentalized migration. In November 2024, the NATO General Secretary Mark Rutte claimed that "Russia is conducting . . . an intensifying campaign of hybrid attacks across our allied territories, interfering directly in our democracies, sabotaging industry and committing violence."[6]

This chapter introduces the key debates of how war is changing, the rationale and puzzle behind the advent of non-military warfare before presenting the thematic and theoretical chapters of the volume. We first review the debate on the nature of modern war and then discuss factors that have contributed to the increased utility of non-military means. The second section proceeds to investigate the rise of non-military warfare. The third section makes a case for the importance of studying non-military warfare and illustrates how ongoing debates relate to Clausewitz's writings. The point of such an exercise is to compare and contrast Clausewitz's foundational theory of war with the ongoing debate about non-military warfare principally dominated by international relations scholars. The following three sections focus on ongoing, yet parallel, camps about the nature of contemporary war – hybrid war, political war, and how emerging technology affects war. The concluding section presents the plan of the anthology.

The Rise of Non-Military Warfare

The increased utility of non-military means is emphasized in contemporary Western, Chinese, and Russian military thinking alike.[7] While Western states have focused on wars in the greater Middle East the last decades, rising powers such as China that have militarily lagged decades behind the United States have sought to contain great power conflict while catching up. For such weaker actors, non-military means offer a way to both offset U.S. military superiority and avoid escalation.[8]

Similarly, Russian thinkers have focused on asymmetric and non-military means. For instance, the Russian Chief of General Staff Valery Gerasimov argued in 2013 that non-military means were four times as important as military ones. For Gerasimov, "the role of nonmilitary means of achieving political and strategic goals has grown, and, in many cases, they have exceeded the power of force of weapons in their effectiveness."[9] Similarly, in 2016, the long-time President of the Russian Academy of Military Sciences, Makhmut Gareyev, stated:

> The definition of the essence of war must, to some extent, be reviewed. . . . The threat is connected with information and other subversive actions, the creation of controlled chaos to provoke various kinds of unrest, overthrow undesirable power structures from within, and violate the internal stability of states, as was done in Iraq, Libya, and Ukraine.[10]

Likewise, Colonel General Vladimir Zarudnitsky, former head of the Main Operations Directorate and the head of the Military Academy of the General Staff, echoed the same sentiments. He noticed the significant increase in the utility of non-military means and argued that they could "prevent an armed conflict or halt it at various phases of development and transition to the resolution of the situation without the use of military force."[11]

In order to come to terms with the transformation of contemporary war, concepts such as "hybrid war" and "hybrid threats"[12] gained popularity among Western observers. Others have suggested that modern wars take place in the "gray zone" between war and peace.[13] Some believe that wars are already being waged in cyberspace – a new domain elevated alongside land, sea, air, and space.[14] Information operations, propaganda, strategic communication, disinformation, and "active measures" have been deemed to pose risks not only to democracies,[15] but, as the Gareyev quote above reveals, even to autocracies. Others still believe that the key change in contemporary warfare is the notion of increased surrogacy, as both human and non-human proxies increasingly carry the burdens of war.[16] Cyberwar and autonomous military systems guided by artificial intelligence offer good examples of how emerging and disruptive technology is thought to transform war.[17] Technology continues to provide new opportunities and to pose new vulnerabilities.

Nonetheless, if one knows their Cold War history, many concepts related to non-military warfare seem familiar. Aside from modern technology, it is difficult to pinpoint what exactly separates contemporary debates on non-military warfare from Cold War "covert action," "subversion," "shadow warfare," "psychological warfare," "political warfare," "economic warfare," "political action campaigns," and "unconventional warfare,"[18] alongside proxy wars and others. Another important, yet often overlooked fact is that just like Gerasimov,[19] classic counterinsurgency texts also claimed that four-fifths of war consists of political non-military means and the remaining fifth of military means.[20] To quote the 2006 U.S. counterinsurgency doctrine put to use in Afghanistan and Iraq, "Some of the Best Weapons . . . Do Not Shoot."[21] A year later, the then Russian Defense Minister Sergei Ivanov stated that

> [t]he development of information technology has resulted in information itself turning into a certain kind of weapon. It is a weapon that allows us to carry out would-be military actions in practically any theater of war and most importantly, without using military power.[22]

All this suggests that the concepts mentioned seek to come to terms with more general developments that concern contemporary war, which have expanded war from a focus on military violence to encompass even non-military means. This is why this anthology focuses on "non-military warfare." The overarching definition of non-military warfare used in this anthology comes from Jack McDonald (see Chapter 3). He defines non-military warfare as "non-kinetic tools of statecraft employed as means of strategic coercion and effect employed with an intensity and purpose such that the target perceives them as significant escalations and/or commensurable to warfare."

The choice to call this activity "warfare" can, of course, be discussed. As pointed out in a 2018 RAND report, most government entities are not mandated to wage warfare and labeling their activities as such can make them less effective. Nonetheless, the report emphasizes "the ways in which actors are employing a host of non-kinetic tactics and how they form part of the spectrum of conflict."[23]

There are three main reasons for calling non-kinetic activity warfare. The first is that such activities are antagonistic in a way that distinguishes them from both everyday statecraft and international relations, as well as the so-called soft power that emphasizes more passive attraction and legitimacy rather than strategic coercion or subversion.[24] The second reason is that their utility has been perceived to have increased in the 21st century and to have become more potent means of achieving political aims. The third reason is that many of the actors involved call these activities warfare due to the political utility of calling it so and/or the perception of it being warfare. While this decision can be criticized for securitizing everyday international relations and taking a step toward total war, it is already being done by the actors themselves and is thus a starting point for our discussion. We hope that by adopting this terminology we can emphasize the importance of non-military means to better understand both war and peace in the 21st century.

This said, we do not claim that the use of non-military means necessarily crosses the threshold of war. Ultimately, war is what we make out of it, as it is a political decision to call something a war. While non-military means can be used on their own, they are often combined with other military and non-military means. It is especially critical to consider the intensity and outcomes of the means employed. As noted, evident in public statements is that with certain intensity, even the use of non-military means can be considered war. For instance, Ivanov's quote above suggests that if informational or other non-military means have more utility than military ones, then comprehending them is crucial for understanding contemporary war and warfare.

The importance of non-military warfare grows from two broad trends: technological developments and the decline of large-scale violence until the Russian invasion of Ukraine 2022. While wars have certainly continued to be fought elsewhere, Western states have witnessed few wars after World War II (and as is increasingly the case, the wars waged have by and large lacked declarations of war). While much of the Cold War–era competition between the East and the West falls under non-military warfare, the four decades that followed World War II have, with some justification, been called a long peace.[25] It is only more recently the revanchist

Russia and the rise of China have forced us to consider great power conflict anew. Indeed, the 2018 U.S. National Military Strategy noted that "the central challenge to U.S. prosperity and security is the reemergence of long-term, strategic competition" with these two countries.[26]

Facing the unsurpassed military might of the United States, Russian and Chinese military thinkers have, however, recognized the need to wage war primarily under the threshold of war and by non-military means.[27] This forces us and others to follow suit and to reconsider contemporary war and warfare where non-military means play an increasingly important, yet not always fully recognized or understood, role.

Reconsidering Non-Military Warfare

The aim of this anthology is to contribute to the understanding of war of our time through an investigation of the utility of non-military warfare, including which combinations of various means can achieve specific political ends. This includes investigating strategic theory, law, and the empirics of specific domains. While a conceptual discussion about war is to an extent unavoidable, it is not the goal of this anthology. Rather, too much of the debate has focused on solely on concepts. This volume seeks to investigate the advent of non-military warfare and how it has changed contemporary war and strategy for practitioners out there.

There are several reasons why this investigation is necessary. The first reason concerns strategy and strategic theory, much of which traditionally has focused on military means. Despairing over his 19th century contemporaries' relative lack of interest about fighting, Clausewitz emphasized that violence forms the essence of war, but also that combat was a mere means for achieving political ends of war. While many of his contemporaries focused on logistics, maintenance, and marching, Clausewitz saw these merely as supporting functions to the only purpose of military force – fighting.[28]

Clausewitz defined war as "an act of violence in order to force our will upon the enemy."[29] It is difficult to overemphasize his influence for military thinking and the military profession across the globe. During World War II, Harold Lasswell defined soldiers as "specialists on violence."[30] Samuel Huntington later famously built on this definition when he dubbed professional officers "managers of violence."[31]

It would of course be an exaggeration to claim that violence has disappeared or lost its utility. This said, violence was for a long time in relative decline. In 1953, there were 23 battle-related deaths per 100,000 people living, which then continually decreased with a 95% decline to 2 in 2016.[32] Even though contemporary wars in places such as Ukraine, Gaza, and Sudan will end this transition, overall fatalities will stay notably below those of the 1980s. Assuming the continuation of international conflicts, as long as direct violence is low, non-military ways and means will continue to play an increasingly important role. Moreover, they are the main instruments for great power competition. While this is reflected in many of the new concepts about war mentioned above, the consequences are often neglected in contemporary understandings of strategy and war. Increasingly, narrow military strategies to achieve desired political ends are subordinated to broader grand strategies that encompass both military and non-military means available to an actor.

The concept of grand strategy possibly emerged in mid-19th-century France, after which it has received several contending meanings. The main contribution relates to the need to widen the means used in strategy from military to encompass even non-military means, and that these are used both in times of peace and war.[33] As noted, non-military warfare focuses on means and ways used in isolation from or in combination with those that belong to the military sphere. Such broadening will inevitably affect the military profession, which has traditionally narrowly focused on organized violence, and hence the organization and use of military force.[34]

The second reason why it is important to investigate non-military warfare concerns its utility. While the increasing utility of non-military measures has been recognized,[35] it is less clear what can be achieved with these means under various circumstances and how. Ukraine offers a recent case where Russia and its opponents have employed wider means for achieving their political ends – and where the use of both military and non-military means has given rise to views of novel ways to wage and win wars.

As noted, this was especially the case with the mostly bloodless Russian occupation and annexation of the Crimean Peninsula in 2014. Critics have, however, subsequently argued that the circumstances surrounding the occupation and annexation of the peninsula were unique. Even further, it would be outright wrong to claim that military force played little or no role in the matter. Michael Kofman, for instance, argued that "what Russia does best is conventional war, and if a conflict does not start that way, it is how it always ends."[36] Andrew Monaghan concluded that while many Western observers focused on non-military means even in the subsequent War in Donbas, the "Russian ground forces would rely on massive salvoes of precision rocket and artillery fire, targeted by UAVs and cyber and electronic warfare capabilities."[37] By and large, it is difficult to disentangle the relative importance of military and non-military means for the outcome when they are used in tandem.

This difficulty is further emphasized with the Russian invasion of Ukraine in 2022. The decision to invade Ukraine can only be understood as the culmination of a long, and unsuccessful, campaign of non-military warfare against Ukraine.[38] Moreover, the failure of the initial part of the Russian invasion was largely a failure of the non-military part of the plan to deliver the preconditions of the military invasion that aimed to decapitate Ukrainian leadership. Indeed, "Russia's intended defeat mechanism was Ukraine's internal destabilization and disorganization" through non-military means.[39] Only after that did the Russian leadership transition to a primarily military strategy.

The core of the Western response came in the form of stifling Russian subversion, economic sanctions, isolating Russia politically and broadcasting Russian plans while simultaneously bolstering Ukraine's defenses. Ukraine, thus, offers an important case study of non-military warfare: Russia failed to achieve its aims with non-military means, possibly suggesting that Russian thinkers had overestimated their utility.[40] As a result, Russia resorted to a large-scale invasion in 2022 after its non-military deadlock with Ukraine.

The third reason why it is important to investigate non-military warfare is that governments and state officials recognize that the international environment and contemporary war are changing. Therefore, it is important to find best practices for managing this changing environment. The defeat in the two–decade-long counterinsurgency campaign in Afghanistan and the way the West has predominantly relied on indirect and non-military means to assist Ukraine in defending itself against Russia serve to further underline the need to improve our understanding of non-military warfare.

Currently, there are at least three broad camps about the increasing utility of non-military means of war. While there are enough commonalities between the three to warrant considering them together under the heading of non-military warfare, doing so also has the benefit of increasing exchange between the camps.

The first camp concerns hybrid warfare and similar seemingly novel threats that consist of non-military means, or a combination of military and non-military means. This section traces the evolution of an operational-level concept into one that ultimately resembles grand strategy. The second camp concerns political warfare, covert operations, and other concepts of Cold War vintage, which have been dusted off and reconsidered anew due to the return of great power rivalries that some have interpreted as Cold War 2.0. While Russia has rattled its opponents with threats of using nuclear weapons and employed sabotage, great powers have sought to limit escalation even in their employment of non-military warfare. Finally, the third camp concerns emerging and possibly disruptive technologies, and how these affect contemporary war. Technological change is both incremental and accelerating. While many technologies are hugely beneficial, they can simultaneously become vulnerabilities when exploited by our adversaries. These three camps are considered in the sections that follow.

Contemporary War and Hybrid Warfare

The concept of hybrid warfare dates to 1998 when it was used to describe combining conventional regular military forces with special forces.[41] A decade later the concept was broadened to describe war,[42] ostensibly to "nudge" the U.S. military to move from its binary logic of either peace or war to a broader view of warfare.[43] The concept's breakthrough came with the Russian occupation and annexation of the Crimean Peninsula from Ukraine in 2014. Six months later, in September 2014, hybrid warfare was discussed alongside Russian aggression in NATO's summit in Wales.[44] Similarly, the European External Action Service saw in 2015 that "Russia's sophisticated use of large-scale, well-coordinated hybrid warfare tactics has compromised Ukraine's territorial integrity and has strived to destabilise the larger neighbourhood."[45]

As envisaged originally, hybrid war concerned operational-level use of military means and methods during times of war.[46] Subsequently, the concept has evolved to the strategic level and to equal gray zone conflict waged below the threshold of war. The concept suggests that the emphasis of contemporary war lies not in narrow military strategy that focuses on violence but rather on grand strategy that applies all available means.[47] Others have gone further and criticized notions of hybrid warfare as a "quasi-theory of foreign policy,"[48] which includes a wide range of things from lies to conventional military force, or an academic fashion that "offered a ready-made interpretation of the conflict in Ukraine as well as Russia's intentions in and beyond Ukraine, and narrowed what researchers are looking at and how, and, hence, which conclusions they draw."[49]

Indeed, hybrid war, warfare, and gray zone conflict alike have after 2014 been usually used as synonyms for Russian (and to a lesser extent Chinese) action in a way that hinders more general analysis.[50] Ultimately, the evolved concept of hybrid war has questionable analytical utility as it is conceptually difficult to distinguish from grand strategy and foreign policy. This risks making everything – and hence nothing – war.

A more practical challenge with notions of hybrid war – and by corollary all attempts to combine various military and/or non-military means – is that this requires both well-designed strategies and superior coordination capability as various means are almost inevitably possessed by multiple actors. Especially, the involvement of non-state actors – militias, rebels, and private companies, for example – demands the ability to control them. Without control, it becomes difficult to employ means strategically.[51] At worst, lack of control risks strategic blunders.

Bettina Renz has criticized viewing all Russian foreign policy moves as a part of an elaborate strategy as implying a "simply unrealistic" "coherence of effort and level of strategic foresight."[52] Even recent examples from Western-led counterinsurgency operations illustrate how difficult cooperation between state bureaucracies tends to be.[53] The assumption that revisionist states have succeeded in doing so or that our societies are particularly susceptible to the non-military means is common, but difficult to back with hard evidence.

A potential hindrance to combining military and/or non-military means comes from the way militaries are used to taking charge of warfare. As a result, it might be more difficult for them to assume a coordinating, let alone a subordinate, role in non-military warfare. For one historical example, when the United States was constructing its capabilities for political warfare in the second half of the 1940s, these were put under the Central Intelligence Agency (CIA). Nevertheless, it was stipulated that in times of war all covert operations "shall be coordinated" and in "theaters of war where American forces are engaged . . . conducted" under the military.[54] In other words, *real* war – war without prefixes – still belonged to the special remit of the military. Despite the broadening of war to encompass non-military means, this supremacy of the military by and large remains the case even today. The next section moves to investigate this kind of political warfare in more detail.

Contemporary War and Political Warfare

In 1948, George F. Kennan – the American diplomat best known for his influence on U.S. Cold War containment strategy – defined political warfare as "the logical application of Clausewitz's doctrine in time of peace. In broadest terms, political warfare is the employment of all the means at a nation's command, short of war, to achieve its national objectives."[55] While the European Recovery Plan, perhaps the greatest aid project to date better known as the Marshall Plan, offers the prime example of a non-military activity with profound political effects, five percent of the billions spent went to CIA covert activities as the National Security Council (NSC) sought to match Soviet covert activities with its own.[56] At the core of covert action was deniability; covert action was "so planned and executed that any US Government responsibility for them is not evident to unauthorized persons and that if uncovered the US Government can plausibly disclaim any responsibility for them."[57] This kind of non-acknowledgment of intervention allows both demonstrating resolve without escalating to military conflict and creating and exploiting strategic ambiguity.[58]

Covert action is often plausibly deniable. Yet is important to acknowledge that the NSC additionally broadened covert action from propaganda to include more direct intervention through

> economic warfare; preventive direct action, including sabotage, anti-sabotage, demolition and evacuation measures; subversion against hostile states, including assistance to underground resistance movements, guerrillas and refugee liberation groups, and support of indigenous anti-communist elements in threatened countries of the free world.[59]

This greatly widened the means available for political warfare, which ultimately built on the observation that countries may be more tolerant when subjected to ambiguous non-military attacks than direct military attacks.[60] The use of non-military means at the core of political warfare thus offered a way to avert escalation that could have led to nuclear exchange.

While the term "political war" has been useful for describing the actions taken beyond war, Frank Hoffman has pointed out that it is also a contradiction in terms. If war indeed is understood in the Clausewitzian manner as violence, then "there is no warfare as we know it in these political and economic activities, which is why the term is an oxymoron."[61] Almost seven decades prior, Kennan had nevertheless already prepared against such counterarguments. As he wrote, Lenin had managed to synthesize politics and war to the extent that:

> Kremlin's conduct of political warfare has become the most refined and effective in history. We have been handicapped however by a popular attachment to the concept of a basic difference between peace and war, by a tendency to view war as a sort of sporting contest outside of all political context, by a national tendency to seek for a political cure-all, and by a reluctance to recognize the realities of international relations – the perpetual rhythm of struggle, in and out of war.[62]

Just like his Communist adversaries who after Lenin had reversed Clausewitz and viewed politics as a continuation of war by other means,[63] even Kennan thus saw struggle and political warfare as inevitable constants in international affairs. With greater intensity of conflict, the boundary between peace and war blurs.

Following the end of the Cold War and what for some appeared a victory of democracy over totalitarianism, U.S. covert operations too were influenced by stronger emphasis of international law. The emphasis on international law continues to influence the question whether we should wage non-military warfare or simply prepare to defend ourselves against it. The U.S. Senate Select Committee to Study Governmental Operations with Respect to Intelligence Activities – also known as the Church Committee after its chairman Senator Frank Church – took stand on this question in its 1976 final report. According to the committee:

> The United States must not adapt the tactics of its enemies. Means are as important as ends. Crisis makes it tempting to ignore the wise restraints that make men free. But each time we do so, each time the means we use are wrong, our inner strength, the strength which makes us free, is lessened.[64]

Another argument against blurring the boundary between peace and war is that this takes the activity a huge step toward total war. This widening of war even questions distinctions between military and civilian and public and private.[65] The risk is obvious: the merging and militarization of society under state authority.

While the war on terror that followed the 9/11 terror attacks was deemed frightening enough to justify wide measures and was often waged by exploiting legal ambiguities (or gray zones), covert action became more overt and paramilitarized. After Special Forces and CIA covert operators began to conduct most covert action (often alongside local proxies), the line that separated covert action and military operations blurred.[66] This too contributed to blurring the line between war and peace.[67]

In order to stay in the gray zone, means and ways should remain difficult to attribute to a certain actor. Plausible deniability and concealment of evidence of direct intervention may limit unwanted escalation.[68] Ambiguity may, however, be more beneficial in the short term than the long term. While ambiguity allows constructing narratives, it can also lead to mythmaking difficult to control and steer.[69] While there may be tactical benefits of operating in unforeseen ways, these may come at a strategic cost. For instance, after repeatedly lying about its interventions in Ukraine after 2014, Russian official statements were almost invariably met with skepticism.[70] Very few believed Russian justifications for invading Ukraine in February 2022. Even the United States continues to suffer from the false claims about Iraq possessing weapons of mass destruction that led to the War in Iraq in 2003, as well as its interrogation methods that amounted to torture. The detainee program was initially run as a covert operation but soon led to unintended consequences as news about the program began to trickle in.

One thing that has changed covert action and hence political warfare is emerging technology. Michael Warner has noted that scale has always posed a problem for covert action. As he elaborates, covert action to be effective has had to remain plausibly deniable for a crucial time period, and to do so, it has had to remain small. Cyberspace with its promises of (relative) anonymity and its near-instantaneous reach to large numbers of computer users has made it possible to run activities that are indistinguishable from covert actions on a much larger scale.[71]

In other words, technology has altered the scale that previously constrained the utility of covert action. Greater scale has enabled greater intensity of non-military means central to the notion of non-military warfare.

Contemporary War, Technology, and Information

Technology has always been one of the critical factors changing war. This includes everything from constructing bows, arrows, and gunpowder as well as railroad, combustion engines, rocket engines, satellite navigation, and telecommunications. Technology has vastly improved our weapons by which we – by and large – deliver kinetic energy that kills people and breaks things. Nonetheless, technology is also constantly changing. This change encompasses everything from our economy to how we live our lives, how we work, and how we consume information. These changes in turn impact how states build power and influence each other.

The modern discussions on how technology impacts military capabilities have a genesis in the Revolution in Military Affairs (RMA), which gained traction after the 1991 Gulf War. The Gulf War was the first war fought by the United States after the end of the Cold War. It became a watershed moment to illustrate how war had changed. The United States and allies relied on superior C4ISR[72] capabilities together with long-range precision weaponry. These capabilities allowed them to maintain a superior battlefield awareness and to engage Iraqi forces from a safe distance. The numerically superior Iraqi forces were beaten by a small but highly effective contingent of U.S. troops. The Gulf War was interpreted to affirm the RMA and the notion that the future of war will be informatized.

While the RMA was brought into the practice by the United States, its intellectual origins were pioneered in the Soviet General Staff in the 1980s. The Soviet General Staff were observing the capabilities developed by the United States and were the first to place these capabilities in a deeper context and understand how doctrinal and organizational change could revolutionize war. For the Soviet General Staff, the "Military-Technical Revolution" that focused on technological advances, "smart" weapons and sensors that improved battle space awareness and increased striking distances.[73] The U.S. application of these capabilities in the Gulf War became a great worry for Russian military theorists because the Iraqi forces were largely relying on Soviet materiel and doctrine. This set of capabilities is at the core of how the Russian General Staff see military force.[74] The crux of the resulting RMA was that as technology changed warfare, militaries had to adapt to the new realities. Overall, much of the debate was geared toward "future visions of war, future wars, and future technology" in ways that made evaluating claims difficult.[75]

While the RMA was changing the way that military force was applied by allowing armed forces to see better, communicate quicker, and strike farther, recent technologies such as those belonging to the cyber and information domains are shifting the relationship between military and non-military means. Technology changes societies and the way they function.

Technological advances are nevertheless a double-edged sword. With opportunities come threats as our dependency on technology forms vulnerabilities that can be exploited by our adversaries. For instance, digitalization affects most aspects of our existence: our economy, our work, our finances, and our social interactions. As these increasingly exist online, they are also vulnerable to cyber operations. As societies have digitized at breakneck speed, our cybersecurity adoption has not kept the same pace. This offers unique opportunities for not only cyber espionage and cyber criminality (e.g., ransomware) but also cyberattacks to either alter or disrupt our systems.[76] As discussed above, the increasing digitalization of everything is providing a hitherto unforeseen scale to these activities.

The revolution in Information and Communication Technology (ICT) impacts non-military warfare in many aspects. Our ways of seeing the world, power, and legitimacy have increasingly moved from radio and print to online and social media. This in turn means that the "process of collecting and organizing information is now a tremendous source of economic, political and cultural power. Data makes us more malleable, easier to predict, and extremely prone to influence."[77] The winners of the data economy are the Big Tech-companies, whose algorithms are increasingly shaping what we experience online.

The shift from offline to online media has decentralized content production. Or less prosaically, it allows anyone to be a publisher. Previously, platforms with editorial oversight decided what was to be published. Today, anyone can use their social media accounts and blogs to publish their views directly and immediately to a larger following. This is not to bring a nostalgic view of how it was before: the point is rather to emphasize that under section 230 in the U.S. Communications Act, none of the social media companies bear publishing responsibility for the content of their platforms. The lack of responsibility coupled with a business model that favors engagement over veracity have made social media a perfect amplifier for information operations. The revolution in ICT means that there is a significant increase in the role of private actors and companies in geopolitical conflict. The key targets for cyber espionage and cyberattacks are rather companies and critical infrastructure, respectively, than military units (even though they are of course also a target).

Russia has been seen to lead the way in terms of "weaponizing information" and implementing an active information warfare effort. This gained attention after the invasion of Ukraine 2014 and disinformation effort that was connected to the use of military means there.[78] This in particular emphasized the role of the "Troll farm," the Internet Research Agency, which was a place in St. Petersburg where several hundreds, if not thousands, were employed to give the Russian leadership's view in different social media, forums, and commentator fields for newspapers all over the world. Thereafter, the most notable information warfare effort came in connection to the 2016 U.S. Presidential elections, where Russian intelligence services hacked the National Democratic Committee's database and spread and amplified the content.[79] This interference was according to U.S. intelligence agencies directed by President Putin himself in an effort to help former U.S. President Donald Trump and to hinder Hillary Clinton.[80]

This is the backdrop against the attention from Russian strategists to the role of information today. As highlighted above, it has been not only an opportunity for Russia to asymmetrically impact their adversaries but also an immense source of vulnerability. The 2021 Russian National Security Strategy saw that ICT were used to "interfere in the internal affairs of states, undermine their sovereignty and violate their territorial integrity, posing a threat to international peace and security."[81] Moreover, it stated that the United States were working together with transnational companies, consolidating their control over the internet. Both the United States and the transnational companies were attacking Russian values and "exerting an informational and psychological impact on individual, group and public consciousness by disseminating social and moral attitudes that contradict the traditions, beliefs and convictions of the peoples of the Russian Federation."[82] Likewise, Chief of General Staff Gerasimov stated in 2017 that "Western countries are increasing the scale of a tough information war unleashed against Russia."[83] Not only did Russian leaders believe that they are in an information war, but also that the "perceived threat is an existential one."[84]

The increasing salience of the information domain is a large part of what is blurring the lines between war and peace. In order to successfully apply cyber operations in wartime, the malware needs to have found its way into the adversary's computers already in peacetime. The case is the same with the use of social media accounts in times of increasing tensions: the accounts need to be created and built long before the start of the conflict.

Outline

This chapter has introduced the reader the increased relevance of non-military means and different ways to understand its role in modern warfare. Non-military means seem to be used ingeniously with military means in "hybrid warfare," and they form the core of political warfare that is enhanced by changes in our technological and informational systems. This anthology brings together leading experts from various backgrounds and expert areas to scrutinize the phenomenon of non-military warfare and its dynamics, utility, and empirics. Going beyond the simple definitional questions – which have often been the focus in the past debates on the transformation of war – the different chapters produce both theoretical insights and empirical investigations into contemporary war.

The second chapter by Jan Angstrom engages with the foundational question of what victory and defeat is in non-military warfare. It starts by discussing our understanding of victory and defeat in warfare, and how those concepts are applicable to a wider range of non-military phenomenon.

Thereafter, in the third chapter, Jack McDonald investigates how we should think of participation in non-military warfare. While combatants in fighting can be defined in a fairly straightforward manner, the questions of participation in non-military warfare are much more difficult to answer. McDonald proposes a framework for judging whether a propagandist, a hacker or a private company exporting chips can be deemed a combatant.

The fourth chapter by Sanshiro Hosaka delves into Russia's history of non-military warfare against Ukraine. In particular, the chapter investigates the wide range of operations and actors involved in trying to impose Russia's will on Ukraine in the period that led up to the full-scale invasion in 2022.

One of the main reasons for the advent of non-military warfare is how the ICT has changed many aspects of our society and economies. Both the fifth and sixth chapters analyze how this impacts modern war. In the fifth chapter, Benjamin Jensen investigates the role of cyber tools as a substitute for military force in crisis management. Cyber tools offer the possibility to signal resolve, while engaging in operations that are less costly than military violence. In the sixth chapter, Nina Kollars and Michael Poznansky focus on the role of private cybersecurity companies and their foundational role in cybersecurity. They are omnipresent in the broader cybersecurity landscape and have to take decisions that impact global stability. However, private companies have vastly different incentive structures than states. For instance, they need to showcase their victories to gain more clients, whereas a state would want to keep its successful methods secret from an opponent. This public-private split is an inevitable part of non-military warfare today.

In the seventh chapter, Justin Key Canfil investigates the increasing use of law as a tool of conflict, most commonly known as lawfare. Canfil proposes a typology for how categorize everyday legal disputes with the instrumentalization of law for conflictual purposes. Moreover, he shows that lawfare is intrinsically connected to broader perceptions of legitimacy, which is a critical condition in international law. This, in turn, connects the phenomenon of lawfare to information operations and the ability to shape perceptions of specific events.

Thereafter, the eight chapter by Elsa B. Kania discusses when competition in science and technology can be considered a form of non-military warfare. Science and technology have always been a source of power and a way to contest advantage in competition and conflict. However, global economies and supply chains are now more interconnected than ever, which creates opportunities for chokepoint exploitation, such that blocking access to critical technology and components can become tantamount to a form of warfare. The ninth chapter by Audrye Wong is also a consequence of increasing global connectedness. The chapter analyzes the phenomenon of diaspora statecraft, and how states, especially authoritarian ones, seek to use diaspora communities to further their geopolitical goals. They seek to control diaspora beliefs and behavior to influence host states through processes that are plausibly deniable.

This book covers warfare in a large number of domains: law, technology, cyber, politics, and diaspora. Moreover, it focuses on the three great powers of today: the United States, China, and Russia. However, this book is not exhaustive. There could be complementary case studies to investigate countries such as India, Iran, and Brazil. Similarly, more themes could be added, such as economic warfare and algorithmic warfare. While we are mindful of more topics and cases that could be studied, this book gives still a broad illustration and a fairly exhaustive investigation into the topic of non-military warfare. We hope it succeeds in providing a starting point to the debate rather than its conclusion, and thus welcome further complementary and contrasting research.

Notes

1 Carl von Clausewitz, On War (New York: Barnes & Noble, 2004), 18.
2 Clausewitz, 1.
3 Clausewitz, 64.
4 For example, see Zeev Maoz and Azar Gat (Eds.), War in a Changing World (Ann Arbor: University of Michigan Press, 2001); Isabelle Duyvesteyn and Jan Angstrom (Eds.), Rethinking the Nature of War (Milton Park, Abingdon, Oxon: Frank Cass, 2005); Sibylle Scheipers and Hew Strachan (Eds.), The Changing Character of War (Oxford: Oxford University Press, 2011); Rosa Brooks, How Everything Became War and the Military Became Everything: Tales from the Pentagon (New York: Simon & Schuster, 2017).
5 Therese Pettersson and Magnus Öberg, "Organized Violence, 1989–2019," Journal of Peace Research 57, no. 4 (2020): 597–613.
6 Joshua Posaner, "NATO's Rutte Says Russian Attacks on Europe Are 'Intensifying'," Politico, 4. November 2024, https://www.politico.eu/article/natos-rutte-says-russian-hybrid-attacks-on-europe-are-intensifying/.
7 For instance, see Liang Qiao and Xiangsui Wang, Unrestricted Warfare (Beijing: PLA Literature and Arts Publishing House, 1999); Rupert Smith, The Utility of Force: The Art of War in the Modern World (New York: Vintage, 2008); Oscar Jonsson, The Russian Understanding of War: Blurring the Lines between War and Peace (Washington, DC: Georgetown University Press, 2019); Mark Galeotti, The Weaponisation of Everything (New Haven: Yale University Press, 2022).
8 Avery Goldstein, "The Diplomatic Face of China's Grand Strategy: A Rising Power's Emerging Choice," China Quarterly 168 (2001): 835–64; Rush Doshi, The Long Game: China's Grand Strategy and the Displacement of American Power (Oxford: Oxford University Press, 2021).
9 Valery Gerasimov, "The Value of Science Is in the Foresight: New Challenges Demand Rethinking the Forms and Methods of Carrying out Combat Operations," Military Review January-February 2016 (2016): 24.
10 Makhmut Gareyev, "Vyzov Prinyat," Voyenno-Promyshlennyy Kuryer 7, no. 622 (2016).
11 Vladimir Zarudnitsky, "Kharakter i Soderzhaniye Voyennykh Konfliktov v Sovremennykh Usloviyakh i Obozrimoi Perspektive [Character and Content of Armed Conflicts in Modern Conditions and Near Term Perspective]," Voennaya Mysl, 34–44, no. 1 (2021).
12 The two are used synonymously in Frank Hoffman, " 'Hybrid Threats': Neither Omnipotent Nor Unbeatable," Orbis 54, no. 3 (2010): 441–55.
13 United States Special Operations Command, "White Paper: The Gray Zone," 2015, https://info.publicintelligence.net/USSOCOM-GrayZones.pdf; Olga Oliker, "Statement before the Senate Armed Services Committee Subcommittee on Emerging Threats and Capabilities: 'Russian Influence and Unconventional Warfare Operations in the "Grey Zone:" Lessons from Ukraine'" (Center for Strategic & International Studies, 2017), https://www.armed-services.senate.gov/imo/media/doc/Oliker_03-29-17.pdf; Frank Hoffman, "Examining Complex Forms of Conflict: Gray Zone and Hybrid Challenges," PRISM 7, no. 4 (2018): 31–47; Michael J. Mazarr, Mastering the Gray Zone: Understanding a Changing Era of Conflict (Carlisle, PA: Strategic Studies Institute, 2015); Oliver Fitton, "Cyber Operations and Gray Zones: Challenges for NATO," Connections QJ 15, no. 2 (2016): 109–19; Oscar Jonsson, "Russia Is Waging Grey-Zone Warfare," in K. Giles and M. Boulegue (Eds.), Myths and Misconceptions on Russian Military Intent (London: Chatham House, 2022).
14 Richard A. Clarke and Robert K. Knake, Cyber War: The Next Threat to National Security and What to Do about It (New York: Ecco, 2010); David E. Sanger, Perfect Weapon: War, Sabotage, and Fear in the Cyber Age (London: Scribe, 2018); Richard J. Harknett and Max Smeets, "Cyber Campaigns and Strategic Outcomes," Journal of Strategic Studies, 45, no. 4 (2022): 534–67.

15 Thomas Rid, Active Measures: The Secret History of Disinformation and Political Warfare (Farrar, Straus and Giroux: New York, 2019).
16 Andreas Krieg and Jean-Marc Rickli, Surrogate Warfare: The Transformation of War in the Twenty-First Century (Washington, DC: Georgetown University Press, 2019).
17 Kenneth Payne, I, Warbot: The Dawn of Artificially Intelligent Conflict (London: Hurst, 2021).
18 All of these come up in David F. Rudgers, "The Origins of Covert Action," Journal of Contemporary History 35, no. 2 (2000): 249–62.
19 Gerasimov, "The Value of Science Is in the Foresight: New Challenges Demand Rethinking the Forms and Methods of Carrying out Combat Operations," 28.
20 For instance, see David Galula, Counterinsurgency Warfare: Theory and Practice (Westport, CT: Praeger, 1964), 63.
21 Headquarters Department of the Army, FM3–24 Counterinsurgency (Washington, DC: University of Chicago Press, 2006), 1–27.
22 Quoted in Stephen Blank, "Russian Information Warfare as Domestic Counterinsurgency," American Foreign Policy Interests 35, no. 1 (2013): 34.
23 Linda Robinson et al., Modern Political Warfare: Current Practices and Possible Responses (Santa Monica: RAND, 2018), xix, 2.
24 Joseph S. Nye, "Soft Power: The Evolution of a Concept," Journal of Political Power 14, no. 1 (2021): 196–208.
25 John Lewis Gaddis, The Cold War (London: Penguin, 2007).
26 Department of Defense, "Summary of the 2018 National Defense Strategy of the United States of America, 2018," 2018, https://www.defense.gov/Portals/1/Documents/pubs/2018-National-Defense-Strategy-Summary.pdf.
27 For instance, see Qiao and Wang, Unrestricted Warfare; Gerasimov, "The Value of Science Is in the Foresight: New Challenges Demand Rethinking the Forms and Methods of Carrying out Combat Operations"; David Kilcullen, The Dragons and the Snakes: How the Rest Learned to Fight the West (Oxford: Oxford University Press, 2020).
28 Michael Howard, Clausewitz: A Very Short Introduction (Oxford: Oxford University Press, 2002), 36–38.
29 Clausewitz, On War, 1.
30 Harold Lasswell, "The Garrison State," American Journal of Sociology 46, no. 4 (1941): 455–68.
31 Samuel Huntington, The Soldier and the State: Theory and Politics of Civil-Military Relations (Cambridge, MA: Belknap Press of Harvard University Press, 1957), 11.
32 Nygård, Håvard Mokleiv, SDG 16 Data Initiative: 2018 Global Report (Oslo: Peace Research Institute).
33 Thierry Balzacq and Donald R. Krebs, "The Enduring Appeal of Grand Strategy," in Thierry Balzacq and Donald R. Krebs (Eds.), The Oxford Handbook of Grand Strategy (Oxford: Oxford University Press, 2021).
34 Ilmari Käihkö, "The Evolution of Hybrid Warfare: Implications for Strategy and the Military Profession," Parameters 51, no. 3 (2021): 115–27.
35 For instance, see Paul A. L. Ducheine and Frans Osinga (Eds.), Winning Without Killing: The Strategic and Operational Utility of Non-Kinetic Capabilities in Crises, NL ARMS, Netherlands Annual Review of Military Studies (Den Haag: T.M.C. Asser Press, 2017).
36 Michael Kofman, "Russian Hybrid Warfare and Other Dark Arts" (War on the Rocks, March 11, 2016), https://warontherocks.com/2016/03/russian-hybrid-warfare-and-other-dark-arts/; see even Bettina Renz, Russia's Military Revival (Cambridge: Polity, 2018).
37 Andrew Monaghan, "The 'War' In Russia's 'Hybrid Warfare,'" Parameters 45, no. 4 (2016): 73.
38 Jack Watling, Oleksandr V. Danylyuk and Nick Reynolds, "Preliminary Lessons from Russia's Unconventional Operations During the Russo-Ukrainian War, February 2022–February 2023", Royal United Services Institute, Special Report, March 29,

2023, https://rusi.org/explore-our-research/publications/special-resources/preliminary-lessons-russias-unconventional-operations-during-russo-ukrainian-war-february-2022
39 Jack Watling, Oleksandr V. Danylyuk and Nick Reynolds, "Preliminary Lessons from Russia's Unconventional Operations During the Russo-Ukrainian War, February 2022–February 2023", Royal United Services Institute, Special Report, March 29, 2023, https://rusi.org/explore-our-research/publications/special-resources/preliminary-lessons-russias-unconventional-operations-during-russo-ukrainian-war-february-2022
40 Dimitri Minic, "How the Russian Army Changed Its Concept of War, 1993–2022," Journal of Strategic Studies, 47, no. 1 (2024): 29–63.
41 Robert G. Walker, Spec Fi: The Unites States Marine Corps and Special Operations (thesis, Monterey, CA: Naval Postgraduate School, 1998), https://apps.dtic.mil/sti/pdfs/ADA359694.pdf.
42 James N. Mattis and Frank Hoffman, "Future Warfare: The Rise of Hybrid Wars" (Proceedings, 2005), https://www.usni.org/magazines/proceedings/2005/november/future-warfare-rise-hybrid-wars.
43 Max Boot, "How to Respond to Russia's Hybrid War on the West," Washington Post, January 2, 2025, A15.
44 NATO, "Wales Summit Declaration," August 30, 2018, https://www.nato.int/cps/en/natohq/official_texts_112964.htm.
45 European External Action Service, "Food-for-Thought Paper 'Countering Hybrid Threats,'" 2015, 2, https://www.statewatch.org/media/documents/news/2015/may/eeas-csdp-hybrid-threats-8887-15.pdf.
46 Frank Hoffman, Conflict in the 21st Century: The Rise of Hybrid Wars (Arlington, VA: Potomac Institute for Policy Studies, 2007).
47 Käihkö, "The Evolution of Hybrid Warfare: Implications for Strategy and the Military Profession," 120.
48 Bettina Renz and Hanna Smith, "Russia and Hybrid Warfare – Going beyond the Label" (Aleksanteri Institute, 2016), https://helda.helsinki.fi/bitstream/handle/10138/175291/renz_smith_russia_and_hybrid_warfare.pdf.
49 Chiara Libiseller, " 'Hybrid Warfare' as an Academic Fashion," Journal of Strategic Studies, 46, 4, 875–76.
50 Käihkö, "The Evolution of Hybrid Warfare: Implications for Strategy and the Military Profession," 118.
51 Ilmari Käihkö, "Slava Ukraini!" Strategy and the Spirit of Ukrainian Resistance, 2014–2023. Helsinki: Helsinki University Press (Helsinki: Helsinki University Press, 2023).
52 Bettina Renz, "Russia and 'Hybrid Warfare,'" Contemporary Politics 22, no. 3 (2016): 297; see even Robinson et al., Modern Political Warfare: Current Practices and Possible Responses, 83–85.
53 Lene Ekhaugen, "Whole-of-Government Coordination for Complex Operations: Symbolic Politics or Coordination Failure?," Journal of Strategic Studies 45, no. 5 (2022): 668–91.
54 Office of the Historian, "292. National Security Council Directive on Office of Special Projects" (Department of State, June 18, 1948), https://history.state.gov/historicaldocuments/frus1945-50Intel/d292.
55 George F. Kennan, "The Inauguration of Organized Political Warfare (Redacted Version)" (Wilson Center, 1948), 1, https://digitalarchive.wilsoncenter.org/document/114320.pdf.
56 Galeotti, The Weaponisation of Everything, 131.
57 Office of the Historian, "292. National Security Council Directive on Office of Special Projects."
58 Rory Cormac and Richard J. Aldrich, "Grey Is the New Black: Covert Action and Implausible Deniability," International Affairs 94, no. 3 (2018): 477–94.
59 Office of the Historian, "292. National Security Council Directive on Office of Special Projects."

60 Gary D. Brown, "State Cyberspace Operations: Proposing a Cyber Response Framework" (RUSI Occasional Paper, 2020), 39.
61 Frank Hoffman, "On Not-So-New Warfare: Political Warfare vs Hybrid Threats" (War on the Rocks, 2014), https://warontherocks.com/2014/07/on-not-so-new-warfare-political-warfare-vs-hybrid-threats/; The same of course goes for many applications of hybrid warfare, which as noted has evolved from its narrow original meaning to mainly concern the use of non-military means. See Käihkö, "The Evolution of Hybrid Warfare: Implications for Strategy and the Military Profession."
62 Office of the Historian, "269. Policy Planning Staff Memorandum" (Department of State, April 30, 1948), https://history.state.gov/historicaldocuments/frus1945-50Intel/d269.
63 Jacob W. Kipp, "Lenin and Clausewitz: The Militarization of Marxism, 1914–1921," Military Affairs 49, no. 4 (1985): 184–91.
64 quoted in Gregory F. Treverton, Intelligence for an Age of Terror (Cambridge: Cambridge University Press, 2009), 235.
65 Jan Willem Honig, "People's War or Professional War: Which Has the Future in Europe?" in Tim Sweijs and Jeffrey Michaels (Eds.), Beyond Ukraine: Debating the Future of War (London: Hurst, 2024): 125–46.
66 Treverton, 228, 234.
67 Brooks, How Everything Became War and the Military Became Everything: Tales from the Pentagon, 105.
68 Austin Carson, "Facing Off and Saving Face: Covert Intervention and Escalation Management in the Korean War," International Organization 70, no. 1 (2016): 103–31.
69 Cormac and Aldrich, "Grey Is the New Black: Covert Action and Implausible Deniability," 491–93.
70 Lawrence Freedman, The Future of War: A History (New York: Public Affairs, 2017).
71 Michael Warner, "A Matter of Trust: Covert Action Reconsidered," Studies in Intelligence 63, no. 4 (2019): 33.
72 Command, Control, Communications, Computers, Intelligence, Surveillance, Reconnaissance.
73 Dima Adamsky, "Through the Looking Glass: The Soviet Military-Technical Revolution and the American Revolution in Military Affairs," Journal of Strategic Studies 31, no. 2 (2008): 257–94.
74 Valery Gerasimov, "Vektory Razvitiya Voyennoy Strategii [Development Vectors of Military Strategy]," Krasnaya Zvezda, 2019, http://redstar.ru/vektory-razvitiya-voennoj-strategii/.
75 Jan Angstrom, "Introduction: Debating the Nature of Modern War," in Isabelle Duyvesteyn and Jan Angstrom (Eds.), Rethinking the Nature of War (Milton Park, Abingdon, Oxon: Frank Cass, 2005), 16.
76 Thomas Rid, "Cyber War Will Not Take Place," Journal of Strategic Studies 35, no. 1 (2012): 5–32.
77 Sara M. Smyth, "The Facebook Conundrum: Is It Time to Usher in a New Era of Regulation for Big Tech?," International Journal of Cyber Criminology 13, no. 2 (2019): 578–95.
78 Katri Pynnöniemi and Andras Rácz (Eds.), Fog of Falsehood: Russian Strategy of Deception and the Conflict in Ukraine (Finnish Institute for International Affairs, 2016), https://www.fiia.fi/wp-content/uploads/2017/01/fiiareport45_fogoffalsehood.pdf.
79 Department of Justice, "Report on the Investigation into Russian Interference in the 2016 Presidential Election," 2019, https://www.justice.gov/storage/report.pdf.
80 US Senate Select Committee on Intelligence, "Russian Active Measures Campaigns and Interference in the 2016 U.S. Election," 2017, https://www.intelligence.senate.gov/sites/default/files/documents/Report_Volume4.pdf.
81 President of the Russian Federation, "Strategiya Natsionalnoy Bezopasnosti Rossiyskoy Federatsii (National Security Strategy of the Russian Federation)," 2021, II.87, http://publication.pravo.gov.ru/Document/View/0001202107030001.

82 President of the Russian Federation, "Strategiya Natsionalnoy Bezopasnosti Rossiyskoy Federatsii (National Security Strategy of the Russian Federation)," 2021, II.87, http://publication.pravo.gov.ru/Document/View/0001202107030001.
83 Valery Gerasimov, "'Tomagavki' Na Nizkom Starte ['Tomahawks' on a Low Start]," Voyenno-Promyshlennyy Kuryer 17, no. 681 (2017), https://www.vpk-news.ru/articles/36547.
84 Keir Giles, Handbook of Russian Information Warfare, Fellowship Monograph, No. 9 (NATO Defence College, 2016), https://www.ndc.nato.int/news/news.php?icode=995, p. 37.

Bibliography

Adamsky, D. (2008). Through the looking glass: The Soviet military-technical revolution and the American revolution in military affairs. *Journal of Strategic Studies*, *31*(2), 257–294.

Ångström, J. (2005). Introduction: Debating the nature of modern war. In I. Duyvesteyn & J. Ångström (Eds.), *Rethinking the nature of war* (pp. 1–27). Frank Cass.

Balzacq, T., & Krebs, D. R. (2021). The enduring appeal of grand strategy. In T. Balzacq & D. R. Krebs (Eds.), *The Oxford handbook of grand strategy*. Oxford University Press.

Blank, S. (2013). Russian information warfare as domestic counterinsurgency. *American Foreign Policy Interests*, *35*(1), 31–44.

Boot, M. (January 2 2016). How to respond to Russia's hybrid war on the West. *Washington Post*, p. A15.

Brooks, R. (2017). *How everything became war and the military became everything: Tales from the Pentagon*. Simon & Schuster.

Brown, G. D. (2020). *State cyberspace operations: Proposing a cyber response framework* [Occasional Paper]. RUSI.

Carson, A. (2016). Facing off and saving face: Covert intervention and escalation management in the Korean War. *International Organization*, *70*(1), 103–131.

Clarke, R. A., & Knake, R. K. (2010). *Cyber war: The next threat to national security and what to do about it*. Ecco.

Clausewitz, C. (2004). *On war* (J. J. Graham, Trans.). Barnes & Noble.

Cormac, R., & Aldrich, R. J. (2018). Grey is the new black: Covert action and implausible deniability. *International Affairs*, *94*(3), 477–494.

Department of Defense. (2018). *Summary of the 2018 National Defense Strategy of the United States of America*. https://www.defense.gov/Portals/1/Documents/pubs/2018-National-Defense-Strategy-Summary.pdf

Department of Justice. (2019). *Report on the investigation into Russian interference in the 2016 presidential election*. https://www.justice.gov/storage/report.pdf

Doshi, R. (2021). *The long game: China's grand strategy and the displacement of American power*. Oxford University Press.

Ducheine, P. A. L., & Osinga, F. (Eds.). (2017). *Winning without killing: The strategic and operational utility of non-kinetic capabilities in crises*. T.M.C. Asser Press.

Duyvesteyn, I., & Ångström, J. (Eds.). (2005). *Rethinking the nature of war*. Frank Cass.

Ekhaugen, L. (2022). Whole-of-government coordination for complex operations: Symbolic politics or coordination failure? *Journal of Strategic Studies*, *45*(5), 668–691.

European External Action Service. (2015). *Food-for-thought paper 'countering hybrid threats.'* https://www.statewatch.org/media/documents/news/2015/may/eeas-csdp-hybrid-threats-8887-15.pdf

Fitton, O. (2016). Cyber operations and gray zones: Challenges for NATO. *Connections: The Quarterly Journal*, *15*(2), 109–119.

Freedman, L. (2017). *The future of war: A history*. Public Affairs.

Gaddis, J. L. (2007). *The Cold War*. Penguin.

Galeotti, M. (2022). *The weaponisation of everything*. Yale University Press.
Galula, D. (1964). *Counterinsurgency warfare: Theory and practice*. Praeger.
Gareyev, M. (2016). Vyzov prinyat. *Voyenno-Promyshlennyy Kuryer*, *7*(622).
Gerasimov, V. (2016). The value of science is in the foresight: New challenges demand rethinking the forms and methods of carrying out combat operations. *Military Review* (January-February), 23–29.
Gerasimov, V. (2017). 'Tomagavki' na nizkom starte ['Tomahawks' on a low start]. *Voyenno-Promyshlennyy Kuryer*, *17*(681). https://www.vpk-news.ru/articles/36547
Gerasimov, V. (2019). Vektory razvitiya voyennoy strategii [Development vectors of military strategy]. *Krasnaya Zvezda*. http://redstar.ru/vektory-razvitiya-voennoj-strategii/
Giles, K. (2016). *Handbook of Russian information warfare* [Fellowship Monograph, No. 9]. NATO Defence College. https://www.ndc.nato.int/news/news.php?icode=995
Goldstein, A. (2001). The diplomatic face of China's grand strategy: A rising power's emerging choice. *China Quarterly*, *168*, 835–864.
Harknett, R. J., & Smeets, M. (2022). Cyber campaigns and strategic outcomes. *Journal of Strategic Studies*, *45*(4), 534–567.
Headquarters Department of the Army. (2006). *FM 3-24 counterinsurgency*. University of Chicago Press.
Helberg, J. (2021). *The wires of war: Technology and the global struggle for power*. Simon & Schuster.
Hoffman, F. (2007). *Conflict in the 21st century: The rise of hybrid wars*. Potomac Institute for Policy Studies.
Hoffman, F. (2010). 'Hybrid threats': Neither omnipotent nor unbeatable. *Orbis*, *54*(3), 441–455.
Hoffman, F. (2014). On not-so-new warfare: Political warfare vs hybrid threats. *War on the Rocks*. https://warontherocks.com/2014/07/on-not-so-new-warfare-political-warfare-vs-hybrid-threats/
Hoffman, F. (2018). Examining complex forms of conflict: Gray zone and hybrid challenges. *PRISM*, *7*(4), 31–47.
Honig, J. W. (2024). People's war or professional war: Which has the future in Europe? In T. Sweijs & J. Michaels (Eds.), *Beyond Ukraine: Debating the future of war* (pp. 125–146). Hurst.
Howard, M. (2002). *Clausewitz: A very short introduction*. Oxford University Press.
Huntington, S. (1957). *The soldier and the state: Theory and politics of civil-military relations*. Belknap Press of Harvard University Press.
Jonsson, O. (2019). *Russian understanding of war: Blurring the lines between war and peace*. Georgetown University Press.
Jonsson, O. (2020). *The evolution of Russian hybrid warfare: EU/NATO*. Center for European Policy Analysis. https://cepa.org/wp-content/uploads/2021/01/CEPA-Hybrid-Warfare-1.28.21.pdf
Jonsson, O. (2022). The "Russia is waging grey-zone warfare". In K. Giles & M. Boulegue (Eds.), *Myths and misconceptions on Russian military intent*. Chatham House. https://www.chathamhouse.org/2022/06/myths-and-misconceptions-around-russian-military-intent/myth-1-russia-waging-grey-zone
Käihkö, I. (2021). The evolution of hybrid warfare: Implications for strategy and the military profession. *Parameters*, *51*(3), 115–127.
Käihkö, I. (2023). *"Slava Ukraini!" Strategy and the spirit of Ukrainian resistance, 2014–2023*. Helsinki University Press.
Kennan, G. F. (1948). *The inauguration of organized political warfare* [Redacted Version]. Wilson Center. https://digitalarchive.wilsoncenter.org/document/114320.pdf
Kilcullen, D. (2020). *The dragons and the snakes: How the rest learned to fight the West*. Oxford University Press.
Kipp, J. W. (1985). Lenin and Clausewitz: The militarization of Marxism, 1914–1921. *Military Affairs*, *49*(4), 184–191.

Kofman, M. (2016, March 11). Russian hybrid warfare and other dark arts. *War on the Rocks*. https://warontherocks.com/2016/03/russian-hybrid-warfare-and-other-dark-arts/
Krieg, A., & Rickli, J. M. (2019). *Surrogate warfare: The transformation of war in the twenty-first century*. Georgetown University Press.
Lasswell, H. (1941). The garrison state. *American Journal of Sociology*, *46*(4), 455–468.
Libiseller, C. (2023). 'Hybrid warfare' as an academic fashion. *Journal of Strategic Studies*, *47*(4), 858–880.
Maoz, Z., & Gat, A. (Eds.). (2001). *War in a changing world*. University of Michigan Press.
Mattis, J. N., & Hoffman, F. (2005). Future warfare: The rise of hybrid wars. *Proceedings*. https://www.usni.org/magazines/proceedings/2005/november/future-warfare-rise-hybrid-wars
Mazarr, M. J. (2015). *Mastering the gray zone: Understanding a changing era of conflict*. Strategic Studies Institute.
Monaghan, A. (2016). The 'war' in Russia's 'hybrid warfare.' *Parameters*, *45*(4), 65–74.
NATO. (2018, August 30). *Wales summit declaration*. https://www.nato.int/cps/en/natohq/official_texts_112964.htm
Nye, J. S. (2021). Soft power: The evolution of a concept. *Journal of Political Power*, *14*(1), 196–208.
Office of the Historian. (1948a, April 30). *269. Policy planning staff memorandum*. Department of State. https://history.state.gov/historicaldocuments/frus1945-50Intel/d269
Office of the Historian. (1948b, June 18). *292. National security council directive on office of special projects*. Department of State. https://history.state.gov/historicaldocuments/frus1945-50Intel/d292
Oliker, O. (2017). *Statement before the Senate Armed Services Committee Subcommittee on Emerging Threats and Capabilities: 'Russian Influence and Unconventional Warfare Operations in the "Grey Zone:" Lessons from Ukraine'*. Center for Strategic & International Studies. https://www.armed-services.senate.gov/imo/media/doc/Oliker_03-29-17.pdf
Payne, K. (2021). *I, warbot: The dawn of artificially intelligent conflict*. Hurst.
Pettersson, T., & Öberg, M. (2020). Organized violence, 1989–2019. *Journal of Peace Research*, *57*(4), 597–613.
President of the Russian Federation. (2021). *Strategiya Natsionalnoy Bezopasnosti Rossiyskoy Federatsii (National Security Strategy of the Russian Federation)*. http://publication.pravo.gov.ru/Document/View/0001202107030001
Pynnöniemi, K., & Rácz, A. (Eds.). (2016). *Fog of falsehood: Russian strategy of deception and the conflict in Ukraine*. Finnish Institute for International.
Qiao, L., & Wang, X. (1999). *Unrestricted warfare*. PLA Literature and Arts Publishing House.
Renz, B. (2016). Russia and 'hybrid warfare.' *Contemporary Politics*, *22*(3), 283–300.
Renz, B. (2018). *Russia's military revival*. Polity.
Renz, B., & Smith, H. (2016). *Russia and hybrid warfare – Going beyond the label*. Aleksanteri Institute. https://helda.helsinki.fi/bitstream/handle/10138/175291/renz_smith_russia_and_hybrid_warfare.pdf
Rid, T. (2012). Cyber war will not take place. *Journal of Strategic Studies*, *35*(1), 5–32.
Rid, T. (2019). *Active measures: The secret history of disinformation and political warfare*. Farrar, Straus and Giroux.
Robinson, L., Helmus, T. C., Cohen, R. S., Nader, A., Radin, A., Magnuson, M., & Migacheva, K. (2018). *Modern political warfare: Current practices and possible responses*. RAND.
Rudgers, D. F. (2000). The origins of covert action. *Journal of Contemporary History*, *35*(2), 249–262.
Sanger, D. E. (2018). *Perfect weapon: War, sabotage, and fear in the cyber age*. Scribe.
Scheipers, S., & Strachan, H. (Eds.). (2011). *The changing character of war*. Oxford University Press.
Smith, R. (2008). *The utility of force: The art of war in the modern world*. Vintage.
Smyth, S. M. (2019). The Facebook conundrum: Is it time to usher in a new era of regulation for Big Tech? *International Journal of Cyber Criminology*, *13*(2), 578–595.

Treverton, G. F. (2009). *Intelligence for an age of terror*. Cambridge University Press.
United States Special Operations Command. (2015). *White Paper: The gray zone*. https://info.publicintelligence.net/USSOCOM-GrayZones.pdf
US Senate Select Committee on Intelligence. (2017). *Russian active measures campaigns and interference in the 2016 U.S. election*. https://www.intelligence.senate.gov/sites/default/files/documents/Report_Volume4.pdf
Walker, R. G. (1998). *Spec fi: The Unites States Marine Corps and special operations* [Thesis, Naval Postgraduate School]. https://apps.dtic.mil/sti/pdfs/ADA359694.pdf
Warner, M. (2019). A matter of trust: Covert action reconsidered. *Studies in Intelligence*, *63*(4), 33–41.
Watling, J., Danylyuk, O. V., & Reynolds, N. (2023, March 29). *Preliminary lessons from Russia's unconventional operations during the Russo-Ukrainian war, February 2022–February 2023* [Special Report]. Royal United Services Institute. https://rusi.org/explore-our-research/publications/special-resources/preliminary-lessons-russias-unconventional-operations-during-russo-ukrainian-war-february-2022
Zarudnitsky, V. (2021). Kharakter i soderzhaniye voyennykh konfliktov v sovremennykh usloviyakh i obozrimoi perspektive [Character and content of armed conflicts in modern conditions and near-term perspective]. *Voennaya Mysl*, *1*, 34–44.

2 The Meaning of Victory and Defeat in Non-Military Warfare

Jan Angstrom

Introduction

Determining victory and defeat in war has always been fraught with inherent difficulties. On top of obvious methodological issues, for example, the extent to which it is possible to find verifiable data on casualties (e.g., Mueller, 1995), there are also problems related to whether or not casualties are relevant for victory and defeat to begin with. Scholars have highlighted these problems over the past decades, when analyzing the wars in, for example, Afghanistan, Iraq, and the greater Middle East (e.g., Hobbs, 1979; Bond, 1998; Mandel, 2005; Angstrom & Duyvesteyn, 2007; Martel, 2007; Johnson & Tierney, 2006; Larsen, 2014; Sullivan, 2012). Understanding and determining victory and defeat in the so-called non-military warfare are, arguably, even more difficult. How do we know that it is a certain information (or disinformation) operation that influences individuals' preferences and not "normal" shifts in political preferences? How can we distinguish between legitimate foreign investments and economic efforts to exercise power and when do we know such investments are successful in changing the behavior of the target state? And how can we properly assess what particular combination of military and non-military means and ways lead to a victory and defeat?

Unfortunately, efforts to more precisely reflect on the meaning of victory and defeat in non-military war is near non-existent. Indeed, most scholars (e.g., Rid, 2020) seem to take success as a given *a priori* in non-military warfare. Disinformation campaigns, for example, are almost always assumed to have an impact on the target state as long as they are carried out. This is problematic since important questions on the utility of force are left out. For the purposes of analysis, this chapter follows McDonald (in this volume) and defines non-military warfare as "non-kinetic tools of statecraft employed as means of strategic coercion and effect employed with an intensity and purpose such that the target perceives them as significant escalations and/or commensurable to warfare" (cf. Galeotti, 2022). This definition highlights the non-kinetic means of warfare, but it should also be pointed out that actual war, following Jonsson (2019), will in all likelihood include information warfare, cyber warfare, economic warfare, political subversion, as well as limited use of or threat to use actual military force.

DOI: 10.4324/9781003616993-2

I will argue that non-military war is associated with both low costs and limited aims as compared to regular war. These low costs can still yield substantial payoffs, and success is thus not only relatively easy to come by but also – correspondingly – easily lost. Non-military war, as compared to regular war, is by the nature of the ends, means and ways involved, highly fluent, transient, and elusive. Understanding victory and defeat in non-military warfare is thus best captured by the so-called cost-benefit understanding of victory. This has profound implications on how to think about strategy in non-military war as compared to regular war. For example, while casualties and the destruction of military assets are usually understood to be an important currency in determining the outcome of regular war, non-military war is waged more directly against the broader domestic support of the opponent's population. Hence, whereas it has long been recognized within strategic thought that the political will of the enemy needs to be broken, it is usually thought to be broken by incurring military costs on the opponent. In non-military war, however, the political will is thought to be influenced via a weakening support of the incumbent government from the population.

The chapter proceeds as follows. First, I will briefly outline different schools of thought on victory and defeat in war to create an analytical tool with which to analyze non-military warfare. Second, in the main analytical part, I will differentiate and elaborate on the problems of possibilities of determining victory and defeat in information warfare, cyber warfare, economic warfare, political subversion, and the limited use of or threat to use actual force, respectively. I conclude this section by also discussing the problems related to balancing, that is how can we understand or determine victory and defeat if one is successful in economic warfare, but not in the other domains. Third, the chapter concludes with a set of challenges for analysts of non-military war in order for the field to develop.

Schools of Thought on Victory and Defeat in War

Perhaps surprisingly considering their importance for the vast literature on use of force, the concepts of victory and defeat remain undeveloped and highly contentious. In essence, there are four schools of thought of how to conceptualize victory and defeat in war (Angstrom & Widen, 2015).[1] When approaching non-military warfare, however, the two most commonly used (although implicitly) by scholars seem to be *end-state understandings* and *cost-benefit understandings*. In this chapter, I will focus on these two.

First, the *end-state understanding* of victory and defeat is by far the most common conceptualization of victory in war. At its core, this school of thought holds that if one reaches one's goals – be they at the strategic, operational, or tactical level of war – one has achieved victory. Not reaching one's goals, correspondingly, constitutes defeat. This is, of course, intuitively reasonable: being successful is related to reaching your aims. Among practitioners, this understanding is even institutionalized in doctrines, field manuals, and planning procedures stating that the military in their planning should identify end-states and then plan how to reach them. Among scholars, Clausewitz is certainly one who approaches war through the political aims of the war even concluding that war is "an act of force to compel our enemy to do our will" (von Clausewitz, 1993, p. 83). As such, von Clausewitz (1993, pp. 99–100; cf. Echevarria, 2007; Herberg-Rothe et al., 2011; Strachan, 2008) emphasized that war is a means with which political aims can be realized. Accordingly, the political aim of the war is central when evaluating how war progresses or how it ended. Furthermore, much of the literature on limited war defines success as the fulfillment of political aims *without* further escalation into all-out war (e.g., Bakich, 2014, p. 20).

The end-state understanding is an essentially rationalist notion that implies that before military action ensues, an end-state has been formed and the progress of the war can be measured against it. It assumes rational actors since such actors ought to set aims and evaluate different courses of action before acting (Hollis & Smith, 1991; Simon, 1955). The idea assumes, furthermore, that war then can be understood to be a tool to reach political aims (e.g., von Clausewitz, 1993; Pillar, 1983; Powell, 2004; Reiter, 2009; Schelling, 1966; Sullivan, 2012). It is precisely the assumed rationality and actors with pre-set preference structures that make it possible to understand victory and defeat in terms of what the combatants try to achieve through the use of force (Betts, 2000, pp. 40–43). The rationalism involved in end-state understandings also make it perfectly compatible with measuring victory and defeat in terms of objective results. This does not necessarily mean that rationalism and perceptions are incompatible (see, e.g., Jervis, 1976), but the whole point of end-state understandings of victory and defeat is that they can be reached and that the actors can measure and evaluate their operations continuously throughout the war. This does not mean that actors measure things correctly, make correct estimations or interpret evidence correctly, but it means that commonly they try to. It does not even suggest that rational planning guarantee success even if it is carried out well. Naturally, there are also a number of empirical examples that indicate that strategic planning – rational and long term – can be flawed in wars. Ricks (2006) has demonstrated that planning for post-war operations in Iraq in 2003 was nearly absent. Similarly, the German General Ludendorff infamously exclaimed before the 1918 spring offensive on the Western front that the aim was to "punch a hole into [their line]. For the rest we shall see" (Parker, 2008, p. 289).

End-state understandings, however, are perfectly compatible with the voluminous literature on deterrence and threats to use force, that is the use of force that does not involve direct violent behavior. Understood as end-states, thus, victory and defeat can be used fruitfully in cases of use of force that does not involve direct violence. It is noteworthy, though, the literature on deterrence and coercion usually do not talk of victory in such cases, instead referring to outcomes in more moderate terms, such as "success" and "failure." Successful deterrence, for example, is often understood as preventing an attack from an opponent, thus measured as continued absence of war (e.g., Freedman, 2004; Morgan, 1983, 2010; Schelling, 1966). Following this logic, both protagonists in a militarized crisis can claim to be victorious if both parties tried to avoid war. Failed deterrence only occurs when a challenger launches an attack despite efforts to deter. Meanwhile, deterrence also operates *within* war (e.g., Stone, 2011). As such, successful deterrence in war could be understood as efforts that succeed in preventing further escalation or preventing the opponent to choose a line of attack particularly non-preferable. For example, on the tactical level of war, deterring a flanking attack on one's forces while they are stretched could be done by using mines. If successful, the minefield effectively deters an attack from that direction.

It is important to recognize that a strict end-state understanding of victory and defeat does not take into account the costs involved. Hence, victory can be declared regardless of the costs of war – in either casualties or economic costs. If we take a strict end-state position, in short, it would be victory if the German Army in the spring of 1940 had lost 20 armored divisions and tied another 20 to drawn out counterinsurgency operations when occupying Norway, despite that such a development would have postponed or made impossible a campaign against France and the Low Countries later the same spring. This is also the nucleus of the main criticism against end-state interpretations. They tend not to regard other aims, other plans, or other uses of military force, and in most situations regarding life or death, peace and war, there is more than one objective at stake.

Measuring victory as war aims versus war outcomes, moreover, struggles to accommodate that end-states are inherently ambiguous and fluctuating (Mandel, 2005). Political end-states tend to be vague in order to create freedom of maneuver for the decision-makers to broker a future peace. Although laudable, this ambiguity also implies problems of accountability as well as problems related to creating a clear-cut "score-card" with which to evaluate the military operations. As war progresses, furthermore, end-states can be modified and changed, leading to what is usually called "mission creep." War aims at all levels of warfare can also be difficult for analysts to uncover. Although it makes sense for political and military leaders to convey to their own forces what the aim of war is, it does not necessarily mean that the actors are interested in publicly announcing these aims. Sometimes, and perhaps especially for the operational and tactical levels of war, actors keep their objectives secret since it would be advantageous for the opponent if it knew the main political and military objectives of the other side. Withholding "private information" (Fearon, 1995, pp. 390–401), that – would it be shared – logically would make the opponents better at assessing each other's preference structures and degrees of credible commitment, could unnecessarily prolong the war and make the process of strategic assessment more imprecise than necessary.

Regardless, if the war aims change or are ambiguous, there are also issues of trade-offs between war aims. For example, in 2003, then-U.S. president George W. Bush outlined 53 (!) different motives to launch the war against Iraq. If we take these 53 as war aims, it makes it incredibly difficult to achieve all of the 53 and what constitutes victory in such circumstances? Is reaching 27 of the aims sufficient or do we need to somehow assess whether or not some war aims are more important than others? If so, can reaching one out of 53 war aims be sufficient for victory? Aggregating war aims in coalition warfare makes this even more difficult. Considering that some states may have caveats in their conduct of operations and that these caveats will look differently (Auerswald & Saideman, 2014), it creates difficulties on how to measure victory and defeat. What happens, for example, if one coalition partner is prevented from partaking in a particular form of operations in an area that later turns out to be critical to the war effort? Does this mean that the entire coalition is victorious or only those allies that took part in the decisive operations?

There are trade-offs at the operational level of war. As Gray (2002, p. 11) notes: "A decisive victory in one theater might be offset by a decisive defeat in another." Empirical examples are abundant, and we only have to think of Germany's continuous balancing between preventing defeat on the Eastern Front against the prospect of success on the Western Front during World War I. At the tactical level, finally, trade-offs between objectives are equally frequent. For example, at the high end of tactics, Soviet offensives following the battle of Stalingrad effectively stretched Soviet lines and exposed them to German counteroffensives that ended with the capture of 90,000 Soviet soldiers during the third battle of Kharkiv in 1943. The continuous balance between offense and defense thus presents clear trade-offs at the tactical level. At the lower end (but no less important), moreover, protection of own forces led allied commanders to incur a 15,000-foot bombing ceiling during the 1999 Kosovo War, making target identification more difficult, but decreasing the risk of casualties (Lambeth, 2001, pp. 140–141).

In sum, end-state understandings of victory and defeat are by far the most common understanding among both practitioners and scholars. Nevertheless, it suffers from a range of problems both for analysts and for practitioners. Most importantly, war aims at all levels of warfare can be fluctuating, ambiguous, secret, and difficult to aggregate and measure. End-state understandings easily accommodate outcomes at different levels of war, and they are frequently used also for outcomes of non-violent use of force or threats to use force.

The second most influential understanding of victory and defeat suggests that victory can be understood in terms of a *cost-benefit* calculus (Freedman, 2013; Mandel, 2005, pp. 7–9). Here, it becomes important to evaluate reaching the political end-state relative to the costs associated with reaching it. Schelling (1966, p. 31) implied as much when arguing that the important effect of the military was "the bargaining power that comes from its capacity to hurt." Hence, we can have situations where the political end-state was reached, but the costs were so over-bearing that it would constitute strategic defeat anyway. The critical question for victory, accordingly, is not if we achieved what we set out to do, but rather whether it was worth it?

The great advantage of cost-benefit understandings over end-state interpretations is that the former takes into consideration the costs (material and immaterial) of war. Hence, rather than just fixing the reference object at the outcome, cost-benefit understandings emphasize that one can be defeated even though one reaches one's war aims if the price for achieving them is too high. It is probably not a coincidence that this type of understanding of victory and defeat was prominent in theorizing about the use of nuclear weapons during the Cold War (e.g., Brodie, 1959; Freedman, 2003; Kahn, 1960; Kaplan, 1983; Schelling, 1966). Should a nuclear war have broken out between the United States and the Soviet Union, the ensuing peace with both countries devastated and in all likelihood millions of dead and more dying of cancer would hardly amount to victory even if political aims were reached in a nominal sense. A second analytical advantage of the cost-benefit understanding is that by assuming that war is a political process involving a counterpart, it avoids problems related to the timing of war endings. It does so by embracing that not only are war aims set by political interests, but what costs that are acceptable is also an inherently political issue. According to this understanding, then, it does not matter if war aims are ambiguous when starting out, what they effectively mean is decided once war has progressed anyway.

According to the cost-benefit understanding, it is beneficial to differentiate costs and benefits at various levels of warfare. One could easily imagine situations where the costs of war at the strategic level, for example, in terms of destroyed industrial capacity surpass any potential gains at the strategic level. This logic operates also on the operational and tactical levels of war. If an army corps, for example, succeeds in reaching its operational level war objectives by cutting the communication and logistics lines of the opponent's army corps, but in doing so effectively loses its capability to maneuver, the army corps is essentially a sitting duck for an enemy counter-attack. Furthermore, if at the tactical level the unit reaches its tactical aims, but in doing so disintegrates due to heavy losses, it can be understood as a tactical defeat regardless of aims achieved.

Understanding victory and defeat as cost-benefit calculus makes the issue of translation of costs at one level of war into benefits at another level crucial. To some extent, much of the costs of war are attached to the tactical level, since it is at the tactical level of war that death and destruction occurs most frequently. However, the benefits of the tactical costs should ideally be translated into strategic level victory. Rather than refraining from mixing the levels of war, thus, a cost-benefit understanding of the outcome of war and the use of force makes it necessary to assess how costs and benefits play out at all levels of warfare. It is precisely this logic that is visible in the concept of Pyrrhic victory. After a victorious battle against the Roman army, Pyrrhus of Epirus exclaimed that "one other such victory would utterly undo him" since even though Roman losses were greater, the Romans total military superiority would defeat Pyrrhus in the long run anyway.

This understanding also relies upon a rationalist framework, and it does so by stressing that military operations consist of a series of related and continuous assessments and decisions where efficiency becomes important and not only effectiveness. Measuring victory by weighing issue salience (benefit) against spent resources (costs) and duration (resolve) implies that you win if you punch above your weight and thus that strategy "is fluid and flexible, governed by the starting point and not the end point" (Freedman, 2013, p. xi). Hence, if you can hurt the opponent (benefit) without costs (or lower costs than your opponent) it amounts to victory, while situations in which your costs exceed gains constitute defeat. For example, by accumulating tactical and operational victories first in the Western theater, then in the Carolinas and Georgia, before launching a two-pronged attack against Virginia, Union commander Ulysses Grant finally broke Confederate resistance over a period of two years during the American Civil War (Gwynne, 2019). By recognizing that the Union was stronger in absolute terms, Grant and then-President Lincoln could sustain losses over a longer period of time (show more resolve) despite the loss of union casualties and resources (costs) to finally break the will of the Confederate states (benefit). Hence, the result of a series of tactical victories accumulated or "traveled" to strategic level victory.

Although it has advantages of considering spent resources and, therefore, enables a far more nuanced understanding of victory, this school of thought still encounters some problems. The most significant of these is that creating an analytical framework to measure victory is, as politically laden, as end-states are ambiguous. It involves at the strategic level, by necessity, making estimates as to whether one's post-war benefits in inherently difficult areas to measure, such as political reputation, freedom, independence, resolve, or power, have improved compared to pre-war estimates and the costs of war. Even at the tactical level, where death and destruction make things seemingly straightforward, estimates of gains of sunk ships or seizing territory can be notoriously difficult to assess both in the short and long term. How would one, for example, value seizing and defending a particular bridgehead in the area next to the division's main operational objectives? It could not only be extremely important if the main effort fails, but it could also be utterly redundant if the main effort succeeded. It could even be impossible to make correct assessments since by seizing the nominally less relevant bridgehead, it divided enemy forces and made breakthrough in the direction of the main effort possible at all.

Estimating cost-benefit calculus when results are measured at different levels of war also makes this highly complex. Consider the case of the battle of Jutland in 1916. At the tactical level, its results were seemingly straightforward: The Imperial German Navy's High Seas Fleet met the Grand Fleet of the Royal Navy in Skagerrak, and the British suffered 14 sunk ships against the German six and some 6,000 British sailors lost their lives as compared to German losses up to 2,000. In Britain as well as in Germany at the time, the battle was received as a German victory. However, since the German high command concluded that further losses at the same level would cripple the German navy it ordered the German High Seas Fleet to stay in harbor for the remainder of the war. Hence, there are sound arguments to be made that the Battle may have been lost at the tactical level for the British but won on the strategic level (Till, 2009, pp. 161–162). Indeed, the Battle may even have increased pressure to wage unrestricted submarine warfare in the Atlantic – a decision that eventually led to U.S. intervention in the war and Germany's subsequent untenable war effort.

Moreover, conflicting or ambiguous war aims make it difficult to assess benefits and while costs seem straightforward, they are anything but. Furthermore, these estimates must be addressed either in absolute terms or in relative terms toward the opponent. In a zero-sum, absolute game, these estimates may be straightforward. In retrospect, one can argue that Finland, for example, won the Winter War against the Soviet Union in 1939–1940, since even though it lost territories in the east, it still hindered an occupation of the entire country and thereby also avoided a later enforced membership in the Warsaw Pact and Soviet domination. At the time, though, these obvious post–World War II benefits were not clear. In relative terms, it is even more difficult to determine whether or not one state is better off after a war in comparison with the opponent.

Furthermore, it is sometimes difficult to reduce matters of life and death – as battle, campaigns, and wars – to a cost-benefit calculus. In empirical terms, when Spartan King Leonidas made a stand against Xerxes' advancing Persian army at Thermopylae in 480 BC, he and his soldiers were killed. Even if the Spartan bravery and the toll they took on the Persian army had a strategic effect insofar as several other city-states in Greece formed an alliance with the Spartans that eventually defeated the Persians, it is very difficult to talk of costs and benefits for King Leonidas since the ultimate cost were already paid. For the victorious Persians, too, it makes more sense to talk of a defeat since Thermopylae left some 20,000 Persian soldiers dead as well as providing the Greeks with a narrative of Greek heroism. Similarly, suicide terrorism is also difficult to approach from a strict cost-benefit understanding at the tactical level. Instead, the benefits in terms of government secessions are reaped at the organizational or strategic level, much in the vein of King Leonidas (Pape, 2005, pp. 27–29). Proposals of how to counter suicide terrorism, however, clearly relies on cost-benefit logic. Pape (2005, p. 238), for example, mentions as operational measures of victory as defeating the current crop of terrorists, while simultaneously discouraging others to join the ranks of terrorist organizations. Hence, victory relies on a dual understanding of extracting costs, while also potentially offering benefits to discourage further recruitment of suicide terrorists. Hence, although intuitively attractive, understanding victory and defeat in terms of a cost-benefit calculus encounters problems of conflicting values.

Moreover, according to cost-benefit calculations, we need to consider if we are to measure victory by absolute or relative costs and gains. In the case of the original Pyrrhic victory above, Roman losses were higher in absolute terms, but still comparably favorable for the Romans. At the battle of Chancellorsville during the American Civil War, moreover, the Confederate Army despite winning on the day (Union casualty figures numbered 18,000 against the Confederates 13,000) could not sustain such losses in the long run. Hence, even if the Union Army suffered more losses, it could replenish its ranks quicker and thereby emerge victorious when the war dragged out in time. This also means that for the cost-benefit school of thought, victory and defeat are not necessarily zero-sum games (cf. Kolenda, 2021). Both sides can claim victory, and scholarly analyses can fairly easily conclude that wars, campaigns, or battles can have winners on all sides. If, for example, one side punches above its weight and reaches gains for minor costs that it would not have even dreamed about before the war began, it can claim to be victorious. The other side involved in the same war, moreover, can also claim victory if the costs, however dear they appear, still are valued less than the benefits of military action. Similarly, to end-state understandings, cost-benefit interpretations of victory can also include perceptions, that is, it may not be the case that certain benefits objectively are worth more than a particular set of costs. It could just as easily be related to perceptions of how the war progresses and things that are perceived to be important may later prove to be less valuable than originally thought.

As in the case of end-state understandings, cost-benefit understandings are perfectly compatible with the literature on deterrence and coercion. To some extent, cost-benefit is even a more perfect match for deterrence and threats of the use of force since the logic of the latter so clearly relies on preference structures and efforts to influence the opponent's costs in order to reach one's political goals. Influential definitions of deterrence, for example, stress: "the persuasion of one's opponent that the costs and/or risks of a given course of action he might take outweigh its benefits" (George & Smoke, 1974, p. 11). Accordingly, outcomes emanating from threats to use force can be fruitfully understood from cost-benefit calculations.

In summary, cost-benefit calculations improve our understanding of victory and defeat by also incorporating costs in the assessment of victory and defeat. Even though it therefore provides a more nuanced understanding of victory and defeat, it still faces a few problems for analyses. How do we, for example, create operational definitions of a whole set of political values that defy precision? How do we evaluate costs in human lives and untold suffering against tactical, operational, or strategic gains? How do we assess alternative ways of using force in order to gauge alternative costs when we have to rely on counterfactuals as evidence? Still, cost-benefit understandings are frequently used to describe outcomes at different levels of war, and they are compatible outcomes of non-violent use of force. When continuing the analysis below, I will elaborate on victory and defeat in non-military warfare using both cost-benefit calculus and the traditional end-state understanding.

The Meaning of Victory and Defeat in Non-Military Warfare

Instead of the traditional differentiation of victory and defeat in terms of the levels of the war, I will analyze the concepts through the dynamics of micro and macro levels or individual versus collective dimensions. Arguably, one of the shortcomings of much of present research on non-military warfare assumes that, for example, disinformation spreads through society like wildfire. However, there are also significant and natural firebreaks in society. If we, for example, assume that information bubbles are key to transmitting disinformation, it also means that there are natural boundaries to the spread of such disinformation. This is an example where a particular logic seems convincing on the micro level, while it is less so on the macro level. Throughout the analytical parts, I will draw upon the case of the war in Ukraine and elaborate what victory and defeat would look like according to end-state and cost-benefit understandings, respectively.

END-STATES AND NON-MILITARY WAR

End-state understandings of victory and defeat seemingly sit easy with non-military war. Across the spectrum of non-military warfare, it is possible to identify ends according to which one can gauge victory and defeat. However, these ends are all quite limited by nature. At large, they deal with intelligence gathering, disruption of societal needs, as well as creating or increasing societal unrest. They are not primarily about seizing territory, political control, or death and destruction as traditional, military means. This means that assessments of victory and defeat in non-military warfare to a higher extent than "proper war" is time sensitive and suffers from more immediate analytical problems of "when to measure" success or failure.

Assessing victory and defeat in traditional war, of course, also entails difficult analytical assessments related to time so it is difference in grade, not in kind. Drawing upon the war in Ukraine to illustrate this point, it is quite clear that disinformation campaigns since 2004 have contributed to create divisions within Ukraine (see Chapter 4 by Hosaka). However, these divisions were clearly not enough to pacify the Ukrainians or paralyze its response to the large-scale invasion in February 2022. Hence, if one were to measure the effect of disinformation in 2013, one would probably conclude that they are a resounding success, but if one measures them in 2025 – when Ukraine has fought through nearly three years of open war – one would in all likelihood conclude that the effects of disinformation are quite limited.

Moreover, if you can identify, target, and isolate the effectiveness of cyber operations, then end-state understanding of victory is useful. Some have suggested that cyber warfare can accomplish three distinct effects: It can disrupt, it can gather intelligence, and it can destroy (Valeriano et al., 2018). First, disruption is potentially the least ambitious and most easy to accomplish of these effects. Using a cyberattack nearly always entails the opportunity to rebuild the object of the attack. For example, on multiple occasions, since the war in Ukraine started in 2014, Russia has used cyberattacks to disrupt parts of the electric grid in Ukraine. Although certainly annoying, it or its potential follow-on effects in terms of temporarily failed communications or loss of productivity in the industry have most often not lasted more than days. The concept of victory when it comes to intelligence gathering suffers from not only the covert nature of cyberattacks but also the covert nature of intelligence. If one has succeeded in accumulating critical pieces of intelligence, it does rarely make sense to divulge what you know and how you came about that knowledge since it disincentives future HUMINT (human intelligence) sources or it makes it obvious to your counterpart where and how his cyber defense must be reinforced. When it comes to destruction, it is also clear that cyber warfare is effective, but mostly in the short term. Cyber warfare seems to be better-suited for intelligence purposes and covert operations, than destruction (Maschmeyer, 2021). Nonetheless, cyber warfare can be a substitute for military force (as shown by Jensen in Chapter 5).

Even if it is clear that cyber warfare is effective, it is also quite clear that the aims it can reach separately are quite modest and the effects are transient. Even if one manages to destroy the computer system of a power plant, the computer can be replaced, wires can be redrawn, and appliances reprogrammed. In technical terms, the same logic applies, of course, to financial systems, transportation, and so on. It is also possible to create systems that are independent of outside connection to protect one's military secrets. Instead, perhaps the one more lasting effect that cyberattacks can create is distrust in societal functions. If cyberattacks disrupt basic societal functions, such as transportation, energy, health care, and supply lines, *repeatedly*, there may be a longer-lasting effect on the population in a breakdown of the basic societal social contract of the citizen surrendering his independence to a sovereign ruler in exchange for societal services.

As cyber warfare, information warfare does not recognize a clear and firm demarcation between war and peace. Disinformation campaigns can, in theory, create changes in the political system of the target state, changes in individuals' belief in the value of the political system of the target state as well as changes in individuals' trust in official state information, including the dissolution of truth. As above, however, far more than traditional war, information war suffers from difficult analytical problems partly due to secrecy, partly due to "normal" shifts in political preferences, and partly due to issues related to time.

In order to reach the ends of information campaigns, it is imperative to maintain secrecy of the transmitter of the information. It is difficult to fathom that anyone initially would be convinced by a message if one knew that the message came from the opponent. Once trust in traditional media outlets or governmental information has begun to erode; however, secrecy may be of lesser importance. Indeed, late in a campaign, it can even be openly framed as "the other side of the story," implying the part that is necessary to make the information complete, that is your government does not tell the whole story and you need *Russia Today* to complete the picture. The problem with the initial covert information campaigns in terms of victory and defeat, of course, is that it is difficult to assess – even in the aftermath of the operation – what the intended ends of the campaign was. If one cannot credibly assess the ends of warfare, it makes the idea of an end-state understanding impossible.

End-state understandings of victory and defeat of economic warfare are seemingly more straightforward. Sanctions, boycotts, interfering with trade, and tolls, after all, ought to have a measurable effect on your opponent's economy. Again, however, while some forms of economic warfare occur in the open (and are thus easier to isolate the effect of), others occur covertly, thus making it difficult to use end-state understandings. Open economic warfare, such as Russia stopping the export of gas and crude oil to Ukraine multiple times since 2014, is visible to us all and relatively easy to measure the effects of. Of course, making your opponent's economy suffer will also – as in the logic of all wars – increase the incentives for your opponent to develop countermeasures or alternative sources of energy, thus rendering your initial upper hand less important. In Ukraine, the past ten years have witnessed on several occasions regular trade wars between Russia and Ukraine. While suffering supply chocks on the energy side, Ukraine has several times stopped buying Russian weaponry and tried to replace old Soviet or Russian weapons systems with Western. In doing so, of course, the Ukraine has also started to sever the ties with Russia in domains where it is hurting Russian industry.

More problematic, however, is that economic warfare can occur behind a façade of private companies. The obvious problem is that it may be impossible to differentiate between "legitimate" foreign private investments or land purchases closely located to critical infrastructure and "illegitimate" attempts to create an interdependency that your opponent can use against you. In this way, you may not only be hurt economically, but such economic warfare also fundamentally undermines Western understandings of the relationship between public and private. The West may wrongly expect private firms to be private, while they are in fact covert state institutions.

In terms of end-states, however, the utility of economic warfare presumes that you have transactions with the target state or leverage and ability to influence it. Without such interactions, it is of course difficult to punish the target with economic means. This inevitably means that there are costs involved when using economic warfare, thus making a "pure" end-state approach to victory and defeat less appealing.

Political warfare has the potential to reach a number of effects on both the individual and societal levels. Hence, political warfare is easily compatible with an end-state understanding of victory and defeat. For purposes of simplicity, we can categorize these effects either as electoral interference and subversion or as effects relating to changing the entire political system. First, electoral interference ranges from influencing the political preferences of individuals to measures that benefit one candidate over others. In Ukraine, Russia has interfered in elections since 2004, long before the war began (Wilson, 2005). However, perhaps the political warfare was more successful during the first few post-independence elections since pro-Russian candidates were elected. In 2004, the main attempt to create favorable conditions for Kremlin's candidate Viktor Yanukovych consisted of attempts to scare the population to vote for Yanukovych with made-up stories of U.S. forces amassing in Poland and Hungary to invade Ukraine. The only way to avoid being occupied was to vote for him since with the aid of Russia, the Americans could be deterred. By continuously reinforcing the image of ethnic Russians in Eastern Ukraine being discriminated against by the government in Kyiv ably encouraged by the West, Russia has also managed to reinforce the divide in the country. In the elections, in 2004, voting patterns were clearly split along the language line of west and east.

Even if electoral interference is far from a modest aim and it is certainly attainable, its importance should not be overestimated. Elections come and go and to reap the rewards, one has to continuously wage political warfare since electoral patterns between elections can be fickle. Russian political warfare has not stopped at tampering with the elections either. More fundamentally, there have been attempts to coach elected political leaders to pursue policies to radically change the political system of Ukraine. Turning Ukraine into a new Belorussia – effectively a subaltern to Russia – would have been a more lasting effect. However, those attempts have so far failed on a large scale.

However, on a smaller scale, Russia has managed to alter the political landscape of Ukraine. Handing out 50,000 Russian passports in Eastern Ukraine created a minority and therefore a "legitimate" cause for Russian interference with the domestic politics of Ukraine. In this sense, Russia pulled a basic racketeering scheme. It first created a demand for "protection" over a long period of time, then handed out passports to reinforce the notion that there existed a Russian-speaking minority in Ukraine that was threatened, and it then had created the political space for agency in the domestic affairs of its neighbor and could then be the supplier of "protection."

Despite that we often associate non-military war with non-military means, the concept does entail the threat to use and use of limited military force. Limited war in comparison with total war is usually understood to have less ambitious aims, involve less numerous troops and resources, demand less coordination on behalf of the state, and involve less width of potential means. In the late 19th century, for example, it was fairly common, and understood to be perfectly legitimate, that states solved unpaid debts to their private companies by using their navies to block a harbor to hold the opponent hostage and force the company to pay its debts (Finnemore, 2003). When war was outlawed for purposes other than self-defense first in the Kellogg-Briand Pact in 1928 and then again in the UN Charter at the end of World War II, such limited war also became problematic. As such, end-state understandings of victory and defeat also from the perspective of limited use of force is problematic as it is even more difficult than in "war proper" to gauge a lasting effect of military force.

LOW COSTS, BUT LIMITED BENEFITS IN NON-MILITARY WAR

Although compatible, in a strict sense, with end-state understandings of victory and defeat, non-military war lends itself better to cost-benefit understandings. One cannot completely avoid sticky, difficult methodological issues with cost-benefit either, but the nature of the means involved in non-military war means that ends are indeterminate, fluid, transient, and long term. Approaching such strategic issues as a balance between benefits wrought and costs attached makes for a more careful analysis. Instead of asking questions about effectiveness, cost-benefit understandings make us ask questions about efficiency. From this perspective, we can more easily grasp that victory and defeat in non-military war are fleeting since the costs are low, while the benefits are small in comparison to war proper. This creates a strategic interaction in which the benefits are short-lived, thus creating incentives to try to reach them continuously, while the costs are low so there are no significant obstacles for trying to get at the benefits.

Even if cyber warfare is effective for a range of less ambitious purposes, how efficient is it? If one approaches cyber warfare from the cost-benefit perspective, it is obvious that cyber warfare also yields relatively easy victories. There are several reasons why costs are low. Because there are huge difficulties with attributing blame in cyberattacks, it is virtually impossible to attach costs to an attacking actor. Any state using cyberattacks – for whatever purpose – will always have an easier time maintaining their innocence or at least create ambiguity or deniability since one can always blame individuals operating from within your country, but not on the behest of the state. Moreover, costs for cyberattacks will be low since cyber forces are relatively cheap in comparison with standing armies, air forces or navies that all require much heavier investments on development of platforms as well as costs associated with maintaining and using them.

The one potential cost attached to cyber warfare is that since it occurs in the shadows, it is impossible to create a deterrent effect among peers. Simply put, if no one learns, and you do not acknowledge, that you have certain capabilities, you cannot expect that counterparts – friendly or not – will be deterred from taking a particular set of actions against you. The logic of deterrence in cyberspace thus plays out partly differently from "regular" interstate deterrence (Mandel, 2017). Hence, by their covert nature, the results of cyberattacks do not travel and solve other political ends indirectly. It should also be pointed out, of course, that sometimes it is not the cyberattack in and of itself that delivers the political effect. As a case in point, Russian cyber warfare against the Ukraine has during the past eight years mostly been carried out at the same time as information warfare and political warfare. Already in January 2014, before the annexation of the Crimea, Russia created fake accounts on social media that were supposed to work as "evidence" in a larger information campaign designed to spread the image that Kyiv since the *Euromaidan* uprising in November 2013 was controlled by Nazis and acted under direction of CIA. In and of itself, this may not have convinced that many neutrals, but it served to reify an already deep suspicion of the West held in Eastern Ukraine.

If one introduces costs to the calculus, the analysis of victory and defeat in information warfare becomes more nuanced. As in many other forms of non-military warfare, the costs are relatively low, which implies that victory is comparatively easy to come by. However, in contrast to goods and services in the economy, which require storage and supply lines, information and ideas can move freely. In this sense, giving voice to an idea intended to weaken your opponent internally is not dissimilar to opening Pandora's box since the idea can also haunt yourself. In the Ukrainian case, Russian demands for increased autonomy for the Russian-speaking minority in Eastern Ukraine have also created an inconsistency in its own policies toward minorities in Russia proper, where autonomy is not granted. Still, the low-cost, low-yield impact is clearly visible in other information campaigns. For example, Russia has amplified the message that Sweden and Finland are Russophobic. At first, it creates a dilemma for the target states, since if they complain about the description, they confirm the image, and if they do not complain, Russia's claim is not contradicted. However, at second glance, it also begs the question why these states at all are afraid of Russia and it therefore reminds everyone of Russia's wars against these states. Although not as firmly set in stone as monuments of past wars, constant presence of discourse about security fears also reminds the target continuously that Russia at least in the past has been a security threat.

Moreover, introducing the idea that truth does not matter seems to be an excellent method to undermine trust in the media, in the state or in other institutions that have earned a high degree of trust in society, such as professions, but it also has other indirect costs. Most importantly, since one cannot guarantee that the ideas spread back to your own society, you will also suffer the consequences of disappearing truth and increased distrust. Furthermore, because information warfare as other forms of non-military war are not decisive, it requires continuous attention and continuous efforts. Hence, even if the costs are relatively low, waging information warfare introduces constant costs and not only a constant sense of being beleaguered but also a constant sense of awareness against attempted disinformation.

Even if economic warfare can be effective, how efficient is it? The major drawback to a number of methods in economic warfare is that they also punish your own economy. A boycott of certain products from a country, for example, can incur everything from decreasing economic growth (in particular, if the boycotted products are important for your economy) to weakened consumption of other products. An embargo, meanwhile, could also imply a weaker economy since your foreign trade will diminish, unless you can find a new market for your products without lowering the prices. Other forms of trade interference – such as tolls or just introducing more cumbersome customs procedures – can also backfire since your opponent can respond tit for tat. Throughout the Ukraine war, Russia has used economic warfare and varied its approach, including not only trade interferences but also more open boycotts and embargoes.

The costs of economic warfare are also obvious insofar as Ukraine has responded in kind. When Russia already in 2014 introduced an embargo on diesel and gas to Ukraine, the latter introduced one on Russian weapons technology. This meant that parts of the Russian arms industry came to a standstill since it – already during Soviet times – was dependent on some parts that were manufactured in Ukrainian factories. Divorcing its arms industry from Russia, moreover, served other purposes for Ukraine. It could then start to import Western arms and in that strengthen the bonds to the West.

Still, Russian economic warfare also incurs indirect costs on the West. If we only approached victory from the perspective of end-states, we would fail to see that potentially the real problem of Russian (and Chinese) economic warfare is that they contribute to dissolving a firm difference between public and private. In a kleptocracy such as Russia, whomever that holds government can enrich himself and his cronies. Although the West views this form of enrichment as something on steroids, dissolving public and private interests presents a problem for the West since it is difficult to distinguish legitimate foreign investments from illegitimate ones.

Approaching political warfare from a cost-benefit perspective has similar consequences as other forms of non-military warfare, that is the costs are small, and therefore it is relatively easy to win political warfare. First, investments for conducting political warfare can vary between direct or indirect campaign funding for the preferred candidates, direct bribes, and setting up bot factories to manage and direct information flows in accounts on social media. Even if these are costly, they are still cheaper than equipping armed forces. Second, as in cyber warfare, there are problems of attribution involved, where states conducting political warfare can do it without being detected since the nature of political warfare is covert. Being open about the political pressure could be counterproductive as it easily could be construed as illegitimate outside influence on the sovereignty of the people voting. There is no deterrent effect in political warfare either since the costs of pursuing these ends are marginal and problems of attribution makes it difficult to attach costs.

Cost-benefit understandings also allow us to identify the costs of secrecy involved in political warfare. Since operations by definition need to be kept secret in sensitive things such as election tampering, it is difficult to demonstrate one's capabilities to ward off other potential challengers. Political warfare, thereby, also ramps up indirect costs of having to repeat behavior toward all other targets. Moreover, if made public, one can expect a stronger, more long-term backlash and increased suspicion from the target state. Although discord will certainly be sown within the target state if election tampering or other forms of political warfare were made public, the costs of that can be high in terms of other later attempts to influence the target.

Waging limited war has periodically been a central concern in Western strategic thought over the last few centuries (Heuser, 2010). It is easily compatible with an end-state understanding of victory, but it is not as easily compatible with the idea of war being decisive. When it comes to non-military war, the limited use of military force is a mirror image not only of these criticisms but also of the advantages of approaching limited war by using costs and benefits. If we were to understand limited use of military force only by its end-states, we would quickly realize that the results are transient. For example, Russia has with varying intensity and with varying troop commitments supported East-Ukrainian separatists since 2014. From special forces to regular armed units have fought a low-intensity, limited campaign for eight years without seemingly reaching the desired end-state of Donbass becoming part of Russia proper. If we instead approach limited use of military force with the concepts cost and benefit, we can more readily see that the benefits must be balanced against the costs. Hence, actually conquering Donbass from the Ukraine would have cost too much early on in the conflict, before more long-term benefits of weakening NATO and the European Union could be accomplished. Maintaining a low-level conflict in Ukraine also contributed to making it more difficult for Ukraine to integrate with the West – to a low cost.

BALANCING VICTORY AND DEFEAT IN NON-MILITARY WAR: FOREVER STRATEGY AND THE ACCUMULATION OF COSTS

The Prussian military theorist Clausewitz famously suggested that "in war, the results are never final" (1993). Although certainly true in regular war, it is an even more poignant observation for non-military warfare. As has been discussed throughout the analysis above, non-military warfare tends to be cheap, relatively easy, and effective, but for quite limited gains. Results are transient. This represents a fundamental challenge to Western strategic thought in several ways, but most fundamentally, it challenges the idea of *finality* or of warfare being decisive (cf. Lewis, 2010). As such, it fully embraces the notion that strategy is not preconditioned upon the distinction between war and peace. Strategy carries on regardless of whether there is war or peace. Instead of approaching strategy from the perspective of applying military means to reach a political goal and once that goal is reached, the war is over, strategy in non-military war even more so than war proper is about muddling through. As such, reiterating Lawrence Freedman, strategy is not about end-states but about the starting point and staying in the game for the next iteration. One iteration at a time. And since such iterations occur constantly and without recognizing a sharp difference between peace and war, strategy becomes ever-present.

Non-military warfare not only challenges Western notions of civil and military but also public and private (see Chapter 6 by Kollars and Poznansky). In a way, predominant configurations of these two conceptual pairings have been a precondition for well-functioning economies and democracies and thus been at the heart of Western strength the last two centuries. However, non-military warfare is the martial art developed to turn these strengths against the West itself. By spreading disinformation, Russia has forced Western societies to start to question the free exchange of ideas – so central to business, innovation, and landmark achievements in science, technology, and the humanities, but also of course these societies' capabilities to wage war. By creating discord, polarization and distrust of state institutions, combinations of cyber, information, and political warfare make the open society an enemy of itself. By not recognizing the distinction between public and private, economic warfare corrupts the free market and prevents it from distributing wealth efficiently, while at the same time changes our idea of the state. By not recognizing civil and military as distinct and mutually excluding categories, non-military warfare dissolves the boundaries between war and peace, thus making war a constant presence.

Even if these developments sound unsettling, we should not forget that target states also have a say in the matter. It is easy to make gloomy predictions about the end of democracy and the Western way of life when faced with non-military war. However, it is also the case that Western societies have proven resilient and although certainly useful, the tools of non-military war also carry substantial reputational and indirect political costs for the aggressor. The dissolution of truth, for example, also has an impact on political, economic, and social interactions in the aggressive state. Even Ukraine, hardly the most socially cohesive of states to begin with, and subject to non-military attacks with varying intensity since the early 2000s, did not crumble as Russia increased its pressures in late 2021. Hence, although some have understood the Russian escalation in February 2022 as a sign of increasing self-confidence and arrogance, it can be understood as an attempt to reach one's goals with overwhelming military power in frustration of non-military means not having reached these goals for the past decade. Neither of the non-military warfare operations fundamentally changed Ukraine's behavior 2014–2021: It tried to form a closer bond with NATO and the European Union throughout the period. Hence, full-scale war was perceived necessary. In isolation, Russian reputation costs among peers did not suffer massively 2014–2021 (sanctions were relatively mild and soon forgotten or partly disbanded), while the February 2022 invasion made Russia an international pariah in large parts of the world. Paradoxically, Western early sanctions could even have strengthened Russia's resolve to create an economy not dependent on the West, thus minimizing the effects of the later sanction regime.

Conclusions: Why Victory in Non-Military War Is Fluent and Short-Lived

If outcomes in non-military warfare are transient, is there a place for victory and defeat as meaningful analytical categories to describe its outcomes? The school of thought that is best equipped to deal with fluent outcomes is the cost-benefit understanding of victory and defeat. Transient outcomes do not make it easier to value costs and benefits, but it allows for a more nuanced understanding of victory and one that already from the start does not approach the concept binary. Non-military warfare, moreover, as Jack McDonald eloquently points out in his contribution to this volume (see Chapter 3) – still cannot escape the close links to war proper. This means that the link to politics is still prevalent and that politics, intended or unintended, therefore permeate also non-military warfare. Even if the fortunes of war also vary during a long, traditional war and even if the results are not final, non-military warfare suffers from being even more fickle and left to chance than events on the battlefields. It is precisely the political nature of war – military and non-military alike – that makes the cost-benefit understanding of victory more meaningful. Politics is about not only determining the aim of war but also addressing what costs are acceptable.

The cost-benefit understanding is also better equipped to deal with the particular dynamics of non-military war since it recognizes and allows for trade-offs between war aims. Instead of being fixed at measuring the achieved end-states, cost-benefit understandings can more easily accommodate changes and trade-offs between end-states. For example, if we are fixed upon certain outcomes, such as Russian enforced regime change in Kyiv, we would clearly judge the Russian invasion as a failure, but if we instead were to understand regime change as a preferred outcome if the costs are low, then we would start to realize that mounting costs can change the end-states.

The transient nature of outcomes also has implications for the strategy of non-military warfare. Most obviously, one needs to stop approaching strategy as if there is a final end-state to be reached and embrace the idea that strategy is equally much about creating force, than it is about using force. Striking a balance between staying in the game and preparing for the next iteration, rather than spending your force to reach an outcome is therefore the major role of the strategist in non-military war. Translating tactical victories into strategic victories is a major difficulty in war proper too, but in the case of non-military warfare, it is even more difficult. It is easy and relatively cheap to accomplish inroads, but achieving long-term systemic changes in political, economic, or social order is extremely difficult. Even after near-seismic events, such as introduction of democracy, revolutions, or the creation of the Internet, there are several possible contingencies for societies and resisting sweeping changes is always an option. Instead of pursuing a theory of immediate, decisive victory, strategy has become the art of muddling through in non-military warfare.

Given that outcomes are transient, we should also ask ourselves when the band starts playing, the parade starts and what conquered artifacts do you parade in victory? In regular interstate war, this is usually associated with cease-fires, peace treaties, and enemy prisoners or other embodiments of enemy strength such as flags. In non-military warfare, bands, parades, and artifacts are still relevant, but instead of being public manifestations of achievements, in which the achievements can become shared among the wider population, they essentially confirm that the costs have been acceptable for the polity. This means that they do not have to be public but instead manifest themselves in bureaucratic advancement within the state. This also means that a defeated party never has to admit or recognize that it has been defeated. Non-military war can therefore go on.

Note

1 In this chapter, I will focus more narrowly on victory and defeat in war, whereas, for example, Beatrice Heuser has correctly highlighted the overlap between some conceptions of peace and victory within the just war tradition. A central tenet of the latter is that a durable peace is not possible without an overlap of justice and force (see Heuser, 2013).

Bibliography

Angstrom, J., & Duyvesteyn, I. (Eds.). (2007). *Understanding victory and defeat in contemporary war*. Routledge.

Angstrom, J., & Widen, J. J. (2015). *Contemporary military theory: The dynamics of war*. Routledge.

Auerswald, D. P., & Saideman, S. M. (2014). *NATO in Afghanistan: Fighting together, fighting alone*. Princeton University Press.

Bakich, S. D. (2014). *Success and failure in limited wars: Information and strategy in the Korean, Vietnam, Persian Gulf and Iraq wars*. University of Chicago Press.

Betts, R. (2000). Is strategy an illusion? *International Security*, *25*(2), 5–50.

Bond, B. (1998). *The pursuit of victory: From Napoleon to Saddam Hussein*. Oxford University Press.

Brodie, B. (1959). *Strategy in the missile age*. Princeton University Press.

Echevarria, A. (2007). *Clausewitz and contemporary war*. Oxford University Press.

Fearon, J. D. (1995). Rationalist explanations for war. *International Organization*, *49*(3), 379–414.

Finnemore, M. (2003). *The purpose of intervention*. Cornell University Press.

Freedman, L. (2003). *The evolution of nuclear strategy* (3rd ed.). Palgrave Macmillan.

Freedman, L. (2004). *Deterrence*. Polity Press.

Freedman, L. (2013). *Strategy: A history*. Oxford University Press.

Galeotti, M. (2022). *The weaponisation of everything*. Yale University Press.

George, A., & Smoke, R. (1974). *Deterrence and American foreign policy*. Columbia University Press.

Gray, C. S. (1979). Nuclear strategy: The case for a theory of victory. *International Security*, *4*(1), 54–87.

Gray, C. S. (2002). *Defining and achieving decisive victory*. Strategic Studies Institute.

Gwynne, S. C. (2019). *Hymns of the republic: The story of the final year of the American Civil War*. Scribner.

Herberg-Rothe, A., Honig, J. W., & Moran, D. (Eds.). (2011). *Clausewitz: The state and war*. Franz Steiner Verlag.

Heuser, B. (2010). *The evolution of strategy: Thinking war from antiquity to the present*. Cambridge University Press.

Heuser, B. (2013). Victory, peace, and justice: The neglected trinity. *Joint Forces Quarterly*, *69*, 6–12.

Hobbs, R. W. (1979). *The myth of victory: What is victory in war?* Westview Press.

Hollis, M., & Smith, S. (1991). *Explaining and understanding international relations*. Oxford University Press.

Jervis, R. (1976). *Perception and misperception in international politics*. Princeton University Press.

Johnson, D., & Tierney, D. (2006). *Failing to win: Perceptions of victory and defeat in international politics*. Harvard University Press.

Jonsson, O. (2019). *The Russian understanding of war*. Georgetown University Press.

Kahn, H. (1960). *On thermonuclear war*. Princeton University Press.

Kaplan, F. (1983). *The wizards of Armageddon*. Stanford University Press.

Kolenda, C. D. (2021). *Zero-sum victory: What we're getting wrong about war*. University Press of Kentucky.

Lambeth, B. (2001). *NATO's air war in Kosovo: A strategic and operational assessment*. RAND.

Larsen, K. (2014). *Substitute for victory: Performance measurements in Vietnam, the Gulf War, and Afghanistan*. Københavns Universitet.

Lewis, J. D. (2010). *Nothing less than victory: Decisive wars and the lessons of history*. Princeton University Press.

Mandel, R. (2005). *The meaning of military victory*. Lynne Rienner.

Mandel, R. (2017). *Optimizing cyberdeterrence: A comprehensive strategy for preventing foreign cyberattacks*. Georgetown University Press.
Martel, W. C. (2007). *Victory in war: Foundations of modern military policy*. Cambridge University Press.
Maschmeyer, L. (2021). The subversive trilemma: Why cyber operations fall short of expectations. *International Security*, *46*(2), 51–90.
Morgan, P. M. (1983). *Deterrence: A conceptual analysis*. SAGE.
Morgan, P. M. (2010). *Deterrence now*. Cambridge University Press.
Mueller, J. (1995). The perfect enemy: Assessing the Gulf War. *Security Studies*, *5*(1), 77–117.
Pape, R. (2005). *Dying to win: The strategic logic of suicide terrorism*. Random House.
Parker, G. (2008). *The Cambridge illustrated history of warfare*. Cambridge University Press.
Pillar, P. R. (1983). *Negotiating peace: War termination as a bargaining process*. Princeton University Press.
Powell, R. (2004). Bargaining and learning while fighting. *American Journal of Political Science*, *48*(2), 344–361.
Reiter, D. (2009). *How wars end*. Princeton University Press.
Ricks, T. E. (2006). *Fiasco: The American military adventure in Iraq*. Penguin.
Rid, T. (2020). *Active measures: The secret history of disinformation and political warfare*. Farrar, Straus & Giroux.
Schelling, T. C. (1966). *Arms and influence*. Yale University Press.
Simon, H. (1955). A behavioral model of rational choice. *The Quarterly Journal of Economics*, *69*(1), 99–118.
Stone, J. (2011). *Military strategy: The politics and technique of war*. Continuum.
Strachan, H. (2008). *Clausewitz's on war: A biography*. Grove Press.
Sullivan, P. (2012). *Who wins? Predicting strategic success and failure in armed conflict*. Oxford University Press.
Till, G. (2009). *Seapower: A guide for the twenty-first century* (2nd ed.). Frank Cass.
Valeriano, B., Jensen, B., & Maness, R. C. (2018). *Cyber strategy: The evolving character of power and coercion*. Oxford University Press.
von Clausewitz, C. (1993). *On war* (M. Howard & P. Paret, Trans.). Everyman's Library.
Wilson, A. (2005). *Virtual politics: Faking democracy in the post-Soviet world*. Yale University Press.

3 Who Are Participants in Non-Military Warfare?

Jack McDonald

Introduction

Non-military warfare, defined in this volume as *non-kinetic tools of statecraft employed as means of strategic coercion and effect employed with an intensity and purpose such that the target perceives them as significant escalations and/or commensurable to warfare*,[1] covers a significant range of activities, from subversion and information warfare, through to economic warfare. This wide variety of hostile activities are linked in that they do not include the use of physical force and violence and are implicitly hard to define as warfare itself.[2]

How can, and should, states respond to non-military warfare? States have recently struggled to respond to hybrid warfare, "best described as covert or deniable activities, supported by conventional or nuclear forces, to influence the domestic politics of target countries."[3] In particular, these activities exploit "domestic weaknesses via non-military means (such as political, informational, and economic intimidation and manipulation)" while backed by the threat of military force.[4] As a concept, non-military warfare recognizes that spheres of everyday civilian activity can be transformed into tools of statecraft. In this sense, we are concerned with activities that might be included in hybrid warfare (or beyond its scope) absent the specific threat of military force. In this regard, non-military warfare examines the strategic use of civilian means to achieve strategic effects and/or to coerce states. It therefore complicates the usual division of politics into civilian and military spheres.[5] Though these have historically been unstable categories, they currently reflect the evolution of international thought regarding war and peace in international politics.[6]

Responding to non-military warfare therefore entails securitizing elements of civilian life, and using legal or extra-legal means to counter hostile statecraft. For this reason, responses to non-military warfare require careful calibration. Any response is likely to have knock-on effects for a state's political climate and economy. Depending on the form of response, overreactions could harm a state's own economy by driving investment and businesses away, or threaten its own values by chilling free speech.

Even though the concept is hard to define, responding to non-military warfare in a proportionate way is therefore important. Proportionate responses are likely to be directed at those involved in non-military warfare, either as participants or as

DOI: 10.4324/9781003616993-3

state-sponsors. Delineating between those whose participation legitimizes action against them, and those whose participation does not will frame any response to non-military warfare. This chapter addresses the issue of how to conceptualize participation in non-military warfare. In particular, this chapter highlights the problems states face in distinguishing between the large number of individuals and organizations that could be considered participants in one way or another, and the smaller number of each that are sufficiently responsible participants to warrant acting against them.

Defining participation is a precondition for distinguishing between those that can, and cannot, be harmed or targeted for their activities. The principle of distinction underpins the law of armed conflict, and many of the contemporary issues with this principle arise from civilians and non-combatants contributing to military operations in direct or indirect ways.[7] Still, as noted above, we cannot directly transpose concepts that regulate war to non-military warfare, due to the latter's primarily civilian character.

As such, one of the problems that follows from the concept of non-military warfare is defining what counts as meaningful participation in such activities. Below the level of the state directing the activities, participation might mean owning or financing a company that is used to conduct election interference, or information warfare. A good example of this is the United States indicting Yevgeny Prigozhin,[8] a Russian oligarch, for funding the Internet Research Agency, which interfered in the 2016 U.S. presidential election.[9] It might equally be working for an organization – be it a formal company or informal hacking group – that conducts operations like cyberattacks.[10] Importantly, many kinds of non-military warfare involve using individuals or groups to achieve an operational goal, such as promoting disinformation. If we were to consider the people circulating disinformation produced by a disinformation campaign, we arrive at a far wider number of participants.

The range of potential activities that could fall under the heading of non-military warfare means that the scope of participation is wider than any existing definition of direct participation in (military) warfare. Furthermore, this wide scope implies a substantial challenge for conceptualizing any single definition of participation, as it would have to cover a huge range of different everyday means, methods, and modes of participation.

The policy problem that this raises is that state responses to non-military warfare require them to identify individuals and organizations as participants prior to distinguishing between those whose actions justify some form of hostile state response and those that do not. Furthermore, what kind of response does an individual or organization's participation justify? Since non-military warfare is often conducted by non-military organizations and persons, this is a sensitive task. It is also one that could open a Pandora's box of tit-for-tat responses between states. For example, states routinely kick out intelligence officers of rival states during diplomatic disputes.[11] But the involvement of private companies complicates matters. State responses in the form of sanctions can be selective. For example, the United States also sanctioned Yevgeny Prigozhin in 2019 for his relationship with a variety of companies, famously the Wagner group that operates as a quasi-mercenary force for the Russian state, alongside the Internet Research Agency that was blamed for enabling Russian interference in U.S. elections.[12] But sanctions or legal tools need not be restricted to owners, executives, and corporations – Russia has opened up criminal cases against the employees of private companies such as Meta for complying with U.S. sanctions.[13]

What does it mean to say that an owner, executive, corporation, or employee is participating in non-military warfare? This chapter provides a framework for evaluating participation in non-military warfare. It does so by drawing upon existing concepts from legal and moral debates about the loss of non-combatant immunity to explain the implications and limitations of using such concepts to evaluate non-military warfare.[14] The policy implications of this are twofold. First, it provides a means of thinking through the problem of identifying participants in non-military warfare. Second, states will need to be selective about which participants they are willing to harm in response to non-military warfare. This is because the set of individuals and entities that could be considered participants is likely to be very wide, and the consequences of taking action against an individual or entity can therefore vary significantly. Nonetheless, if states are to develop countermeasures to non-military warfare, they will need processes and policies to identify participants and then select those it is both practical and legitimate to act against.

This chapter first outlines some of the key issues of defining non-military warfare that are relevant to defining participation in such activities. It then examines how organizations and individuals can be conceptualized as participating in non-military warfare by examining different ways of analyzing participation in warfare, such as organizational ties, individual acts of participation, and continuous functional participation.[15] I then highlight the key problem of identifying individual participants in warfare, which is particularly pertinent in domains. such as cyber warfare and cybersecurity, more generally. Finally, this chapter examines what can be done to participants in non-military warfare, given the non-military nature of their activities, and highlights the key issue of how states need to collaboratively define arbitrary rules to define the limits of legitimate responses to participation in non-military warfare.

Defining Non-Military Warfare

Warfare fundamentally refers to the organized use of physical violence for political purposes. As an activity, it is characterized by the myriad roles that members of professional militaries and organized armed groups fulfill in order to conduct military operations. Whereas the international system features basic shared conceptual and legal definitions of war and warfare, non-military warfare – "non-kinetic tools of statecraft employed as means of strategic coercion and effect employed with an intensity and purpose such that the target perceives them as significant escalations and/or commensurable to warfare" – is largely a matter of state interpretation at this point. This means that state arguments that (for example) a set of economic sanctions are akin to warfare, raise the prospect of retaliatory behavior, or even the resort to force. The implication of equating non-military warfare to warfare is therefore that states might seek to justify the use of force either in response to non-military warfare or in response to a state's non-violent countermeasures for non-military warfare.

Non-military warfare lacks the central cohesion of military operations as an organizing fulcrum for organizational activity. This is because many different forms of strategic coercion and effect fall under the umbrella of non-military warfare. As outlined in the introduction to this volume, this ambiguity and lack of cohesion is a principal reason for engaging with the concept of non-military warfare – despite the lack of conceptual clarity, many different forms of state activity, or state-directed activity, appear to form a cluster of statecraft tools that are increasingly used to achieve strategic effects that might have once required military intervention and operations to achieve.

The second issue with defining non-military warfare is that many of the activities are continuous with "peaceful" tools of statecraft. This is not so different from "normal" war and warfare as one might initially consider. As Quincy Wright noted in his landmark *A Study of War*, the intense military activity of battles and campaigns within war are to a certain extent continuous with the arms races and preparations for war and normal levels of military activity.[16] Wright's study, which precedes the formation of disciplines, such as security studies in the mid-late 20th century, contained a looser definition of war than one might find in contemporary international relations – he pointed out that from the perspective of belligerent powers, war meant "an extreme intensification of military activity, psychological tension, legal power, and social integration."[17] From this perspective, non-military warfare bears many of Wright's hallmarks – it is characterized by intense psychological tension (the notion of mutual enmity or antagonism between parties), abnormal legal power (often the use of expansive legal powers under national security laws, or extra-legal means), and what Wright called "intense political integration" – abnormal direction of civilian areas of life by the government to achieve its aims. There are many areas where non-military warfare is discontinuous with Wright's ideas. For instance, Wright held that these conditions were likely to occur where material power was relatively balanced between two sides,[18] whereas non-military warfare appears to be a means by which states can overcome imbalances of material and military capability. That said, in my view, the general point remains the same: non-military warfare is in part defined by its relative intensity and abnormality against a standard of normal peaceful competition and interaction between states.

Still, as other chapters in this volume point out, the character of non-military warfare is markedly different between tools and contexts. What states can do with cyber power differs from what they can achieve with economic statecraft, or information warfare during election cycles. Fundamentally, we must recognize that non-military warfare – in this volume's framing – covers operations and campaigns on different time-scales. Specific instances of election interference, political warfare, or even sanctions and cyberattacks sit alongside long-term diplomatic, economic and technological competition, and battles for narratives.[19] For present purposes, it is more important to focus upon differentiating non-military warfare using a contextual approach, rather than a tool-centric approach.

Political context is a useful heuristic to differentiate between what is "normal" and where the intense use of tools of statecraft might be discernable or definable as non-military warfare. I use four contexts as a guide for working through this issue: everyday competition, antagonistic competition, crisis, and armed conflict. By "everyday competition," we recognize that "peace" in international relations is characterized by states engaging in a wide variety of activities that could be construed as hostile – intelligence collection (including computer network exploitation), economic competition, and political signaling over salient claims. I distinguish this from antagonistic competition in that states engaged in long-term antagonistic competition tend to focus their resources upon their long-term competitors and engage in the activities that this volume terms non-military warfare as a core component of their everyday interactions with their key competitors.[20] This long-term everyday hostility can be distinguished from crisis periods, what Michael Brecher and Jonathan Wilkenfeld refer to as a "foreign policy crisis," involving a perceived "threat to one or more basic values, along with an awareness of finite time for response to the value threat, and a heightened probability of involvement in military hostilities."[21] These are situations where states engage in intense campaigns of political and military threats and coercion that are markedly heightened from both everyday competition and antagonistic competition, risking escalation into armed conflict. Finally, armed conflict is a key context for non-military warfare in that these means can be used alongside military operations to achieve political and strategic goals.

For present purposes, the importance of these distinctions is that a person or organization might be participating in non-military warfare inside and outside armed conflict and may be doing so with or without a connection to a specific political crisis. The importance of this distinction is that the further we move from a salient armed conflict or political crisis, the harder it becomes to justify coercive measures aimed at participants on the basis of national security concerns. Furthermore, this frames the key policy problem of responding to non-military warfare: it might be easy to identify state-based hostile activity as non-military warfare in the context of armed conflict, but state characterizations of intense statecraft as warfare outside the context of an armed conflict or crisis are likely to face pushback from other states and international civil society. Since responses to non-military warfare are likely to rely upon coalitions of like-minded states, or promoting specific norms of behavior in international society, this is a key problem. Defining a given activity as non-military warfare is a form of securitization, and therefore we should expect the efforts of states to define activities as non-military warfare to be contested by other stakeholders through acts of resistance to these labels.[22] At the domestic level, the social, political, and economic linkages inherent in a globalized world mean that responses to non-military warfare are also likely to have knock-on effects for diaspora or migrant communities that share some cultural tie to those targeted in response to non-military warfare.

Defining Participation in Non-Military Warfare

In order to respond to non-military warfare, states need to be able to define a class of activity as non-military warfare, identify individuals and entities participating in that activity, and then identify the subset of participants whose actions have made them liable to be harmed in response. This effort of distinction is important because it is a precondition for any organized response to non-military warfare that goes beyond purely defensive actions. The main focus of this chapter is how states and scholars approach the issue of distinction between individuals and entities that may be harmed in war, and those who for whatever reason should not be harmed.

Typically, scholars differentiate between the legitimacy of harm caused in self-defense during armed conflict and the legitimacy of harm caused due to a person's role or actions in an armed conflict.[23] If states perceive non-military warfare as commensurable to warfare due to its potential effects, then the question of what harm can be legitimately inflicted upon persons and organizations due to their participation in non-military warfare is a central issue. Prior to considering whether it would be legitimate to respond to economic sanctions or election interference with physical force, it is first necessary to outline the discussions and debates about participation in warfare, before moving to outline how these might inform our understanding of participation in non-military warfare.

What Constitutes Participation in War

The two main frameworks by which people assess the legitimacy of force in international politics, international law and just war theory, both differentiate between the legitimacy of resorting to force (*jus ad bellum*) and the legitimate use of force within an armed conflict (*jus in bello*).[24] Here I constrain my analysis to *jus in bello* principles, for the reason that these are where participation matters as a reason for harming someone, or protecting them from harm.

The distinction made between persons (or objects) that should be protected from harm in war and persons (or objects) that are permissible targets of attack in war goes to the heart of the legal and moral frameworks that constitute the international regulation of warfare. Typically, the legitimacy of harm is made with reference to three factors – whether a person may be permissibly harmed either directly or indirectly, whether it is proportional to do them harm (and the proportionate level of harm that may be inflicted upon them), and whether it is necessary to cause them harm. Military necessity is increasingly balanced against the principle of humanity[25] – taking account of the fact that the harms of war should be minimized where possible and that it is impermissible to harm people in some inhumane ways.[26]

Participation in war matters because it generally legitimizes other participants in war to harm a person or object. While individuals and objects can be harmed in self-defense, conceptually the rationale for harming individuals and objects is their relationship to, or use to, an armed organization (either a state's military or an organized armed group). Typically, individuals or groups can become participants due to their status (e.g., membership of armed forces) or their conduct (e.g., civilians picking up weapons and attacking combatants). Parties to an armed conflict are meant to restrict their direct attacks to one another, prohibiting direct attacks on non-participants. Given that war necessarily involves physical violence, participation in armed conflict therefore exposes an individual to lawful premeditated killing. Given the potentially extreme consequences of participation in armed conflict, there is tension between those who seek to regulate warfare to enable its conduct and those who seek to regulate warfare to reduce the harms of warfare itself.

So, who can be harmed in war? International humanitarian law seeks to balance military necessity and the principle of humanity. For this reason, it is a permissive legal framework where many forms of violence are lawful so long as they are necessary to achieve lawful objectives, but it also constrains parties to an armed conflict in order to reduce and ameliorate the harms of war. There are some generally non-controversial areas of agreement. Armed forces consist of military personnel classed as combatants – therefore subject to attack owing to this status – as well as non-combatants (e.g., medics) who are protected from attacks so long as they perform their lawful role and do not engage in combat. Civilians are similarly protected against premeditated attack, unless they take up arms by joining an organized armed group, or similar.[27]

Importantly, there is a difference between active participation in armed conflict and passive participation. International law, for example, recognizes that civilians are protected unless their actions make them liable to attack.[28] In a war between two states, each will be fighting using militaries that are supported by taxation paid by their respective citizens. At the same time, this form of participating and enabling warfare is generally held to be too insignificant to warrant legitimizing physical force against taxpayers.[29] Therefore, when considering participation in war, we are generally examining what active forms of participation and occupation render a person liable to attack. Higher thresholds of participation are required to legitimize lethal force than the relatively low bar of paying taxes.

Non-participants in armed conflicts can still be legitimately harmed, but the point of distinction is that they cannot be made the direct focus or targets of attack. A military can legitimately attack a military target knowing that it will cause significant harm to non-participants (civilians) nearby, so long as the focus of the attack is not upon the civilians, and the harm inflicted upon them is proportional to the military advantage that is perceived to be gained by the attack.[30]

But what about civilians that actively participate in armed conflicts? The history of warfare is replete with this problem, from the *levée en masse* of 19th-century wars, through to insurgent fighters in contemporary conflicts.[31] This problem is exacerbated by the fact that civilians might be sporadically engaged in a conflict, sometimes active participants, sometimes bystanders. This is the most pertinent issue for the consideration of non-military warfare, since one of the central problems of non-military warfare is that it can be conducted by civilians who are, it seems, protected from direct attacks, or might only engage in operations sporadically.

International humanitarian law contains the concept of "direct participation in hostilities" (DPH) in non-international armed conflicts. This holds that civilians are liable to direct attack "unless and for such time as they take a direct part in hostilities." This term is undefined in international law, and subsequently, there has been much debate regarding the scope of this term.[32] Opinions differ regarding the kinds of actions that result in civilians losing their protection against attack, as well as the point at which this loss of protection starts and ends.[33] The International Committee of the Red Cross (ICRC), the humanitarian organization that has played a huge role in the development of international humanitarian law, has a restrictive view that hinges upon the fact that a "revolving door" notion of protection applies – that a farmer can pick up a gun and participate in an armed conflict, but that they regain the protection of their civilian status at the point that they put the weapon down and cease to participate in fighting.[34] More expansive views, such as those held by the United States, hold that a person counts as a participant for so long as they remain engaged in a conflict through their repeated activity, rejecting the notion of revolving door protections.[35]

Key to this debate, however, is the notion of hostilities itself – the actions and activities that constitute warfare. As the conduct and character of war changes, so, too, do the actions that constitute DPH. The question, then, is, "What use can we make of these concepts to define participation in non-military warfare, and differentiate between levels of involvement in non-military warfare?"

What Constitutes Participation in Non-Military Warfare?

The concept of distinction is a useful basis for thinking through the construction of categories of participation in non-military warfare. The notion of participation in non-military warfare is necessarily built atop the existing conceptual architecture of international politics, such as international law. The question, therefore, is first how we might distinguish someone as a participant in non-military warfare in the same manner that someone might be distinguished as a combatant or non-combatant.

One aspect of international humanitarian law that is unlikely to be much help in this regard is status-based accounts of participation. The notion of combatant in international humanitarian law reflects centuries of convergence and debate between states as to the character of legitimate and authorized involvement in warfare. While in the long term such convergence may happen regarding non-military warfare, it is unlikely to occur in the short term. That said, the fact that a person is a member of a state institution, or a third-party contractor, is likely to matter in the calculation of a state's response to what it perceives as non-military warfare. For example, imprisoning a state's citizens on spurious charges (in what is known as hostage diplomacy) is likely to be less escalatory than the detention of state officials (such as diplomats).

Here I argue for a four part test, similar to that of DPH, but dissimilar due to the varied contexts in which non-military warfare can occur, or state activities can be defined as such. A useful test for defining a person as participating in non-military warfare is:

- They are directly and actively engaged in non-military warfare operations
- Their actions are connected to an organized campaign of strategic effect
- That campaign has a state nexus
- The campaign is abnormal, according to the context

One reason to adopt these tests is that they enable us to differentiate between the large number of persons that can be defined as participants in non-military warfare from the smaller group of people that are responsible for such activities. Such distinction is necessary since we are likely distinguishing between two different classes of civilians (be they individuals, organizations, or objects). A second reason is that we may not be able to distinguish based upon individual or organizational actions. When considering something as tricky as disinformation, for example, a large number of people might be involved in propagating disinformation without being part of any campaign, even while a small number of people or organizations may be dedicated to promoting the disinformation that is circulating. Depending upon the activity itself, there may be a far greater set of persons engaged in activities that are adjacent to, or contribute to, non-military warfare who are in a sense innocent. In practical terms, it is likely to be a waste of time to target such people. In terms of legitimacy, it is likely to be disproportionate to do so.

Active and Direct Participation in Non-Military Warfare Operations

Operations refers to the organized activity required to achieve a specific form of strategic effect. While political warfare or economic warfare might both be used for coercive effect as a threat, or weaken an adversary in practice, they can be distinguished on the basis of the activities required to successfully undertake them. A wide understanding of participation would include anyone connected to an organization that was undertaking non-military warfare operations and anyone that might be used by them to achieve their goals.

Ideally, selectively targeting specific individuals would be the ideal allocation of resources and is easiest to justify. For example, we can usefully distinguish between a cybersecurity professional whose day-to-day role is maintaining the tools required to break into computer networks from administrative personnel that might be responsible for paying their salary. One is central to the organization's ability to sustain operations, and the latter is likely replaceable.

In practice, an individual's participation in an organization is likely to be a more efficient means of identifying those whose actions justify targeted responses. Membership of a ransomware group, or similar organization, even if only in a supporting role, means that an individual is an active and direct participant in the group's operations.

In this sense, distinguishing between those who know their actions are likely to be part of, or support, an operation and those who lack such knowledge is important. We might consider an administrative figure a participant, but a minor contractor a non-participant unless they continue to support a group after being supplied with information about that group's activities. Even then, ancillary individuals and organizations are likely replaceable and therefore might not be worth targeting.

Given that many forms of non-military warfare now involve the manipulation of social networks, it makes sense to distinguish between active participants that are directly responsible for disseminating disinformation and individuals that might wittingly or unwittingly repeat disinformation. In the case of disinformation, this distinguishes, say, the individual responsible for creating and disseminating disinformation, or paid to run accounts on a "troll farm," from, say, ultranationalist civilians who willingly repeat that same message on an internet platform. To do otherwise (in the case of disinformation) would run the risk of widespread infringements of free speech rights. More widely, non-military warfare usually plays upon pre-existing political divisions within societies, and policy responses to non-military warfare are unlikely to change or heal those divisions.

Third, using the term "direct" to denote participation here eliminates swathes of persons, and organizations that might be unwillingly made accessories to such activities. Non-military warfare mediated by digital networks requires the transmission of information via neutral third party computer networks and infrastructural layers of the internet. To include these in the concept of participation would, in my view, make the concept of participation automatically scale to encompass the internet, and again, render it useless.

Actions Are Connected to an Organized Campaign of Strategic Effect

The second criterion is that a person's activities or role are connected to an organized campaign of strategic effect. This places three further constraints on defining a person as participating in non-military warfare.

First, the notion of "campaign" implies an identifiable and coherent sequence of activities that are sustained over time. In this sense, while a person is working on that campaign, they can be considered a participant. The problem of temporal "stickiness," as described above in the debates about continuous participation and DPH presents itself here.[36] For example, if someone logs off for the day, are they no longer a participant? In legal debates regarding the notion of DPH, some argue that direct participation ends when an individual's specific actions cease (e.g., an insurgent sympathizer puts down a gun and return to farming), others argue that for as long as a person retains something akin to an active organizational tie (e.g., joins an insurgent movement and does not then actively distance themself from it), they have a "continuous combat function."[37] This latter concept, while contested in international law, provides a good explanation of how and why a person should be considered a participant in non-military warfare due to their pattern of actions and activities tied to a specific campaign.

Second, the notion of an organizational nexus is important. The damaging/dangerous forms of non-military warfare are in many senses the exact same activities that can be performed by "pick up groups" of like-minded activists. To make non-military warfare meaningful, an organizational nexus – even if the organization is only inferred from a standard pattern of activity – is important. In this sense, we seek to distinguish, say, a state-backed group of computer hackers from non-state activists that occasionally manage to deface web pages.

Third, the campaign is classified as a campaign of strategic effect. Following from the definition of non-military warfare above, the use of strategic effect here encompasses both coercive campaigns and those that are used to weaken an adversary without the clear signaling that is usually associated with coercion.[38] The intention is to distinguish between low-level signaling behavior and the kinds of campaigns that seek to coercively alter state behavior or capability.

The Campaign Has a State Nexus

The problem of non-military warfare is states leveraging a range of non-military operations to achieve strategic effects. However, a range of groups and organizations in domestic and international politics use some of these means to achieve their own goals. In effect, focusing upon the actions and operations associated with non-military warfare without considering their relationship to a state would draw in a wider variety of political groups and non-state actors.[39]

Using a test of whether the campaign has a state nexus is a test that sharply reduces the number of potential persons and organizations involved. In essence: is a person or group's activities being directed by a state or its direct affiliates? A second variant of this test provides a useful counterpoint. A repressive state is likely to ask: does a person or group's activities benefit another state?

By focusing upon direction or immediate affiliation, we include the substantial majority of individuals and organizations involved in the direction and operation of non-military warfare campaigns. At the same time, we exclude social movements that might use the same means or methods, and domestic political agents that are likely to take advantage of such campaigns. This avoids a number of sensitive political issues. A good example of this is that we would be focused on the individuals and groups conducting hack-and-leak operations rather than those political parties or social movements that take advantage or benefit from them. Of course, the distinction between the two can be complicated by collusion, most notably that of the Republican political operative Roger Stone with WikiLeaks, prior to WikiLeaks releasing hacked emails and disinformation to interfere with the 2016 U.S. Presidential election.[40]

In doing so, we exclude a number of individuals and groups that are an important element of non-military warfare, but we do so because the alternative is unworkable in democratic states. In particular, evaluating whether or not a domestic political group or social movement is engaged in non-military warfare, because it is involved in such operations without knowledge of a state's direction is a recipe for repression. For disinformation campaigns, it would mean categorizing all kinds of domestic political groups, media organizations, or similar, as engaged in non-military warfare for repeating information promulgated online.

Typically, this criterion is likely to draw upon a society's definition of what constitutes an agent of a foreign power. It means living with what are colloquially referred to as "fellow travelers" – ideologues unconnected to a foreign state's government. Maintaining such a distinction is important, because widening the scope of non-military warfare to include domestic agents, or unconnected political agents in international politics, would unnecessarily include a large number of persons and organizations that are, fundamentally, not the primary problem.

The Campaign Is Abnormal, According to the Context

Following from the contextual approach outlined at the start of this chapter, the final criterion for determining whether a person is participating in non-military warfare is to consider the campaign in its political context.

In this sense, the approach taken here is to constrain the notion of non-military warfare to a time-limited period of high intensity and to distinguish it from background activities that are part and parcel of statecraft. We might say that a civilian fishing vessel fishing in disputed waters is fundamentally not an act of non-military warfare, but a group of fishing vessels swarming into a disputed zone to block the fishing activities of others (or prevent the passage of naval vessels more generally) in the context of a political crisis is.

A second reason for such a criterion is that it requires consideration of baseline activities, which are context specific. For example, we might expect a certain level of hostile public messaging during periods of intense competition, crisis, or armed conflict. But organized campaigns of disinformation might be considered abnormal, depending upon the states involved.

Responding to Non-Military Warfare

How can, or should, states respond to non-military warfare? For states to respond to non-military warfare, they need to be able to identify participants and also have effective policy tools to use against them. The remainder of this chapter seeks to address the issue of state responses to non-military warfare by highlighting the range of issues involved in identifying participants in non-military warfare and suggests a series of criterion for calibrating state responses to participation in non-military warfare.

One of the central problems that non-military warfare poses to states is that it is often difficult to identify participant groups and individuals and also to attribute their actions to state-directed campaigns. As discussed above, many of the individuals and organizations involved in the conduct of non-military warfare are civilians. Importantly, these individuals and organizations do not necessarily need to be situated within traditional state bureaucracies. Indeed, something that is often observed is the use by states of third parties and private contractors to achieve strategic aims using non-military means. This outsourcing appears to be one of the key features of successful non-military warfare.

The issue of participation in warfare is inseparable from the consequences of participation. This goes to the central pillar of the regulation of war itself: who (or what) can be harmed, and who (or what) must be protected from harm during armed conflict? However, while some form of responsibility via participation is necessary to legitimize a response, the actual form and limits of the response are likely a function of the threat posed by an individual or their group's actions. For example, while we might imagine a state turning to extreme, perhaps even violent, responses to a campaign of sustained and effective cyberattacks on critical national infrastructure and responses to technological espionage are likely to be less extreme. Therefore, these theoretical questions must, in the real world, be turned into practical guidance for policymakers.

Participation is, therefore, more than a conceptual issue. Concepts of participation in warfare delineate who can, or cannot, be directly or indirectly harmed in war, yet there are substantial practical issues associated with identifying participants in war that cannot be wished away. In analyzing participation in contemporary warfare, analysis tends to focus upon edge cases that raise important substantive questions. Academics spend their time evaluating the margins (whether and when, e.g., performing a specific support function for a terrorist group renders an individual liable to attack) than substantively settled issues (whether or not a member of a non-state armed group can be targeted during an insurgent attack). This is because these edge cases test the conceptual boundaries of concepts such as "direct participation in hostilities" and similar concepts that are core to the principle of distinction in war.

The conceptual and practical problems of defining, or identifying, participation in war structure create practical dilemmas that are addressed by policy choices and decisions. In order for participation to be treated in a systematic way, states and their armed forces make both general and specific policy choices, ranging from state-level interpretations of relevant legal rules, through to targeting directives that may explicitly render some targets as protected.

Policy choices about determining participation in war are both bounded and arbitrary. Not only they are bounded in that they draw from shared international frameworks for regulating war and warfare, but they are also arbitrary in that they incorporate a range of factors in every decision that cannot be justified from the first principles of, say, law. They are informed by professional military experience, as well as political conditions, as much by the neatness of conceptual categories derived from international law. A good example of this is the way in which military professionals and military lawyers have worked to translate international law concepts and rules into explicit guidance related to cyberattacks in the Tallinn Manual processes.

In terms of policy responses, there are therefore two issues that differentiate participation in warfare from participation in non-military warfare. First, policy choices related to non-military warfare are relatively unbounded. They are of course shaped by international law, but there are no comparable specific and explicit frameworks that address participation in non-military warfare to the same extent that international humanitarian law addresses participation in armed conflict. Second, there are no agreed international baselines for conceptualizing non-military warfare, or its constituent activities. Therefore, the range of activities that could result in someone being identified as a participant in non-military warfare is going to be arbitrary.

Identification: How Can a State Know Someone Is a Participant?

Identification problems are likely to constrain the ability of states to understand exactly who is operating against them using non-military warfare. States and their agents must usually infer from secondary evidence and intelligence whether or not a person or organization is a participant in non-military warfare. For example, the complicated ties between Yevgeny Prigozhin and the Russian government have been explained by both the U.S. Treasury Department[41] and the open-source investigative organization Bellingcat,[42] yet we must assume that the U.S. government's understanding of his relationship with the Russian government also relies upon non-public intelligence.

Following our criteria above, states seek to identify individuals or organizations that are responsible for non-military warfare campaigns with a state nexus. There has been considerable work done on attribution in recent years, notably in connection to cyberattacks[43] and deterring or countering hybrid warfare.[44] Here we are primarily concerned with internal identification and attribution processes rather than the public attribution of non-military warfare operations or campaigns.

As in normal military operations, the form of non-military warfare will fundamentally enable or constrain the ability of states to identify individual participants. Again, we can look to the law of armed conflict for the multiple challenges facing military commanders in terms of distinction. For some forms of non-military warfare (say, cyberattacks and information warfare), it is likely that states will first have to identify the organizations conducting such campaigns and then identify individual participants based upon their links to these organizations. Other forms, for example, swarming fishing vessels, are in theory more easy to observe. In both cases, however, individuation – separating an organization into targetable individuals – is a resource-intensive task.[45] For this reason, there may be no reasonable way for a state to understand the precise degree of participation of a person, except in a backwards-looking manner as intelligence is collected and analyzed.

These are not new problems – they would be familiar to any military commander facing an insurgency or rebellion. Absent self-identification (uniforms), identifying many forms of civilian participation in irregular warfare requires substantial surveillance and intelligence work. At the same time, there are specific activities that constitute hostilities which are reasonably identifiable as such, and commanders have a relatively high degree of latitude in determining if someone poses a physical threat. Government departments and civil servants tasked with responding to non-military warfare, therefore, have less established precedent and organizational knowledge to guide them in their decisions.

In a practical sense, it is not enough to ask "who is a participant," instead it is "who can be identified as a participant, and what can be done to them?" Whereas military commanders have the option of using physical force, this is likely to be unavailable in many situations of non-military warfare without significant escalation problems, and backlash. Furthermore, in non-military warfare, following from the participation criterion above, states are seeking to differentiate between different degrees of civilian participation, rather than between combatants and civilians.

A central problem with non-military warfare is that a large variety of the activities that constitute it are intangible. In armed conflict, any military operation is a combination of intangible work (e.g., communication and information processing) and tangible work (e.g., physically fighting and killing people). The balance of the two in non-military warfare appears tilted toward intangible work. Therefore, the observability of individual contributions is likely to be low.

For this reason, it is far likelier that states will be able to identify states or groups engaged in non-military warfare, rather than be able to identify participants on an individual-by-individual basis.

Legitimate Harm: What Can Be Done to Participants in Non-Military Warfare?

The second problem in responding to non-military warfare is the question of what kinds of legitimate harm may be inflicted upon participants. Again, armed conflict legitimates the use of lethal force, perhaps the clearest example of harm, against individuals that participate in it. From the above analysis, the large majority of participants in non-military warfare are likely to be civilians. What, then, can states do in response to their activities?

First, it is necessary to consider that the current architecture of international law permits states to resort to the use of force in self-defense, or when the United Nations Security Council (UNSC) authorizes the use of force.[46] Given that the main states engaged in non-military warfare are permanent members of the UN security council, we can safely park the question of UNSC authorizations for the use of force.

In this case, lethal force is therefore only permitted in individual or collective self-defense, which would require non-military warfare to rise to the level of an armed attack. Some have argued, and states have claimed, that some forms of non-military warfare – e.g., cyberattacks on critical national infrastructure – could rise to this threshold.[47] The 2007 cyberattacks on Estonia that led to widespread disruption of government and banking services drove a significant amount of research and policy work into how to translate cyberattacks (in particular) into existing categories of international law.[48] This has led to intergovernmental processes such as the Tallinn Manual process to identify areas of agreement regarding international law[49] and policy statements by major states such as the United Kingdom on how it views the application of international law to cyberattacks.[50]

In the contextual approach outlined above, physical force is only likely to be a legitimate option where something like a crippling cyberattack occurs in any context, or where non-military warfare activities can be construed as DPH during an ongoing armed conflict. This, however, means that it would be illegitimate to respond to most of what we might consider to be non-military warfare with physical force.

Furthermore, outside the context of armed conflict, human rights law applies.[51] Therefore, states are bound by their human rights obligations in their responses to non-military warfare.

The concept of participation fits into this, because states can either respond to non-military warfare openly or in secret. Typically, secret responses are usually visible to some states, although not publicly acknowledged.[52] For open responses, however, states need to construct a public narrative for their actions that justifies the measures that they take. Such narratives are open to interpretation by the public,[53] but official articulations and justifications are central elements of national security policymaking.[54]

For public responses, the question of "who is a participant in non-military warfare?" is likely to be replaced with "who can be blamed for non-military warfare?" Responsibility for non-military warfare can be leveled at states, organizations, and individuals. The attribution of responsibility to states for the actions of organizations operating on their behalf is a particularly important area of contemporary research in international relations. In cybersecurity, the "attribution problem" is a fundamental issue, because states may have to infer responsibility of another state from the logs and technical workings of computers.[55] There is a fundamental difference between this kind of policy-led attribution and attribution processes related to international law. Here, arguing that a state is responsible for something remains a policy issue, but one guided by the structure and technicalities of international law.[56] However, as Florian Egloff and Max Smeets note, "public attribution is a means towards an end," and therefore depending upon a state's policy aim, public attribution might be counterproductive.[57]

Following on from the issues of identification above, one issue here is that narratives of participation are usually supported by public releases of evidence, but the degree of information released will be constrained by political and technical considerations. Dumping information obtained via intelligence agencies into the public sphere risks revealing intelligence sources and methods.

This limitation aside, in the past decade, public open-source organizations have generated substantial evidence sets about hybrid warfare and non-military warfare.[58] This points to a future where states no longer have the monopoly on calling out such activity, nor the monopoly on generating the evidence required to substantiate their claims. In the 2022 Russo-Ukrainian War, the United States and the United Kingdom took to publishing detailed intelligence assessments of the non-conventional elements of the Russian invasion plan in order to limit the ability of the Russian government to use ambiguity to its advantage.[59]

Policy Responses: What Should Be Done to Participants in Non-Military Warfare?

What, then, should states do to those that participate in non-military warfare? Any state response is going to be a policy decision that incorporates multiple competing demands (e.g., whether to escalate or de-escalate, or even to respond at all). Where do definitions of participation fit into this? What kinds of standards of defining participation are likely to enable effective state responses to non-military warfare?

Academics naturally tend toward identifying theoretically coherent standards of participation. In reality, defining and determining participation requires rules or principles that can either be ambiguous (leaving responsibility for identification to individuals) or strict (creating specific thresholds). Defining participation in policy terms requires balancing a variety of competing concerns to define a set of people as participants and to calibrate actions that can be taken against them. Of course, there is likely to be asymmetry between the actions that a state condemns when performed by their adversaries and similar actions performed by their allies.

In this sense, participation is an arbitrary definition in policy terms. States will select from a pool of organizations and connected persons those that they deem important enough to take action against. It is, however, a form of bounded arbitrariness, as it must take into account existing standards of international law noted above.

Participation in policy terms is, therefore, about selecting what matters, and what does not – what are the resources that are available to be directed against participants, and which ones are worth acting against.

Policy participation therefore involves conceptual definitions of participation, practical constraints on determining participants, resource concerns about the effectiveness of action against different kinds of participation, strategic concerns about the consequences of targeting participants, and political concerns related to escalation and legitimacy. Here, participation is about identifying the important persons to target, with an eye on the available means to do so. Following the logic of this chapter, we move from the conceptual questions of "Who is a participant and what could be done to them?" to the practical questions of "Given our policy goals, what should we do to these participants, and do we have the resources to do so?"

In policy terms, the key issue in responding to non-military warfare is the problem of escalation. Non-military warfare is usually calibrated so as to place its target in a dilemma, where responding with military force (vertical escalation) would be counterproductive, but failing to respond, or bargaining to cease the campaign of non-military warfare are both unacceptable in the long term.

Second, there is the issue of *tu quo que* – by responding with measures that might be used by the opponent, the target state risks giving their opponents greater latitude to harm them in future.

The problem for states is therefore how to define participation publicly so as to legitimate responses that avoid both perceived escalation and do not substantially widen an opponent's range of policy tools in the process. The smaller the defined range of participants are, the easier it will be to articulate their responsibility and legitimate action against them. Doing so also reduces the collateral harms that might be inflicted by state responses on individuals or groups that are not responsible for non-military warfare campaigns. In this sense, we can see public definitions of these patterns of activity as part of state responses to non-military warfare itself. State claims about responsibility, and identifying participants, are part and parcel of this interaction. Notably, selecting individuals for public identification, and even prosecution (even though such prosecutions are not ever likely to occur) is part of this. For instance, the U.S. government's actions against a hacking group tracked as APT41 led to indictments for five Chinese nationals who are unlikely to ever be extradited.[60] Similarly, the four persons held responsible for the downing of MH-17 over Ukraine in 2014 are being tried in absentia, with only one of them sending lawyers to engage with the trial.[61]

Outside extremely intensive campaigns and armed conflict, state responses to individual participation take the form of public shaming, criminal investigations and potential prosecutions, and arbitrary sanctions. Targeting organizations responsible for non-military warfare with financial sanctions restricts their ability to operate, as does preparing criminal cases against individuals involved where the law enables a state to do so. These are forms of soft sanctions that rely upon the connectedness of the global economy. It typically means that individuals who are identified as operational figures in non-military warfare campaigns cannot travel to countries where they are liable to extradition, significantly curtailing their future individual freedom. Ultimately, framing non-military warfare as criminal behavior might be the most practical way to punish individuals and organizations for their participation in such campaigns. By increasing the potential organizational or individual costs for engaging in such activity, it could deter it to an extent, but the state-backed nature of these campaigns means complete deterrence is unlikely.

Conclusion

This chapter has outlined a way of evaluating a person or organization's participation in non-military warfare. It has left the definition of non-military warfare itself intentionally vague, identifying it as an abnormally intense campaign of strategic coercion using non-military means, relative to its political context. My argument here is that given the potentially expansive nature of who or what could be considered a participant in non-military warfare, it is necessary to create a relatively tight definition of participation in non-military warfare. The principle of distinction, in this case, is used to differentiate between different two classes of civilian – those who are defined as participants and those whose participation is relatively ancillary to a given campaign of strategic effect or coercion.

We cannot draw concepts straight from international humanitarian law to define participation in non-military warfare. This is because international humanitarian law is specific to a single context (armed conflict), whereas a key feature of non-military warfare is that it occurs in a variety of different situations in international politics. Second, international humanitarian law is built upon a shared acceptance of legitimate engagement in military operations (combatant status) which is absent in non-military warfare. Still, if states are faced with non-military warfare, some of the debates regarding civilian participation in armed conflict have parallels to these campaigns, as do the problems faced by states in identifying civilians taking direct part in hostilities.

The four-part test developed in this chapter – that participation in non-military warfare is defined by direct and active engagement in non-military warfare operations, connected to an organized campaign of strategic coercion that has a state nexus and is abnormal, according to the context – is designed to constrain participation to persons and organizations involved in an operational or command role. The reason for constraining participation in this way is to make the concept amenable to translation into state practices of identification and attribution that will be integral to any response to non-military warfare and also to restrict it to persons who are most responsible for non-military warfare itself. This, I have argued, makes it easier to build shared understandings of what does, or does not, constitute non-military warfare and also networks of interstate cooperation in response to non-military warfare itself.

Notes

1 Ilmari Käihkö and Oscar Jonsson, "Introduction: A War of Our Time," in Oscar Jonsson and Ilmari Käihkö (Eds.), Non-Military Warfare: A War of Our Time (Routledge, 2025).
2 Ulrik Franke, War by Non-Military Means: Understanding Russian Information Warfare (Försvarsdepartementet, 2015); Martin C. Libicki, "The Specter of Non-Obvious Warfare," Strategic Studies Quarterly 6, no. 3 (2012): 88–101; Andrew Mumford and Jack McDonald, "Ambiguous Warfare," Report Produced for the DCDC, 2014.

3 Andrew Radin, "Hybrid Warfare in the Baltics: Threats and Potential Responses" (RAND Corporation Santa Monica United States, 2017), 5.
4 Julio Miranda Calha, "Hybrid Warfare: NATO's New Strategic Challenge," Report to NATO Parliamentary Assembly 7 (2015): 1.
5 Jan Ångström and Sofia Ledberg, ""Civil and Military" as a Constitutive Categorization of the Study of War and Politics," in Oxford Research Encyclopaedia of Politics (Oxford University Press, 2021), http://urn.kb.se/resolve?urn=urn:nbn:se:fhs:diva-9910.
6 Beatrice Heuser, War: A Genealogy of Western Ideas and Practices (Oxford and New York: Oxford University Press, 2022); Jens Bartelson, War in International Thought (Cambridge: Cambridge University Press, 2017), https://doi.org/10.1017/9781108297707.
7 Emily Crawford, Identifying the Enemy: Civilian Participation in Armed Conflict (Oxford University Press, 2015).
8 Neil MacFarquhar, "Yevgeny Prigozhin, Russian Oligarch Indicted by U.S., Is Known as 'Putin's Cook'," The New York Times, February 2018, https://www.nytimes.com/2018/02/16/world/europe/prigozhin-russia-indictment-mueller.html.
9 Neil MacFarquhar, "Inside the Russian Troll Factory: Zombies and a Breakneck Pace," The New York Times, February 2018, https://www.nytimes.com/2018/02/18/world/europe/russia-troll-factory.html. The indictment was subsequently dropped to prevent national security information being revealed in court, see Spencer S. Hsu, "Justice Dept. Abandons Prosecution of Russian Firm Indicted in Mueller Election Interference Probe," Washington Post, March 2020, https://www.washingtonpost.com/local/legal-issues/us-justice-dept-abandons-prosecution-of-russian-firm-indicted-in-mueller-election-interference-probe/2020/03/16/5f7c3fd6-64a9-11ea-912d-d98032ec8e25_story.html.
10 Andy Greenberg, "Feds Allege Destructive Russian Hackers Targeted US Refineries," Wired, 2022, https://www.wired.com/story/triton-berserk-bear-russian-hackers-doj-indictment/.
11 Julian E. Barnes, "Countries Have Expelled 120 Russian Spies Since the War in Ukraine Started," The New York Times, April 2022, https://www.nytimes.com/2022/04/06/world/europe/russian-diplomats-expelled.html.
12 Donna Borak and Nicole Gaouette, "US Sanctions Putin's 'Chef' for Attempting to Influence 2018 Midterm Elections," CNN, 2019, https://www.cnn.com/2019/09/30/politics/us-yevgeniv-priogzhin-sanctions/index.html.
13 Interfax, "Russian Investigative Committee Opens Extremism Case Against Meta Employees," Interfax, 2022, https://interfax.com/newsroom/top-stories/76426/.
14 Jeff McMahan, "The Ethics of Killing in War," Ethics 114, no. 4 (2004): 693–733; Jeff McMahan, Killing in War (Oxford University Press, 2009); Michael N Schmitt, "Deconstructing Direct Participation in Hostilities: The Constitutive Elements," New York University Journal of International Law and Politics 42 (2009): 697; Kenneth Watkin, "Opportunity Lost: Organized Armed Groups and the ICRC Direct Participation in Hostilities Interpretive Guidance," New York University Journal of International Law and Politics 42 (2009): 641.
15 Bill Boothby, "And for Such Time as: The Time Dimension to Direct Participation in Hostilities," New York University Journal of International Law and Politics 42 (2009): 741.
16 Quincy Wright, A Study of War (The University of Chicago Press, 1942), 685–91.
17 Wright, A Study of War, 698.
18 Wright, A Study of War, 698.
19 See chapters 6, 7, 8, 9, 10, 11, 12 of this volume.
20 This is most visible in "great power competition" between leading states vying for regional or global hegemony, however all states can, at least in theory, compete with their peers in this manner.
21 Michael Brecher and Jonathan Wilkenfeld, A Study of Crisis (The University of Michigan Press, 2000), 3.

22 What security studies scholars refer to as "counter-securitisation", see Holger Stritzel and Sean C Chang, "Securitization and Counter-Securitization in Afghanistan," Security Dialogue 46, no. 6 (December 2015): 548–67, https://doi.org/10.1177/0967010615588725; Thierry Balzacq (Ed.), Securitization Theory: How Security Problems Emerge and Dissolve (Routledge & CRC Press, 2011), https://www.routledge.com/Securitization-Theory-How-Security-Problems-Emerge-and-Dissolve/Balzacq/p/book/9780415556286.
23 McMahan, "The Ethics of Killing in War."
24 Seth Lazar, "Just War Theory: Revisionists Versus Traditionalists," Annual Review of Political Science 20, no. 1 (2017): 37–54, https://doi.org/10.1146/annurev-polisci-060314-112706.
25 Michael N. Schmitt, Military Necessity and Humanity in International Humanitarian Law: Preserving the Delicate Balance, Michael N. Schmitt (Ed.) (The Hague: T.M.C. Asser Press, 2012), 89–129, https://doi.org/10.1007/978-90-6704-740-1_3.
26 The typical distinction being that it is permissible to kill via high explosive shells or causing them gruesome injuries in innumerable ways, but to do so via asphyxiating gases is, for some, distinct and wrong as a means of waging war.
27 Emily Crawford and Alison Pert, International Humanitarian Law, Higher Education from Cambridge University Press, 2nd ed. (Cambridge University Press, 2020), https://doi.org/10.1017/9781108635448. Chapter 4.
28 Nils Melzer, "Interpretive Guidance on the Notion of Direct Participation in Hostilities Under International Humanitarian Law" (ICRC, 2009).
29 Cécile Fabre, "Guns, Food, and Liability to Attack in War," Ethics 120, no. 1 (2009): 36–63.
30 William H. Boothby, The Law of Targeting (Oxford and New York: Oxford University Press, 2012), 94–97.
31 Crawford, Identifying the Enemy, Ch. 2.
32 Melzer, "Interpretive Guidance on the Notion of Direct Participation in Hostilities Under International Humanitarian Law."
33 See e.g. Schmitt, "Deconstructing Direct Participation in Hostilities."
34 Melzer, "Interpretive Guidance on the Notion of Direct Participation in Hostilities Under International Humanitarian Law," 70.
35 United States Department of Defense Office of General Counsel, Department of Defense Law of War Manual (Department of Defense, 2015), 258.
36 W. Hays Parks, "Part IX of the ICRC 'Direct Participation in Hostilities' Study: No Mandate, No Expertise, and Legally Incorrect," New York University Journal of International Law and Politics 42 (2010): 769–830.
37 Melzer, "Interpretive Guidance on the Notion of Direct Participation in Hostilities Under International Humanitarian Law," 71–72.
38 Richard Ned Lebow, "Beyond Parsimony: Rethinking Theories of Coercive Bargaining," European Journal of International Relations 4, no. 1 (March 1998): 31–66, https://doi.org/10.1177/1354066198004001002.
39 Mark Galeotti, The Weaponisation of Everything: A Field Guide to the New Way of War (New Haven: Yale University Press, 2022).
40 Ali Dukakis and Lucien Bruggeman, "Roger Stone Found Guilty on All 7 Counts," ABC News, November 2019, https://abcnews.go.com/Politics/roger-stone-found-guilty-counts/story?id=67015102.
41 U.S. Department of the Treasury, "Treasury Sanctions Russians Bankrolling Putin and Russia-Backed Influence Actors," March 2022, https://home.treasury.gov/news/press-releases/jy0628.
42 Bellingcat Investigations Team, "Putin Chef's Kisses of Death: Russia's Shadow Army's State-Run Structure Exposed," Bellingcat, August 2020, https://www.bellingcat.com/news/uk-and-europe/2020/08/14/pmc-structure-exposed/.
43 Thomas Rid and Ben Buchanan, "Attributing Cyber Attacks," Journal of Strategic Studies 38, no. 1–2 (January 2015): 4–37, https://doi.org/10.1080/01402390.2014.

977382; Nicholas Tsagourias and Michael Farrell, "Cyber Attribution: Technical and Legal Approaches and Challenges," European Journal of International Law 31, no. 3 (August 2020): 941–67, https://doi.org/10.1093/ejil/chaa057; Florian J. Egloff and Max Smeets, "Publicly Attributing Cyber Attacks: A Framework," Journal of Strategic Studies 46, no. 3 (2023): 502–33, https://doi.org/10.1080/01402390.2021.1895117.

44 Attribution as a counter to "low intensity conflict" or "ambiguous warfare" was a theme in the cold war, see George P. Shultz, "Low Intensity Warfare: The Challenge of Ambiguity," Department of State Bulletin 86 (March 1986): 15–18; recent work has been tied to the case of Russia's attack on Ukraine in 2014, See Andrew Mumford and Jack McDonald, "Ambiguous Warfare," Development Concepts and Doctrine Centre, 2014; David Kilcullen, The Dragons and the Snakes: How the Rest Learned to Fight the West (Oxford University Press, 2020), https://doi.org/10.1093/oso/9780190265687.001.0001.

45 Samuel Issacharoff and Richard H. Pildes, "Targeted Warfare: Individuating Enemy Responsibility," New York University Law Review 88, no. 5 (November 2013): 1521–99, https://heinonline.org/HOL/PDFsearchable?handle=hein.journals/nylr88&collection=journals&section=41&id=&print=section&sectioncount=1&ext=.pdf&nocover=&display=0.

46 Ian Hurd, "Legitimacy, Power, and the Symbolic Life of the UN Security Council," Global Governance 8, no. 1 (2002): 35–51, https://www.jstor.org/stable/27800326.

47 Eric Talbot Jensen, "Computer Attacks on Critical National Infrastructure: A Use of Force Invoking the Right of Self-Defense," Stanford Journal of International Law 38, no. 2 (2002): 207–40, https://heinonline.org/HOL/P?h=hein.journals/stanit38&i=217.

48 Scott J. Shackelford, "From Nuclear War to Net War: Analogizing Cyber Attacks in International Law," Berkeley Journal of International Law 27, no. 1 (2009): 192–252, https://heinonline.org/HOL/P?h=hein.journals/berkjintlw27&i=197.

49 Wolff Heintschel von Heinegg, "The Tallinn Manual and International Cyber Security Law," in Yearbook of International Humanitarian Law, Vol. 15 (Cambridge University Press, 2012), 3–18, https://doi.org/10.1007/978-90-6704-924-5_1.

50 Jeremy Wright, "Cyber and International Law in the 21st Century," GOV.UK, 2018, https://www.gov.uk/government/speeches/cyber-and-international-law-in-the-21st-century.

51 The operation of human rights law during armed conflict is a tricky issue, where many scholars and states consider it applicable, but notable states like the US do not. See Gary D. Solis, The Law of Armed Conflict: International Humanitarian Law in War, 3rd ed. (Cambridge University Press, 2021), 19–22.

52 Austin Carson, Secret Wars: Covert Conflict in International Politics (Princeton University Press, 2018).

53 Thomas Colley, Always at War: British Public Narratives of War (Ann Arbor: University of Michigan Press, 2019).

54 Ronald R. Krebs, Narrative and the Making of US National Security, Cambridge Studies in International Relations (Cambridge: Cambridge University Press, 2015), https://doi.org/10.1017/CBO9781316218969.

55 Rid and Buchanan, "Attributing Cyber Attacks."

56 Mark Gibney, "The Downing of MH17: Russian Responsibility?" Human Rights Law Review 15, no. 1 (March 2015): 169–78, https://doi.org/10.1093/hrlr/ngu036; Marieke de Hoon, "Navigating the Legal Horizon: Lawyering the MH17 Disaster," Utrecht Journal of International and European Law 33, no. 84 (2017): 90–119, https://heinonline.org/HOL/P?h=hein.journals/merko33&i=90.

57 Florian J. Egloff and Max Smeets, "Publicly Attributing Cyber Attacks: A Framework," Journal of Strategic Studies 46, no. 3 (2023): 502–33, https://doi.org/10.1080/01402390.2021.1895117. 5065

58 Perhaps the most famous of these is Bellingcat, see Eliot Higgins, We Are Bellingcat: An Intelligence Agency for the People (Bloomsbury Publishing, 2021).

59 Warren P. Strobel, "Release of Ukraine Intelligence Represents New Front in U.S. Information War with Russia," The Washington Post, April 2022, https://www.wsj.

com/articles/release-of-secrets-represents-new-front-in-u-s-information-war-with-russia-11649070001; Carol E. Lee Ken Dilanian Courtney Kube and Dan De Luce, “Bold, Effective and Risky: The New Strategy the U.S. Is Using in the Info War Against Russia,” NBC News, 2022, https://www.nbcnews.com/politics/national-security/us-using-declassified-intel-fight-info-war-russia-even-intel-isnt-rock-rcna23014.

60 United States Department of Justice Office of Public Affairs, “Seven International Cyber Defendants, Including ‘Apt41’ Actors, Charged in Connection with Computer Intrusion Campaigns Against More Than 100 Victims Globally,” September 2020, https://www.justice.gov/opa/pr/seven-international-cyber-defendants-including-apt41-actors-charged-connection-computer.

61 Stephanie van den Berg, “Prosecutors Seek Life Terms in Dutch Murder Trial over Flight MH17,” Reuters, December 2021, https://www.reuters.com/article/uk-ukraine-crisis-mh17-idUKKBN2J10MC.

Bibliography

Ångström, J., & Ledberg, S. (2021). ‘Civil and Military’ as a Constitutive Categorization of the Study of War and Politics. In *Oxford Research Encyclopedia of Politics*. Oxford University Press.

Balzacq, T. (Ed.). (2011). *Securitization Theory: How Security Problems Emerge and Dissolve*. Routledge & CRC Press. https://www.routledge.com/Securitization-Theory-How-Security-Problems-Emerge-and-Dissolve/Balzacq/p/book/9780415556286

Barnes, J. E. (2022, April). Countries Have Expelled 120 Russian Spies Since the War in Ukraine Started. *The New York Times*. https://www.nytimes.com/2022/04/06/world/europe/russian-diplomats-expelled.html

Bartelson, J. (2017). *War in International Thought*. Cambridge University Press.

Bellingcat Investigations Team. (2020, August). Putin Chef’s Kisses of Death: Russia’s Shadow Army’s State-Run Structure Exposed. *Bellingcat*. https://www.bellingcat.com/news/uk-and-europe/2020/08/14/pmc-structure-exposed/

Boothby, B. (2009). And for Such Time as: The Time Dimension to Direct Participation in Hostilities. *NYUJ International Law and Politics*, *42*, 741.

Boothby, W. H. (2012). *The Law of Targeting*. Oxford University Press.

Borak, D., & Gaouette, N. (2019). US Sanctions Putin’s ‘Chef’ for Attempting to Influence 2018 Midterm Elections. *CNN*. https://www.cnn.com/2019/09/30/politics/us-yevgeniv-priogzhin-sanctions/index.html

Brecher, M., & Wilkenfeld, J. (2000). *A Study of Crisis*. The University of Michigan Press.

Calha, J. M. (2015). *Hybrid Warfare: NATO’s New Strategic Challenge*. NATO Parliamentary Assembly.

Carson, A. (2018). *Secret Wars: Covert Conflict in International Politics*. Princeton University Press.

Colley, T. (2019). *Always at War: British Public Narratives of War*. University of Michigan Press.

Crawford, E. (2015). *Identifying the Enemy: Civilian Participation in Armed Conflict*. Oxford University Press.

Crawford, E., & Pert, A. (2020). *International Humanitarian Law* (2nd ed.). Cambridge University Press.

de Hoon, M. (2017). Navigating the Legal Horizon: Lawyering the MH17 Disaster. *Utrecht Journal of International and European Law*, *33*(84), 90–119.

Dilanian, K., Lee, C. E., Kube, C., & De Luce, D. (2022, March 26). Bold, Effective and Risky: The New Strategy the U.S. Is Using in the Info War Against Russia. *NBC News*. https://www.nbcnews.com/politics/national-security/us-using-declassified-intel-fight-info-war-russia-even-intel-isnt-rock-rcna23014

Dukakis, A., & Bruggeman, L. (2019, November). Roger Stone Found Guilty on All 7 Counts. *ABC News*. https://abcnews.go.com/Politics/roger-stone-found-guilty-counts/story?id=67015102

Egloff, F. J., & Smeets, M. (2023). Publicly Attributing Cyber Attacks: A Framework. *Journal of Strategic Studies*, *46*(3), 502–533.

Fabre, C. (2009). Guns, Food, and Liability to Attack in War. *Ethics*, *120*(1), 36–63.

Franke, U. (2015). *War by Non-Military Means: Understanding Russian Information Warfare*. Swedish Defense Research Agency (FOI).

Galeotti, M. (2022). *The Weaponisation of Everything: A Field Guide to the New Way of War*. Yale University Press.

Gibney, M. (2015, March). The Downing of MH17: Russian Responsibility? *Human Rights Law Review*, *15*(1), 169–178.

Greenberg, A. (2022). Feds Allege Destructive Russian Hackers Targeted US Refineries. *Wired*. https://www.wired.com/story/triton-berserk-bear-russian-hackers-doj-indictment/

Heuser, B. (2022). *War: A Genealogy of Western Ideas and Practices*. Oxford University Press.

Higgins, E. (2021). *We Are Bellingcat: An Intelligence Agency for the People*. Bloomsbury Publishing.

Hurd, I. (2002). Legitimacy, Power, and the Symbolic Life of the UN Security Council. *Global Governance*, *8*(1), 35–51.

Hsu, S. S. (2020, March). Justice Dept. Abandons Prosecution of Russian Firm Indicted in Mueller Election Interference Probe. *Washington Post*. https://www.washingtonpost.com/local/legal-issues/us-justice-dept-abandons-prosecution-of-russian-firm-indicted-in-mueller-election-interference-probe/2020/03/16/5f7c3fd6-64a9-11ea-912d-d98032ec8e25_story.html

Interfax. (2022, March 11). Russian Investigative Committee Opens Extremism Case Against Meta Employees. *Interfax*. https://interfax.com/newsroom/top-stories/76426/

Issacharoff, S., & Pildes, R. H. (2013). Targeted Warfare: Individuating Enemy Responsibility. *New York University Law Review*, *88*(5), 1521–1599.

Jensen, E. T. (2002). Computer Attacks on Critical National Infrastructure: A Use of Force Invoking the Right of Self-Defense. *Stanford Journal of International Law*, *38*(2), 207–240.

Kilcullen, D. (2020). *The Dragons and the Snakes: How the Rest Learned to Fight the West*. Oxford University Press.

Krebs, R. R. (2015). *Narrative and the Making of US National Security*. Cambridge University Press.

Käihkö, I., & Jonsson, O. (2025). Introduction: A War of Our Time. In *Non-Military Warfare: A War of Our Time*. Routledge.

Lazar, S. (2017). Just War Theory: Revisionists Versus Traditionalists. *Annual Review of Political Science*, *20*(1), 37–54.

Lebow, R. N. (1998). Beyond Parsimony: Rethinking Theories of Coercive Bargaining. *European Journal of International Relations*, *4*(1), 31–66.

Libicki, M. C. (2012). The Specter of Non-Obvious Warfare. *Strategic Studies Quarterly*, *6*(3), 88–101.

MacFarquhar, N. (2018a, February 16). Yevgeny Prigozhin, Russian Oligarch Indicted by U.S., Is Known as 'Putin's Cook'. *The New York Times*. https://www.nytimes.com/2018/02/16/world/europe/prigozhin-russia-indictment-mueller.html

MacFarquhar, N. (2018b, February 18). Inside the Russian Troll Factory: Zombies and a Breakneck Pace. *The New York Times*. https://www.nytimes.com/2018/02/18/world/europe/russia-troll-factory.html

McMahan, J. (2004). The Ethics of Killing in War. *Ethics*, 114(4), 693–733.

McMahan, J. (2009). *Killing in War*. Oxford University Press.

Melzer, N. (2009). *Interpretive Guidance on the Notion of Direct Participation in Hostilities Under International Humanitarian Law*. ICRC.

Mumford, A., & McDonald, J. (2014). *Ambiguous Warfare* [Report produced for the DCDC]. Development Concepts and Doctrine Centre.

Parks, W. H. (2010). Part IX of the ICRC 'Direct Participation in Hostilities' Study: No Mandate, No Expertise, and Legally Incorrect. *NYU Journal of International Law and Politics, 42*, 769–830.

Radin, A. (2017). *Hybrid Warfare in the Baltics: Threats and Potential Responses*. RAND Corporation.

Rid, T., & Buchanan, B. (2015). Attributing Cyber Attacks. *Journal of Strategic Studies, 38*(1–2), 4–37.

Schmitt, M. N. (2009). Deconstructing Direct Participation in Hostilities: The Constitutive Elements. *NYUJ International Law and Politics, 42*, 697.

Schmitt, M. N. (2012). Military Necessity and Humanity in International Humanitarian Law: Preserving the Delicate Balance. In M. N. Schmitt (Ed.), *Yearbook of International Humanitarian Law* (pp. 89–129). T.M.C. Asser Press.

Shackelford, S. J. (2009). From Nuclear War to Net War: Analogizing Cyber Attacks in International Law. *Berkeley Journal of International Law, 27*(1), 192–252.

Shultz, G. P. (1986, March). Low Intensity Warfare: The Challenge of Ambiguity. *Department of State Bulletin, 86*, 15–18.

Solis, G. D. (2021). *The Law of Armed Conflict: International Humanitarian Law in War* (3rd ed.). Cambridge University Press.

Stritzel, H., & Chang, S. C. (2015). Securitization and Counter-Securitization in Afghanistan. *Security Dialogue, 46*(6), 548–567.

Strobel, W. P. (2022, April 5). Release of Ukraine Intelligence Represents New Front in u.s. Information War with Russia. *The Washington Post*. https://www.wsj.com/articles/release-of-secrets-represents-new-front-in-u-s-information-war-with-russia-11649070001

Tsagourias, N., & Farrell, M. (2020). Cyber Attribution: Technical and Legal Approaches and Challenges. *European Journal of International Law, 31*(3), 941–967.

General Counsel of the Department of Defense. (2015). *Department of Defense Law of War Manual*. Department of Defense.

United States Department of Justice, Office of Public Affairs. (2020, September). *Seven International Cyber Defendants, Including 'Apt41' Actors, Charged in Connection with Computer Intrusion Campaigns Against More Than 100 Victims Globally*. https://www.justice.gov/opa/pr/seven-international-cyber-defendants-including-apt41-actors-charged-connection-computer.

U.S. Department of the Treasury. (2022, March 15). *Treasury Sanctions Russians Bankrolling Putin and Russia-Backed Influence Actors*. https://home.treasury.gov/news/press-releases/jy0628

van den Berg, S. (2021, December). Prosecutors Seek Life Terms in Dutch Murder Trial over Flight MH17. *Reuters*. https://www.reuters.com/article/uk-ukraine-crisis-mh17-idUKKBN2J10MC

von Heinegg, W. H. (2012). The Tallinn Manual and International Cyber Security Law. In *Yearbook of International Humanitarian Law* (Vol. 15, pp. 3–18). Cambridge University Press.

Watkin, K. (2009). Opportunity Lost: Organized Armed Groups and the ICRC Direct Participation in Hostilities Interpretive Guidance. *NYU Journal of International Law & Politics, 42*, 641.

Wright, J. (2018). Cyber and International Law in the 21st Century. *GOV.UK*. https://www.gov.uk/government/speeches/cyber-and-international-law-in-the-21st-century

Wright, Q. (1942). *A Study of War*. The University of Chicago Press.

4 Russian Non-Military Warfare

Active Measures in Ukraine

Sanshiro Hosaka

Introduction

Since February 24, 2022, the salience of Russia's conventional warfare – characterized by tanks, missiles, and trenches in Ukraine – has temporarily diverted attention from its non-military warfare. However, the application of Russia's military means was premised on a strategy of non-military means to ensure a swift and decisive victory. During the first weeks of Russia's full-scale invasion of Ukraine, Russia's Federal Security Service (FSB) was tasked with establishing a puppet government following a planned Russian special operations assault on the Ukrainian leadership. This effort ultimately failed when the Russian forces lost control of Hostomel airfield near Kyiv. Furthermore, the FSB took responsibility for consolidating Russia's occupation regimes and counterintelligence apparatuses with collaborators in the Ukrainian regions seized after February 2022.[1] Russian military and non-military means, especially covert actions and other intelligence operations, are integrated strategically.

This chapter focuses on the political operations that are an integral part of the concept of non-military warfare. They are best understood through the long-standing concept of active measures. It was defined by a U.S. working group in the mid-1980s as encompassing "a wide span of practices including disinformation operations [and] political influence efforts" designed to enhance Soviet influence, usually by tarnishing the image of opponents.[2] Other experts include kidnappings and assassinations as well as paramilitary dimensions, such as sabotage by Spetsnaz and support for foreign guerrilla and insurgency groups.[3] A defining feature of active measures is their reliance on covert means to obscure Moscow's involvement, distinguishing them from overt propaganda.

During the Cold War, Western analysts assessed that Moscow developed active measures with long-term global ambitions and a diverse set of tools. While the U.S. government separates covert actions from overt foreign policy activities, assigning the former to the parcel of the Central Intelligence Agency, Moscow blurs the line between covert and overt activities, with active measures being an integral part of Russian foreign policy.[4]

An investigation into the anatomy of Russian active measures offers insights into the broader phenomenon of non-military warfare, where they both complement and

DOI: 10.4324/9781003616993-4

substitute military means. This chapter highlights Russia's covert actions against Ukraine, paying particular attention to how active measures and other intelligence activities intersect with the military domain.[5] Ukraine is a focal point due to Moscow's significant allocation of military and non-military resources to its neighboring state and the wealth of data on exposed operations.

As argued in this chapter, Russia's active measures can be interpreted in two ways: as standalone efforts, though not excluding parallel military operations, to achieve Moscow's political goals and as auxiliary measures directly supporting its military operations. The dynamic interplay between military and non-military means is evident in Russia's policy toward Ukraine, which Moscow has sought to maintain within its "sphere of influence." In 2013, Russia's active measures to derail Ukraine's pro-European trajectory failed, prompting Moscow to invade Ukraine and illegally annexed Crimea.[6] The Kremlin subsequently pursued the "federalization" of Ukraine and "autonomy" for the Donbas region. Both are euphemisms for Moscow's projects to block Kyiv's pro-Western foreign policy. Through the deployment of its armed forces and paramilitary elements, Moscow installed puppet regimes, such as the "Donetsk People's Republic" ("DPR") and "Lugansk People's Republic," and sought to legalize them in the Ukrainian constitutional system through the Minsk agreements. These efforts aimed to disguise Russia's military invasion as an internal "civil war" in Ukraine.[7] Over eight years, however, Ukraine resisted Russia's violations of its sovereignty and territorial integrity, preventing Moscow's active measures from achieving their objectives. In February 2022, Moscow abandoned the Minsk process and launched a full-scale invasion aimed at imposing a regime change in Kyiv.

The remainder of this chapter is structured as follows. It begins by the introducing the multiple agencies and agents involved in Moscow's active measures in Ukraine and other "near abroad" countries (former Soviet republics). It then analyzes toolkits and covers for operations, and the main executors – political technologists. By mining the leaked Kremlin emails,[8] this chapter provides an in-depth analysis of Moscow's "eco-special status" project in the Zaporizhzhia region from 2014 through 2016. Finally, it explores how Russia's non-military warfare was designed to assist its full-scale invasion. The concluding part discusses the overall effectiveness of Russia's covert actions.

Agencies and Actors in Russian Active Measures

Russia's active measures were planned and implemented by its intelligence agencies, namely the Foreign Intelligence Service (SVR), the FSB, and the Main Intelligence Directorate of the General Staff (GRU).[9] However, Russia's covert operations in Ukraine and other former Soviet countries often include direct involvement from the president and relevant directorates of the Presidential Administration (see Figure 4.1).

President Vladimir Putin, an experienced agent handler, personally managed Ukrainian lawmaker Viktor Medvedchuk, Head of the Presidential Administration under President Leonid Kuchma from 2002 to 2005. Medvedchuk was used to interfere with Ukraine's domestic affairs.[10] As recommended in the Soviet Committee

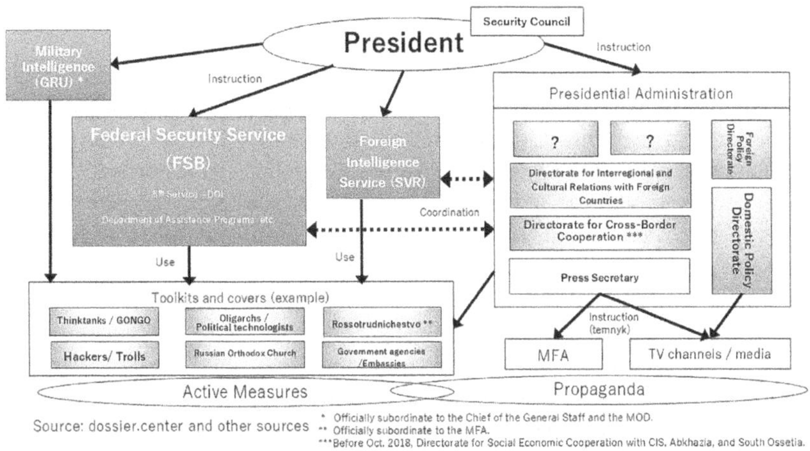

Figure 4.1 The Power Structure and Agencies Involved in Russia's Covert Actions

for State Security (KGB) agent handler regulations, Putin established close relationships with Medvedchuk, even baptizing his youngest daughter in 2004.[11] By 2013, Medvedchuk became a major promoter of Putin's Eurasian integration project in Ukraine. Putin met with Medvedchuk multiple times when Russia implemented the Complex of Measures, a comprehensive plan of active measures authored by Sergei Glazyev, Putin's adviser on Eurasian integration, to derail Ukraine's aspiration to join the European Union. It is perhaps not a coincidence that Putin radicalized his policy toward Ukraine by invading the Crimean peninsula and Donbas when it failed to elevate Medvedchuk's popularity sufficiently, making his return to the main arena of Ukrainian politics unviable at the end of 2013 (Moscow did not trust ostensibly "pro-Russian" Ukrainian president Viktor Yanukovych).[12] Despite these setbacks, Putin and his administration continued to regard Medvedchuk as a key coordinator who lobbied for Russian interests in Ukraine through his political movement "Ukrainian Choice," envisioned as "a basis for a 'separate political project.'"[13] Medvedchuk functioned as an agent-gruppovod (agent group leader), who built up and maintained Russia's agent networks in Ukraine.[14]

Medvedchuk's political comeback in the 2019 parliamentary elections with his political party "Opposition Platform – For Life" (OPZhZ) taking second place with 44 seats, following Zelensky's People's Servant party, was celebrated in the Kremlin. Putin's idea of a full-scale invasion may have consolidated when Ukrainian authorities charged Medvedchuk with treason and placed him under house arrest in May 2021, whereas no previous president removed "Putin's de facto plenipotentiary in Ukraine."[15] Medvedchuk attempted to flee after Russia's full-scale invasion but was arrested by the Security Service of Ukraine (SBU) in April 2022, and later handed over to Russia in a prisoner exchange in September. According to the Ukrainian side, it took back 200 Ukrainian fighters in exchange for Medvedchuk.[16] This exchange, which freed 108 Azov Regiment fighters, caused anger among hawks in

Russia while showing how much Putin cared about Medvedchuk. Putin may retain Medvedchuk as a representative of a future "party of peace" in Ukraine.[17]

Key centers of Russia's covert actions in Ukraine were the Directorates of the Presidential Administration. Although not traditionally viewed as intelligence agencies, at least two directorates were actively engaged in active measures. One is the Directorate for Interregional and Cultural Relations with Foreign Countries, which oversee activities in the countries of the "near abroad." Compared by some experts to the International Department of the Central Committee of the Soviet Communist Party, the Directorate has been led by Russian foreign intelligence generals.[18] SVR General Vladimir Chernov, with the support of reserve and seconded officers from the SVR, FSB, and GRU and political technologists, oversaw a pro-Russian opposition party in Georgia, conducted influence operations with local agents in Armenia and Azerbaijan, and supported the election campaign of Moldovan President Igor Dodon.[19] The Directorate also covered Central Asian and Baltic countries, coordinating covert actions disguised as cultural and public diplomacy with other government agencies, such as Rossotrudnichestvo (Russia's Federal Agency for the Commonwealth of Independent States Affairs, Compatriots Living Abroad, and International Humanitarian Cooperation).[20]

The other directorate engaged in covert actions was the Directorate of Social and Economic Cooperation for the Commonwealth of Independent States (CIS), Abkhazia, and South Ossetia, previously supervised by Presidential Aide Vladislav Surkov. Known as a "grey cardinal" in Russian politics, Surkov, who crafted the "sovereign democracy" ideology to legitimize Putin's autocracy, served in the GRU's special force (Spetsnaz) during the late Soviet times. The term "special forces" typically conjures images of trained guys in camouflage uniforms and exotic face paint, yet GRU Spetsnaz units are specially trained to carry out reconnaissance and sabotage missions behind enemy lines: "If you are cunning, resourceful, and act unconventionally, then you are a special forces soldier."[21] Dmitry Kozak, Deputy Chief of the Presidential Administration and Surkov's successor overseeing this directorate, also received training in the GRU Spetsnaz earlier in his career. Surkov's directorate conducted covert political operations in Ukraine and managed Moscow's puppet "people's republics" in the areas of Donetsk and Luhansk regions temporarily occupied by the Russian armed forces.[22] In 2018, his directorate was renamed the Directorate for Cross-Border Cooperation to focus on "complicated border areas," avoiding the duplication of efforts with Chernov's directorate, which focuses mainly on the CIS countries.[23]

Furthermore, the two key intelligence agencies operating in Ukraine are the FSB and the GRU. Russia's foreign intelligence activities against former Soviet states have primarily been conducted by the FSB, not the SVR. Following the dissolution of the Soviet Union in 1991, Russia agreed with CIS countries not to engage in foreign intelligence against each other; thus, the SVR was unable to target Ukraine and other countries. However, the FSB, which inherited the KGB Second Directorate with domestic counterintelligence as its main function, was legally authorized to carry out foreign intelligence activities since its establishment in 1995. A 2003 law revision established the Operational Information Division ("DOI" in Russian abbreviation) as a specialized foreign intelligence agency within the FSB's Directorate of Operational

Information and International Relations, commonly referred to as the "Fifth Service." Cynically, Russia's counterintelligence agency, the FSB, conducts intelligence operations against CIS countries, despite promises not to engage in such activities. The Fifth Service chief Sergei Beseda and DOI director Anatoly Bolyukh were responsible for intelligence and political operations against Ukraine for many years. For example, both officers were present in Kyiv just days before President Viktor Yanukovych fled to Russia during the Euromaidan protests in February 2014. The DOI's ninth division, in charge of Ukraine, is known to have recruited Ukrainian politicians and oligarchs, especially those who fled to Moscow along with Yanukovych.

Other organizations, often in coordination with Russian intelligence agencies, engage in subversive activities in the "near abroad," such as the Institute of CIS Countries (Institute of Diaspora and Integration) led by Konstantin Zatulin since 1996, Rossotrudnichestvo led by Eleonora Mitrofanova in 2004–2008 and 2017–2020, succeeded by Yevgeny Primakov Jr. and the Russian Orthodox Church led by Patriarch Kirill since 2009. These individuals or their close relatives worked for Soviet intelligence.[24]

Russian oligarchs also provide resources and plausible deniability for active measures. Alexander Borodai, a Russian political technologist, suddenly appeared in eastern Ukraine in the spring of 2014 as the "prime minister" of the self-proclaimed "Donetsk People's Republic" (DPR). Konstantin Malofeev, a Russian oligarch, publicly claimed that Borodai had been a "consultant" of his investment company. However, leaked emails from Surkov's directorate suggest direct contact between his office and Malofeev on the selection of possible candidates for the "DPR" cabinet members.[25] Furthermore, intercepted phone calls published by the Joint Investigation Team on the downing of flight MH17 in eastern Ukraine demonstrate that in July 2014, Surkov and Borodai had a direct phone conversation, during which the former informed the latter that there would be military reinforcements from Russia to support the "DPR."[26]

Political Technology

Moscow's active measures against Ukraine and other "near-abroad" states use various toolkits and covers. Key assets for the Kremlin and Russian intelligence agencies are political technologists – professional spin doctors and media managers.[27] While much of the work by political technologists is not entirely covert, their role in Moscow' active measures is significant. In 2007, Russian political technologist Sergei Markov emphasized to Andrew Wilson the need for political technology in Georgia and Ukraine, disregarding their independence: "We should use political technology internationally in Georgia and Ukraine. I don't think of these countries as independent." Markov argued for using think tanks, roundtables, conferences, affiliated media, and exchanges to elevate new pro-Russian leaders in these countries.[28]

Political technologists often work proactively by formulating and proposing projects according to key parameters given by senior officials and officers, aligning their goals with the Kremlin's overarching tactics and strategy.[29] Their employments are typically project-based and temporary, with technologists shifting from one target country to another based on Moscow's operational priorities and needs.

Leaked Kremlin emails provide insights into how these political technologists contributed to Moscow's efforts to disinform targets and destabilize the situation in Ukraine. Below are some examples.

Aleksandr Kazakov – From Russian Language Advocate to Liaison of Donetsk Puppet Ruler

Aleksandr Kazakov was initially known as a Russian language advocate in Riga and close to former Russian deputy prime minister Dmitry Rogozin. He then served as the director of the Center for Strategic Planning under the "Parliament of Novorossiya" in 2014. Later, he became Surkov's liaison with Aleksandr Zakharchenko, the self-proclaimed "prime-minister of the Donetsk People's Republic," a former smuggler recruited by the Kremlin. Under Surkov's oversight, Kazakov engaged in various tasks and projects, including mythmaking to transform a Donetsk criminal into a respected "Donbas leader." Kazakov arranged the publication of a book by bestselling Russian author Zakhar Prilepin, featuring Zakharchenko as the protagonist. Through a staged press conference of Zakharchenko, Kazakov disseminated disinformation to a Western author and a broader international audience via the magazine *Foreign Affairs*.[30]

After Zakharchenko was killed by a bomb in a Donetsk café in 2018, Kazakov returned to Russia and became the first deputy chairman of Prilepin's party "For Truth." During Russia's full-scale invasion, he actively participated in Russian television shows, making claims about "the genocide of Donbas, to which the West has turned a blind eye for 8 years," aiming to discredit the Ukrainian government and army.[31]

Oleg Bondarenko – Disinforming Ukrainian and Balkan Audiences

Oleg Bondarenko, officially the Director of the Agency for Strategic Communications, was one of the initial members of media expert meetings on Ukraine. These were prepared by Surkov's right hand in charge of information operations Aleksei Chesnakov in February 2014. The meetings, which also included other political technologists such as Evgeny Minchenko and Sergei Mikheev, were held as tensions escalated in Kyiv. They wrote articles and gave interviews to saturate the information space with the Kremlin's preferred narratives about East-West geopolitical competitions and Ukrainian "Banderites." In the second half of 2015, responsibility for media expert meetings on Ukraine was transferred from Chesnakov to Bondarenko, who held monthly meetings mostly with Ukrainian experts who had fled to Moscow. These meetings coordinated the launch and management of anti-Ukrainian information resources.[32] The overt Ukrainian citizenship of the contributors served as a cover, creating the false impression that dissenting voices were emerging from Ukrainians themselves (some of whom later obtained Russian citizenship).

In November 2014, Bondarenko staged an "international election observation mission" in the "DPR." In September 2015, he held a press conference in Luhansk under Russian occupation. The political technologist seized all possible occasions to accuse the Ukrainian government of alleged political repression and restriction on freedom of speech, such as the murder of Ukrainian journalist Oles Buzina and the arrest of the editor of a pro-Russian news site in Odesa Artem Buzila. In 2018, when Surkov's political and influence operations in Ukraine reached an impasse, Bondarenko shifted his focus from Ukraine to the Balkans by launching a new web resource (balkanist.ru), which hosted over 80 local authors. Bondarenko's resource courted Milorad Dodik, president of the Republic Srpska (2010–2018), who was advocating for the self-determination of Bosnian Serbs.

Inal Ardzinba – From G20 Youth Ambassador to Subversion Coordinator in Odesa

In December 2013, Surkov recruited Inal Ardzinba, a former head of the Russian youth delegation to the G20 at a discussion event he hosted at the Moscow State Institute of International Relations (MGIMO).[33] Ardzinba was born in the Abkhaz Autonomous Soviet Socialist Republic and initially looked forward to assisting Surkov with "socio-economic and political problems of Abkhazia and South Ossetia." However, he was subsequently appointed as the Kremlin's curator in charge of "humanitarian assistance to the southeast of Ukraine."[34]

According to the Ukrainian Security Service chief Vasyl Hrytsak, Ardzinba coordinated agent networks in Odesa to conduct a subversive operation. The plan included proclaiming the "Bessarabia People's Republic" in the nine western districts of the Odesa region. This operation involved coordination with Russian troops stationed in the Russia backed de facto-state Transnistria, which were expected to blow up bridges and capture the strip near the Ukraine-Moldova border under the guise of a "militia." This plan was exposed by the Ukrainian counterintelligence service, and Russia's agents were arrested and later exchanged with Ukrainian citizens detained by Russia. According to the Ukrainian counterintelligence chief, an FSB group blamed Surkov's directorate for this debacle.[35] In 2017, leaked emails purportedly belonging to Ardzinba confirmed the allegations. Ardzinba's communications with "Andrei Gold," – likely Dmitry Soin, the ideologue of the Transnistrian project "Breakthrough!" – in May 2015 included proposals for a "future plan of action," which mentioned possible terrorist attacks.[36] Despite this, Ardzinba was then in 2021 appointed "minister of foreign affairs" of the Republic of Abkhazia – a Georgian territory occupied by Russia.[37]

"Eco-Special Status" Project in Zaporizhzhia

In 2014, the Kremlin pursued the "federalization" of Ukraine, aiming to block Kyiv's pro-Western foreign policy orientation by establishing its puppet "people's republics" and staging pseudo-regional autonomy initiatives in Ukraine. The project to create a "special status" for the Zaporizhzhia region was entrusted to Pavel Broide, a political technologist who had worked for Viktor Yanukovych's shadow structure in Zaporizhzhia. Broide was handled by Surkov's subordinate, Inal Ardzinba.[38]

A seemingly unlikely theme was used by the Kremlin as a smokescreen for its covert action in Zaporizhzhia. On October 25, 2014, Broide sent Ardzinba a document titled "Potential Vectors of Activity in the Zaporizhzhia Region to Popularize the Ideas of Decentralization and the Special Status of the Region." This document cautioned that promoting the topic of special status carried "the risk of being accused of the organizers of separatism and forceful suppression of such actions at the very first steps," suggesting it would be "appropriate to start a campaign to popularize this topic from non-political aspects, namely from the environment." Broide argued that the Zaporizhzhia region, which holds the status of an ecological disaster zone, had been promised special status by various politicians, yet no substantive actions had been taken to implement this. The document stated:

> By starting the promotion of the topic of ecology, and on its basis popularizing the topic of special ecological status, one can achieve a significant information resonance, form public opinion and tools (expert groups, initiative groups of citizens, an association of deputies, a public movement, etc.).

According to the plan, once the special ecological status gains public attention, the campaign can then expand to supplement the content of the special status with socioeconomic issues, and then with political aspects. Broide proposed specific action plans to promote the topic, including organizing a rally of 300 people in front of Zaporizhzhia City Council with slogans like "We breathe poison and drink poison, but there is no money in the budget to solve the environmental problems of Zaporizhzhia!" and "We demand a legislative initiative to solve the environmental problems of Zaporizhzhia!" These slogans were used during a staged public meeting the following month.

Broide advised that this rally should be followed by a roundtable discussion with representatives from public organizations and district administrations, who would call for a legislative solution. An initiative group would be created to develop a draft law, which should propose expanding the powers of local self-government bodies beyond those declared in the Ukrainian government's decentralization efforts. The roundtable indeed took place on December 19, 2014, the same day as the street meeting, with both events covered by several local media outlets.[39]

Following the script written by Broide, a presentation of the draft law on the special status of the region was organized by the Council for Eco-Special Status on February 6, 2015. The presentation was covered by local media, and the promotion of the news in more than 50 news sites and groups was reported to Ardzinba.

The Kremlin's next move was to collect signatures in support of the draft law and establish inter-factional local deputy groups. On March 23, Broide reported to Ardzinba that twelve local deputies had entered the group. Broide explained,

> If there is a controlled group in the city council (able to reject a decision or majority), it is possible to block social, political, budgetary and other initiatives of Kyiv at the city level, speaking out against them with appropriate information coverage.

Two days later, a delegation of the Council for Eco-Special Status delivered a petition to the Verkhovna Rada (national parliament) requesting consideration of the bill, which proposed granting the Zaporizhzhia region special environmental status. Several commissions of Zaporizhzhia City Council also supported an appeal of deputies of Zaporizhzhia City Council to the President, Verkhovna Rada and the Cabinet of Ministers to adopt the law.

While the stated aim of the bill was to improve the ecological situation and social welfare in the region, the Kremlin's covert goal was to ensure the region's right to block decisions of the parliament and the government in Kyiv, effectively pursuing the "federalization" of Ukraine under the guise of solving ecological issues. The document Broide sent to Ardzinba argued that "the Zaporizhzhia region receives de facto the powers of a federal subject, endowed with the right of veto." The paper emphasized that "formally the Zaporizhzhia region will not receive the status of autonomy (since such attempts would inevitably lead to accusations of separatism), but in fact, it will receive broader powers than autonomy." The bill would provide local councils with "legal grounds for blocking all draft decisions of Kyiv affecting the Zaporizhzhia region." On June 6, 2016, the draft law, identical to the one Broide sent to Ardzinba, was registered in the Verkhovna Rada by a Zaporizhzhia deputy from the Opposition Bloc. However, the legal expertise of the parliament recommended that the bill be rejected on the grounds that it did not comply with the Constitution and other laws of Ukraine.[40]

In December 2015, the SBU intercepted a congress organized by the NGO "Social Zaporizhzhia," whose participants attempted to demand a special status. The SBU initiated criminal proceedings against the activists of the NGO under Article 110 of the Criminal Code of Ukraine (encroachment on the territorial integrity of Ukraine) on suspicion of separatism under the guise of environmental activities, but the case was closed without finding criminal elements.[41] The Office of the United Nations High Commissioner for Human Rights registered the case as a violation of the right to freedom of opinion and expression.[42] Nevertheless, the activities and members of Social Zaporizhzhia – which later evolved into a interfactional deputy group in the Zaporizhzhia City Council – were also reported by Broide to Ardzinba.

The Kremlin also promoted similar regional autonomy projects in the Odesa region (Porto-Franko project) and the Kharkiv region (Slobozhanshchyna project), although none of these projects produced tangible results for the Kremlin. These cases underscore the challenges faced by the state in promptly identifying influence operations and defining the scope of their participants – most of whom are likely unaware of the instructions coming from a foreign power – and holding them accountable for such actions (see McDonald's Chapter 3).

Interactions Between Covert Actions and Military Operations

In the spring of 2019, Volodymyr Zelensky became elected President of Ukraine. He had earlier criticized his predecessor Petro Poroshenko for making profits from war. In his election campaign, Zelensky pushed for a peace settlement to end the war in eastern Ukraine. However, the Normandy Format summit meeting in Paris, in December, where Zelensky finally met with Putin, did not produce any breakthroughs. The gaps between Ukraine's position on defending its sovereignty and territorial integrity and Russia's policy seeking to limit them were irreconcilable. The negotiations ended in a stalemate and Zelensky's Russia policy grew more similar to his predecessor's hardline stance.

Against this backdrop, in February 2020, Vladislav Surkov was dismissed from his role as Presidential Aide responsible for the Directorate of Cross-Border Cooperation. In an interview, several days after his dismissal, Surkov did not conceal his view of Ukraine, stating, "There is no Ukraine. There is Ukrainian-ness."[43] The dismissal of Surkov, who had overseen Ukraine policy since 2013, signaled a change in Russia's approach towards Ukraine.

According to senior Ukrainian officials, the ninth division (the Ukraine unit) in the FSB's DOI was reorganized as the Ninth Directorate in 2019, with its staff increasing from 30 to between 120 and 200 by the summer of 2021.[44] With the benefit of hindsight, this expansion marked a shift in Russia's objective in Ukraine – from subversion to occupation – although the Ukrainian government initially assessed it as a broadening of Russia's disinformation campaigns. Only later did it become apparent that the Ninth Directorate had developed a detailed map of potential local collaborators and personnel it could influence or coopt when it runs Russia's puppet administrations in the newly occupied territories. The Ninth Directorate also identified Ukrainians who were hostile to Russia and might lead resistance to the Russian occupation.[45] In the autumn of 2021, several Ukrainian individuals suspected by the SBU as Russian agents suddenly went on short trips to Turkey, Cyprus, and Egypt, likely for secret meetings with their FSB handlers.[46]

Penetration Into the Enemy's Camp – Counterintelligence Active Measures

Back in 2014, the FSB recruited senior officials from the Yanukovych government. Thereafter, it broadened its recruitment to targets such as aides and secretaries of members of parliament. For example, an assistant to Ukrainian parliamentarian Yury Zdebsky, a member of Ukraine's ruling party, People's Servant, was arrested in July 2022. The aide was recruited by the Fifth Service during his visit to Moscow in 2018. Receiving $1,500 a month, he passed internal information obtained from the parliament and his colleagues to his FSB handler. The compromised information includes details about Ukraine's anti-terrorist operation, circumstances in border areas, and U.S. military-technical assistance to Ukraine.[47]

Nevertheless, according to the SBU, the FSB's main assets were senior members of the Yanukovych regime who fled to Russia during the Euromaidan revolution in 2014, as well as the Medvedchuk-led OPZhZ party.[48] Furthermore, as with the goal of active measures, the FSB works to build up the position of its agents within the enemy's camp and its surroundings.[49] In Ukraine, the FSB succeeded in infiltrating the most sensitive parts of the target state, namely its counterintelligence agency.

The SBU exposed Major General of the SBU Valery Shaitanov, in 2020, on suspicion of working for the FSB's First Service (Department of Counterintelligence Operations) to assist Russia's terrorist acts and sabotage in Ukraine.[50] However, a more devastating blow to Ukraine's counterintelligence was revealed in 2022. In July, it was announced that Oleg Kulinich, the former head of the SBU General Directorate of the Autonomous Republic of Crimea, had been arrested on suspicion of treason. Kulinich was removed from office in early March 2022, at the beginning of Russia's full-scale invasion. From 1989 to 1994, Kulinich attended the KGB/FSB Academy and served in the Russian Federal Counterintelligence Service (the predecessor to the FSB). Kulinich was also known as the right-hand man of Volodymyr Sivkovich, the former Secretary of the National Security and Defense Council of Ukraine, who served as a KGB officer during Soviet times and fled to Russia after the Euromaidan in 2014. Despite Kulinich's track record, President Zelensky appointed him to lead the SBU Crimea directorate – relocated to Kherson following Russia illegally annexation of Crimea – in October 2020.

Two days after Kulinich's arrest was announced, the SBU head Ivan Bakanov, Zelensky's old friend, was removed from his duties by a decree signed by Zelensky. Bakanov's dismissal was based on Article 47 of the Disciplinary Statute of the Armed Forces of Ukraine, which envisages a "failure to perform service duties, which led to human casualties or other grave consequences, or created a threat of such consequences." According to Zelensky, a total of 60 officials of the General Prosecutor's Office and SBU defected to the Russian side in areas occupied by the Russian troops. Particularly, Kherson and Kharkiv regions had Russian collaborators, and a total of 651 criminal investigations were opened on suspicion of treason.[51]

Three days after Bakanov's dismissal, the National Bureau of Investigation (NBR) of Ukraine, which is responsible for investigating crimes committed by law enforcement officials, issued an indictment against Igor Chumakov, head of the Ninth Directorate (in charge of Ukraine) of the FSB's Fifth Service. According to the indictment, in 2019, Chumakov recruited Sivkovich and Kulinich (codename "Kotygoroshko"). Sivkovich set up a "political office" in Moscow and recruited collaborators in Ukraine under the guise of "political and social activities." The earliest communication between Sivkovich and Kulinich mentioned in the indictment occurred on June 2, 2019, ten days after President Zelensky took office and Bakanov was appointed first deputy head of the SBU.

The investigation established that Kulinich, as a confidant of Sivkovich, actively cooperated with the "political office" and implemented instructions to lobby for the appointment of people of Moscow's preference to various state and law enforcement agencies, and to conduct intelligence and subversive activities against Ukraine. It is alleged that Kulinich obtained intelligence about the plan of the Russian armed forces to carry out an attack from the temporarily occupied territory of Crimea on the mainland of Ukraine but deliberately hid this information from the SBU leadership. In the first hours of the full-scale invasion, according to the indictment, Kulinich blocked any attempts to inform the SBU leadership about the real situation in the region.[52] At the same time, the Russian intelligence service attempted to exert a "special positive influence."[53] Sivkovich instructed Kulinich to persuade the country's leadership to forsake NATO membership in favor of neutrality. They also allegedly discussed the possibility of hijacking the SBU counterintelligence directorate by promoting Andriy Naumov, the head of

SBU's Internal Security Directorate who fled to Serbia several hours before Russia launched full-scale assault, to the SBU first deputy chief.[54]

Earlier journalist investigations revealed that Ivan Bakanov, representing Zelensky's People's Servant Party during the 2019 presidential election, received support for his trip and lobbying efforts in Washington, D.C., from American lobbyist Marcus Cohen through the firm Signal Group Consulting, LLC.[55] In 2020, a Ukrainian journalist speculated that Cohen's funding came from Sivkovich, claiming that Bakanov, who later became SBU chief, reciprocated by promoting Kulinich to a key counterintelligence position.[56] During Kulinich's trial in October 2023, he stated that he had met with Bakanov at the request of American partners and at the invitation of Zelensky's election headquarters. He further claimed that his communication with Sivkovich, in 2019, was not as an FSB agent but as part of Zelensky's election campaign and that Signal Group Consulting was funded by Sivkovich.[57]

Another journalist investigation, in 2025, confirmed the allegation that Sivkovich had hired U.S. lobbyist Cohen. According to the investigation, in 2017, Cohen collected corruption and money-laundering information from U.S. law enforcement agencies and reported it to Sivkovich. In August of the following year, Cohen met Sivkovich and Kulinich in Vienna and passed them a report investigating Ukraine's top anti-corruption prosecutor, who had made it difficult for Moscow to fund influence operations in Ukraine. Cohen further introduced Kulinich to Zelensky's election campaign team, in 2019, and his lobbying for presidential candidate Zelensky was paid for by Sivkovich through shell companies. A senior Ukrainian counterintelligence official compared Kulinich to Kim Philby, who destroyed MI6 from the inside: "Kulinich was the most valued Russian FSB agent in Ukraine."[58]

Russia's penetration was not limited to the SBU. In August 2022, the Prosecutor General's Office of Ukraine disclosed information implicating a major general, who had served as a deputy head of the Main Intelligence Directorate (HUR) of the Ministry of Defense of Ukraine from 2008 to 2010, in suspected acts of treason. According to the investigation, the suspect was recruited, in 2014, and operated in the interests of Russian special services. He allegedly collected and transmitted classified and top-secret information about the military-political situation in Ukraine, the forms and methods used by the Armed Forces and information exposing Ukraine's informer networks in the temporarily occupied territory.[59]

Synchronization of Covert Actions and Combat Operations

The outcome of peace negotiations often depends not only on the battlefield's success but also on the enemy's situational awareness. Russia's maneuvers in political negotiations are closely coordinated with, and backed by, military actions. This strategy of "reflexive control" synchronizes covert actions with combat operations to maximize Russia's political gains at the negotiation table. For example, in September 2014, Russia advanced its regular army deeper into the Ukrainian Donbas near Ilovaisk to combat the Ukrainian armed forces and impose terms on Ukraine in the Minsk talks. Likewise, Moscow intensified the battle for Debaltseve in the run-up to the Normandy Summit meeting in Minsk, in February 2015.[60]

Leaked emails also suggest that, in June 2015, Vladislav Surkov, responsible for Russia's covert actions against Ukraine, received a document titled "Addition to the Package of Measures for the Implementation of the Minsk Agreements in Part Relating to the Establishment of a Demilitarized Zone in the Area of the Settlement of Shirokino," a strategically important city near Mariupol facing the Sea of Azov. The military disengagement map included in the same email demonstrates Surkov's attention to military positions of Russia's puppet "republic," which could be used to exert pressure on Ukraine during the Minsk talks.[61]

In preparation for and during the 2022 February assault on Kyiv, the FSB collaborated with Russian airborne forces. According to Bellingcat, it was the FSB Fifth Service that designed the plan to seize power in Ukraine through a blitzkrieg operation. The Fifth Service reportedly informed the Russian leadership that the Ukrainian people were increasingly dissatisfied with the Zelensky administration, and that upon the formation of a new government with Russia's intervention, the majority of Ukrainians would have no choice but to accept it without resistance.[62] At the end of 2021, the Ninth Directorate of the Fifth Service began testing scenarios together with Russian paratroopers to establish the command-and-control links between intelligence assets and military units in seizing critical infrastructure and government buildings, as well as locating and neutralizing Ukrainian resistance leaders.[63] Agents recruited by the Ninth Directorate also arranged apartments in Kyiv for FSB officers to stay in case of a successful military occupation. According to Ukrainian authorities, Igor Kovalenko, a senior officer of FSB's Ninth Directorate, accompanied Russian forces and was within a few miles of the center of Kyiv. However, the plan was aborted due to Ukraine's resistance and the retreat of Russian forces.[64]

GRU's Work in Ukraine and Competitions Between Agencies

The GRU, Russia's military intelligence agency, is notorious for a series of subversive operations and assassinations in Europe, such as the attempted poisoning of an arms dealer in Bulgaria, in 2015, the failed coup in Montenegro, in 2016, and the assassination attempt of former GRU officer Sergei Skripal in the United Kingdom using the chemical weapon Novichok, in 2018.[65]

In Ukraine, the GRU has complemented and sometimes competed with the FSB Fifth Service in creating agent networks. While the GRU's activities predominantly support military tactics and operations, it has also engaged in political intelligence and conducted active measures (see later).

Typically, during a conflict or shortly before it, GRU special units are formed from reserves that have undergone appropriate training, and they "settle" in the target territory. Alongside these detachments, special forces are deployed behind enemy lines to destroy strategic military targets: missile launchers, headquarters and command posts, army formations, ammunition depots, weapons, airfields, and naval bases. The GRU also conducts reconnaissance of enemy troops from illegal positions and carries out subversive measures behind enemy lines: sabotage, capturing "tongues" (enemy servicemen) and delivering them across the front line, and terrorist attacks against the enemy command and government officials.[66]

At the outset of the full-scale invasion of Ukraine, according to Christo Grozev, the number of GRU officers in charge of Ukraine was estimated at 30 to 40, with over 100 agents.[67] For example, in 2016, the GRU successfully recruited Ukrainian parliamentarian Andrii Derkach. In the early 1990s, Derkach was trained at the Academy of the Federal Counterintelligence Service (the predecessor to the FSB) together with Oleg Kulinich, his close friend and the former head of the SBU General Directorate of the Autonomous Republic of Crimea. Derkach headed Ukraine's state nuclear enterprise Energoatom for seven years, and before the full-scale war, belonged to Medvedchuk's pro-Russian party OPZhZ. In September 2020, Derkach was sanctioned by U.S. authorities as an "active Russian agent" for interfering in the U.S. presidential election. In June 2022, the SBU arrested Derkach's former aide Ihor Kolesnikov, a former employee of the GRU during Soviet times, as a liaison agent between the GRU and Derkach. Both Derkach and Kolesnikov were recruited by Vladimir Alekseyev, the GRU's first deputy director. It is reported that with millions of dollars of funding provided by the GRU, they established "security companies" in Kyiv and other major cities to help GRU Spetsnaz to infiltrate and suppress the Ukrainian resistance.[68] Just before the all-out invasion in February 2022, Derkach fled the country.

According to the SBU and the State Bureau of Investigation, incumbent member of parliament Oleksandr Dubinsky, with the call sign "Buratino," was part of a criminal organization set up by the GRU. The main task of this organization was "shaking the socio-political situation in Ukraine and discrediting the state in the international arena." The GRU financing of this network reached more than 10 million U.S. dollars. Specifically, Dubinsky disseminated disinformation about the higher military and political leadership of Ukraine, for example, the alleged interference of Ukrainian high-ranking officials in the U.S. presidential election, in 2019. These activities were coordinated by Derkach's ex-assistant Kolesnikov.[69]

Russian intelligence agencies are said to be competing with each other in their turf war, which Mark Galeotti calls "competitive intelligence."[70] Presidential Aide Surkov and the FSB were key competitors in Ukraine. Russian media reported conflicts between Surkov's presidential directorate and the security forces responsible for counterintelligence in the Russia-occupied territories.[71]

Moreover, there were overlaps between multiple agencies and agents in planning and conducting tasks in Ukraine. For example, in 2015, Aleksandr Kazakov requested Surkov to clarify why instructions given to other Russian collaborators were interfering with his task. Similarly, in that year, Russian Duma Deputy and Surkov's close associate, Mikhail Markelov, rumored to be working for Surkov's Directorate, informed him about potential duplications of subversive operations in Kharkiv.[72]

However, interagency rivalries and conflicts should not be overstated; they can be resolved with Putin's personal blessing. In 2013, when Putin appointed Surkov as the "curator" for Ukraine, the FSB and another Putin advisor Sergei Glazyev, who were both engaged in subversive activities in Ukraine, had to coordinate their efforts and share resources with Surkov. For example, the FSB's DOI maintained Serhii Kurchenko, a young oligarch and Yanukovych's closest ally who fled to Russia after Euromaidan, as an asset.[73] Leaked emails suggest that, in July 2014,

Surkov met with Kurchenko, who was accompanied by his FSB handler and DOI director Anatoly Bolyukh, supposedly to discuss leveraging Kurchenko's influential Ukrainian media resources to discredit the Ukrainian government and promote Donbas "separatism."[74]

Conclusion

Moscow's combination of active measures and military operations is designed to achieve its overarching goals. Russia employs active measures, either as standalone initiatives in non-military warfare in parallel with military operations or as measures directly supporting its military offensives.

For operational reasons, Moscow conducts active measures on multiple fronts to different aims. When faced with a stalemate in one direction, it seeks other opportunities for destabilization and subversion to have more bargaining chips in potential negotiations with the United States or major European powers that can exert pressure on Moscow's target countries. Thus, timing affects the intensity of using resources for covert actions. For example, in early 2016, as the Kremlin leaks show, liaison agent Aleksandr Kazakov repeatedly requested Surkov for new instructions for "DPR" head Zakharchenko, but Surkov ignored these requests, reflecting a temporary decrease in the Kremlin's interest in Ukraine amidst the military campaign in Syria.[75] According to investigative journalists RISE Moldova, at the beginning of Russia's full-scale invasion, Dmitry Milyutin, deputy head of the DOI, was instructed to develop a plan for a "second front" from Russia-occupied Transnistria.[76]

The effectiveness of Russia's covert actions in Ukraine remains contentious. For example, active measures against Ukraine, in 2013, achieved few of Moscow's strategic objectives. Instead of isolating Ukraine from the West, these efforts inadvertently pushed Russia's "fraternal country" toward European integration. Similarly, the "eco-special status" project in Zaporizhzhia, despite significant resource allocation, yielded no positive outcomes for Moscow. One can even make the case that it was the lack of success from earlier active measures that contributed to Russia's decision to launch the full-scale invasion in 2022. However, it is important to recognize that the invasion itself was not merely a reaction to past failures – it was also grounded in a wide-ranging and ambitious new set of active measures.

Yet, successful operations are often less visible, so one cannot solely base conclusions on exposed cases. Indeed, this presents another complication for understanding victory and defeat in non-military warfare (see Chapter 2). For example, the FSB's penetration into the heart of Ukraine's counterintelligence, showcased by Oleg Kulinich's case, likely aided Russia's successful initial military operations on the southern front during its full-scale invasion. Timely detection and investigation by counterintelligence agencies can disrupt and may deter active measures financed by Moscow, although such an intervention carries a risk of being perceived by international observers as inappropriate state pressure on civil society and freedom of speech.

Russian active measures used against Ukraine have also been applied elsewhere in Europe. In October 2022, *The Washington Post* revealed that the FSB

DOI secretly financed the political campaigns of Ilan Shor, a Moldovan opposition leader in exile, and deployed a team of political technologists to instigate massive protests aimed at derailing Moldova's path toward the European Union.[77] In August 2023, The Insider reported that Vladimir Sergienko, a naturalized German citizen and aide to a Bundestag deputy from the far-right party AfD, was recruited and instructed by the FSB to hinder Germany's provision of Leopard 2 battle tanks by filing a lawsuit to challenge the government's decision. In leaked communications between Sergienko and his FSB handler, the latter explicitly inquired: "By the way on active measures, has something advanced?"[78] According to *Expressen*'s report in April 2023, a leaked document from the Russian Presidential Directorate for Cross-Border Cooperation revealed an ecology project called the "Baltic Platform," aimed at shifting "from discussion of non-political topics to current political plans" and coopting Swedish counterparts.[79]

In sum, the Russian leadership have applied a wide range of active measures in support of its non-military warfare toward Ukraine. These measures are conducted not only through a variety of intelligence and security services but also through the Presidential Administration and more loosely connected entities, such as oligarchs and criminals. The Russian strategy of non-military warfare has sought to both substitute military means and, when that has failed, complement and aid their increase.

Notes

1 Jack Watling, Oleksandr V. Danylyuk and Nick Reynolds, "Preliminary Lessons from Russia's Unconventional Operations During the Russo-Ukrainian War, February 2022–February 2023" (London: The Royal United Services Institute, March 29, 2023), 20–26, https://rusi.org/explore-our-research/publications/special-resources/preliminary-lessons-russias-unconventional-operations-during-russo-ukrainian-war-february-2022.

2 Dennis Kux, "Soviet Active Measures and Disinformation: Overview and Assessment," The US Army War College Quarterly: Parameters 15, no. 1 (July 4, 1985): 19.

3 John J. Dziak, "Soviet Intelligence and Security Services in the 1980's: The Paramilitary Dimension," in Roy Godson (Ed.), Intelligence Requirements for the 1980's: Counter-Intelligence (Washington, DC: National Strategy Information Center, Inc., 1980); Amy Knight, The KGB: Police and Politics in the Soviet Union, Revised (Boston: Unwin Hyman, 1990), 284–85.

4 Sanshiro Hosaka, "Cold War Active Measures," in Rubén Arcos, Irena Chiru and Cristina Ivan (Eds.), Routledge Handbook of Disinformation and National Security (Abingdon, Oxon: Routledge, 2023), 45–58, https://www.taylorfrancis.com/chapters/edit/10.4324/9781003190363–7/cold-war-active-measures-sanshiro-hosaka.

5 This article treats covert actions broadly by including other components of intelligence, such as collection, analysis, and counterintelligence. It is not rare that the same individuals recruited to collect information are in a position to influence events. Roy Godson, "Covert Action: An Introduction," in Roy Godson (Ed.), Intelligence Requirements for the 1980's: Covert Action (Washington, DC: National Strategy Information Center, Inc, 1981), 2–6.

6 Sanshiro Hosaka, "The Kremlin's Active Measures Failed in 2013: That's When Russia Remembered Its Last Resort – Crimea," Demokratizatsiya: The Journal of Post-Soviet Democratization 26, no. 3 (2018): 321–64.

7 Sanshiro Hosaka, "Welcome to Surkov's Theater: Russian Political Technology in the Donbas War," Nationalities Papers 47, no. 5 (2019): 750–73, https://doi.org/10.1017/nps.2019.70.

8 This article used the third tranche of the Surkov leaks, which contains email communications of Vladislav Surkov's first deputy, Inal Ardzinba (1,046 emails from viktor_vinogr@mail.ru). For this leak, see Alya Shandra and Robert Seely, "The Surkov Leaks: The Inner Workings of Russia's Hybrid War in Ukraine," Occasional Paper (London: The Royal United Services Institute for Defence and Security Studies, 2019), 8–9, https://rusi.org/sites/default/files/201907_op_surkov_leaks_web_final.pdf.
9 Mark Galeotti, "Putin's Hydra: Inside Russia's Intelligence Services" (The European Council on Foreign Relations, May 11, 2016), 3, https://ecfr.eu/publication/putins_hydra_inside_russias_intelligence_services/.
10 Hosaka, "The Kremlin's Active Measures Failed in 2013," 351–54.
11 The KGB agent handler regulations encouraged an intelligence officer to establish close relationships with his agent so that the agent would not only accomplish intelligence tasks but also "share the most intimate things from his personal and family life" and "seek good advice and friendly support" from the KGB handler. V. Chebrikov, "Prikaz predsedatelya Komiteta gosudarstvennoi bezopasnosti SSSR № 00140 o vvedenii v deistvie Polozheniya ob agenturnom apparate i doverennykh litsakh organov gosudarstvennoi bezopasnosti SSSR [Order of the Chairman of the State Security Committee of the USSR No. 00140 on the introduction into force of the Regulations on the agent apparatus and proxies of the state security bodies of the USSR]" (Komitet gosudarstvennoi bezopasnosti SSSR, July 4, 1983), 19–20, https://s.kgb.arhivi.lv/public/psrs_vdk_pavele_1983_00140/psrs_vdk_pavele_1983_00140.pdf.
12 Hosaka, "The Kremlin's Active Measures Failed in 2013."
13 Hosaka, 354.
14 The KGB foreign intelligence manual recommends using senior politicians, businessmen, and bureaucrats in the target country as curators to give plausible deniability in the development of intelligence networks. Agents recruited by agents-gruppovod most likely believe they are tasked by senior officials of their own country. Watling et al., "Preliminary Lessons," 5.
15 James Sherr OBE and Igor Gretskiy, Why Russia Went to War: A Three-Dimensional Perspective (Tallinn: International Centre for Defence and Security, 2023), 9, https://icds.ee/en/why-russia-went-to-war-a-three-dimensional-perspective/; see also Owen Matthews, Overreach: The Inside Story of Putin's War against Ukraine (London: Mudlark, 2022).
16 Alona Mazurenko, "215 Defenders of Ukraine Released from Captivity: 200 of Them Exchanged for Medvedchuk," Ukrainska Pravda, accessed November 19, 2023, https://www.pravda.com.ua/eng/news/2022/09/22/7368539/.
17 Konstantin Skorkin, "What the Return of Kremlin Ally Medvedchuk Means for the War in Ukraine," Carnegie Endowment for International Peace, January 20, 2023, https://carnegieendowment.org/politika/88849.
18 Yury Fedorov, "Hybrid War à La Russe" (Kyiv: Center for Army, Conversion and Disarmament Studies, 2016), 98–101.
19 "Kak Kreml' vmeshivaetsya vo vnutrennyuyu politiku sosednikh stran. Chast' tret'ya: general Chernov, kuratory i filbifily [How the Kremlin interferes in the internal politics of neighboring countries. Part Three: General Chernov, curators and philbiphiles]," Dossier, September 2, 2020, https://dossier.center/chernov/.
20 "Chto takoe i zachem nuzhno Rossotrudnichestvo [What Is Rossotrudnichestvo and Why Is It Needed?]," Dossier, February 24, 2021, https://dossier.center/rossotr/.
21 A. I. Kolpakidi and Dmitriĭ Prokhorov, Imperiya GRU: Ocherki istorii rossiiskoi voennoi razvedki., 2 vols., Dos'e (Moskva: OLMA-Press, 2001).
22 Sanshiro Hosaka, "Welcome to Surkov's Theater: Russian Political Technology in the Donbas War," Nationalities Papers 47, no. 5 (September 2019): 750–73, https://doi.org/10.1017/nps.2019.70.
23 Russian media reported that not a single employee in Surkov's directorate engages the CIS countries. "Upravlenie s prigranichnymi vozmozhnostyami [Management

with Cross-Border Capabilities]," Kommersant", October 2, 2018, https://www.kommersant.ru/doc/3759024. This is confirmed by multiple leaked emails and internal reports of Surkov's directorate. The directorate dealt with the breakaway republics in the post-Soviet space, with a renewed focus on Ukraine, starting from the summer of 2013, when there was no signs of Donbas "separatism," Hosaka, "Welcome to Surkov's Theater," 753.
24 The fathers of Zatulin and Mitrofanova were Soviet intelligence officers. Primakov's grandfather was KGB agent and headed the post-Soviet Russian foreign intelligence service. Kirill was a KGB agent.
25 Hosaka, "Welcome to Surkov's Theater," 753.
26 Politie, "Witness Appeal June 2019: Chain of Responsibility in the Russian Federation 4 (8)," 2019, https://www.youtube.com/watch?v=hPGmFJH2ZO8.
27 Some political technologists are suspected to be intelligence officers at the same time.
28 Andrew Wilson, Ukraine Crisis: What It Means for the West (New Haven: Yale University Press, 2014), 34–35.
29 Hosaka, "Welcome to Surkov's Theater," 765.
30 Hosaka, 764.
31 "KAZAKOV Alexander Yurievich – Biography, Dossier, Assets," War and Sanctions, accessed November 17, 2023, https://sanctions.nazk.gov.ua/en/sanction-person/8954/.
32 Hosaka, "Welcome to Surkov's Theater," 761–62.
33 Hosaka, 752–53.
34 Hosaka, 761.
35 Petr Shuklinov, "Direktor Voiny [Director of War]," LIGA.Net, June 22, 2016, http://www.liga.net/projects/gricak_interview/.
36 Mark Krutov, "Oskolki 'Russkogo mira' [Fragments of the 'Russian World']," Radio Svoboda, November 10, 2017, https://www.svoboda.org/a/28846639.html.
37 "Eks-Kurator Ukrainy v Kremle Vozglavil MID Abkhazii [The Ex-Curator of Ukraine in the Kremlin Headed the Foreign Ministry of Abkhazia]," RBK, November 17, 2021, https://www.rbc.ru/politics/17/11/2021/619525fe9a794763d07eee57?from=from_main_1.
38 Hosaka, "The Kremlin's Active Measures Failed in 2013," 336; Hosaka, "Welcome to Surkov's Theater," 762. Since the start of the Russian invasion of Ukraine in 2014, Broide began to work for Surkov and developed plans for various news websites tailored to manipulate the Ukrainian audience, with their ties to the Kremlin sophisticatedly concealed.
39 "Spetsial'nyy status Zaporozh'ya – a pogovorit'?" Novyi den', December 19, 2014, https://nd.zp.ua/zp/66-specialnyy-status-zaporozhya-a-pogovorit.html.
40 "VYSNOVOK na proekt Zakonu Ukrayiny "Pro terytoriyi zi spetsial'nym ekolohichnym statusom" (reyestr. N 4785 vid 06.06.2016 r.)" [CONCLUSION on the draft Law of Ukraine "On territories with special ecological status" (Reg. No. 4785 of 06.06.2016)]. https://ips.ligazakon.net/document/XH3OX00A?an=33.
41 "SBU zakryla delo 'Sotsial'nogo Zaporozh'ya' [SBU closed the case of Social Zaporozhzhia]," Prav.DA!, October 12, 2016, https://pravda.in.ua/news-zaporozhye/sbu-zakrila-delo-sotsialnogo-zaporozhya/.
42 The report states: "In Zaporizhzhia, the 'Social Zaporizhzhia' NGO has faced pressure from the Zaporizhzhia Regional SBU. . . . One of the members told OHCHR that the NGO has suspended all public activity, fearing that they may 'go from being witnesses to becoming the accused'." Office of the United Nations High Commissioner for Human Rights, "Report on the Human Rights Situation in Ukraine 16 February to 15 May 2016," 2016, 31–32, https://www.ohchr.org/sites/default/files/Documents/Countries/UA/Ukraine_14th_HRMMU_Report.pdf.
43 Mike Eckel, "In First Interview Since Departure, Russia's Former 'Gray Cardinal' Questions Existence of Ukraine," Radio Free Europe/Radio Liberty, February 26, 2020, https://www.rferl.org/a/in-first-interview-since-departure-russia-s-former-gray-cardinal-questions-existence-of-ukraine/30456301.html.

44 Jack Watling and Nick Reynolds, The Plot to Destroy Ukraine (London: The Royal United Services Institute, February 15, 2022), 9, https://rusi.orghttps://rusi.org; " 'FSB i GRU konkurirovali v Ukraine'. Kak proval spetssluzhb stoil Rossii 40 000 zhiznei ["FSB and GRU competed in Ukraine," How the Failure of the Special Services Cost Russia 40,000 Lives]," LIGA, July 25, 2022, https://www.liga.net/politics/articles/fsb-i-gru-konkurirovali-v-ukraine-kak-proval-spetsslujb-stoil-rossii-40-000-jizney; Greg Miller and Catherine Belton, "FSB Errors Played Crucial Role in Russia's Failed War Plans in Ukraine," The Washington Post, August 19, 2022, https://www.washingtonpost.com/world/interactive/2022/russia-fsb-intelligence-ukraine-war/?tid=ss_tw.

45 Jack Watling and Nick Reynolds, "The Plot to Destroy Ukraine," 9–10.

46 Watling et al., "Preliminary Lessons," 5.

47 "SBU podozrevaet v rabote na FSB pomoshchnika nardepa iz Slugi naroda – video doprosa [SBU Suspects Assistant to People's Deputy from Servant of the People of Working for the FSB – Video of Interrogation]," LIGA, July 5, 2022, https://news.liga.net/politics/video/sbu-podozrevaet-v-rabote-na-fsb-pomoschnika-nardepa-iz-slugi-naroda-video-doprosa.

48 "SBU rasskazala, na kogo Rossiya delala stavku v Ukraine, planiruya vtorzheniye [The SBU Told Who Russia Relied on in Ukraine When Planning the Invasion]," LIGA, June 29, 2022, https://news.liga.net/politics/news/dehtyarenko-rasskazal-na-kogo-rossiya-delala-stavku-planiruya-vtorjenie.

49 Sanshiro Hosaka, "Repeating History: Soviet Offensive Counterintelligence Active Measures," International Journal of Intelligence and CounterIntelligence 35, no. 3 (2022): 431, https://doi.org/10.1080/08850607.2020.1822100.

50 "SBU razoblachila general-mayora, kotoryy rabotal na FSB RF i planiroval ubiystvo Osmayeva [The SBU Exposed a Major General Who Worked for the FSB of the Russian Federation and Planned the Murder of Osmayev]," ZN.ua, April 14, 2020, https://zn.ua/UKRAINE/sbu-razoblachila-general-mayora-kotoryy-rabotal-na-fsb-rf-i-planiroval-ubiystvo-osmaeva-351140_.html.

51 "Zelenskiii ob"yasnil otstranenie glav SBU i OGP bol'shim kolichestvom predatelei v organakh [Zelensky Explained the Removal of the Heads of the SBU and the OGP by a Large Number of Traitors in the Authorities]," LIGA, July 17, 2022, https://news.liga.net/all/video/zelenskiy-obyasnil-otstranenie-glav-sbu-i-ogp-bolshim-kolichestvom-predateley-v-organah.

52 Derzhavne byuro rozsliduvan', "Kolyshniy nachal'nyk holovnoho upravlinnya SBU v AR Krym Oleh Kulinich postane pered sudom (VIDEO) [Oleg Kulinich, the former head of the main department of the SBU in the Autonomous Republic of Crimea, will appear in court (VIDEO)]," accessed November 13, 2023, https://dbr.gov.ua/news/kolishnij-nachalnik-golovnogo-upravlinnya-sbu-v-ar-krim-oleg-kulinich-postane-pered-sudom.

53 Hosaka, "Cold War Active Measures," 47.

54 "Povidomlennya pro pidozru Chumakovu I.O. [Notice of Suspicion of Chumakov I.O," (Derzhavne byuro rozsliduvan', July 22, 2022), 9–12, https://www.gp.gov.ua/ua/posts/povidomlennya-pro-pidozru-cumakovu-io-original-ta-pereklad.

55 Christopher Miller, "Exclusive: Enigmatic Attorney Says He Hired D.C. Lobbyists for Ukrainian President," Radio Free Europe/Radio Liberty, August 8, 2019, https://www.rferl.org/a/ukraine-zelenskiy-cohen-lobbyist-washington/30100141.html.

56 Volodymyr Boiko, "Bakanov rozplatyvsya z Sivkovychem [Bakanov Paid Off with Sivkovich]," Naspravdi.Today, October 22, 2020, https://naspravdi.today/uk/2020/10/22/bakanov-rozplatyvsia-z-sivkovychem/.

57 Maksym Kamenev, ""Narodzhenyi v SRSR," U Kyevi pochaly sudyty Olehа Kulinicha – kolyshn'oho nachal'nyka kryms'koho hlavku SBU, yakoho zvynuvatyly u derzhavniy zradi ["Born in the USSR," The Trial of Oleg Kulinich, the Former Head of the Crimean SBU Headquarters, Who Was Accused of Treason, Has Begun in Kyiv]," Graty, November 2, 2023, sec. Без категорії, https://graty.me/narodzhenij-v-srsr-u-ki%d1%94vi-pochali-suditi-olega-kulinicha-kolishnogo-nachalnika-krimskogo-glavku-sbu-yakogo-zvinuvatili-u-derzhavnij-zradi-ta-shpigunstvi/.

58 Maxim Tucker, "How the Kim Philby of Kyiv Infiltrated Team Zelensky," The Times, February 8, 2025, https://www.thetimes.com/world/russia-ukraine-war/article/kim-philby-in-kyiv-became-zelenskys-aide-then-helped-putin-invade-m2703gl5c.
59 "Ekszastupnyka holovy HUR pidozryuyut' u derzhavniy zradi [The Ex-Deputy Head of the GUR Is Suspected of Treason]," Radio Svoboda, 15:33 2022, https://www.radiosvoboda.org/a/news-gur-ekszastupnyk-derzhzrada/31975346.html.
60 Sanshiro Hosaka, "Putin the 'Peacemaker'? – Russian Reflexive Control During the 2014 August Invasion of Ukraine," The Journal of Slavic Military Studies 32, no. 3 (July 3, 2019): 345–46, https://doi.org/10.1080/13518046.2019.1646950.
61 Hosaka, 330.
62 Miller and Belton, "FSB Errors Played Crucial Role in Russia's Failed War Plans in Ukraine."
63 Jack Watling and Nick Reynolds, "The Plot to Destroy Ukraine," 10.
64 Miller and Belton, "FSB Errors Played Crucial Role in Russia's Failed War Plans in Ukraine."
65 "The Dreadful Eight: GRU's Unit 29155 and the 2015 Poisoning of Emilian Gebrev," bellingcat, November 23, 2019, https://www.bellingcat.com/news/uk-and-europe/2019/11/23/the-dreadful-eight-grus-unit-29155-and-the-2015-poisoning-of-emilian-gebrev/.
66 Kolpakidi and Prokhorov, Imperiya GRU.
67 "FSB i GRU konkurirovali v Ukraine."
68 "GRU vydelyalo desyatki millionov dollarov Derkachu na zakhvat Ukrainy: detali razoblacheniya [The GRU Allocated Tens of Millions of Dollars to Derkach to seize Ukraine: Details of the Exposure]," LIGA, June 24, 2022, https://www.liga.net/politics/articles/gru-vydelyalo-desyatki-millionov-dollarov-derkachu-na-zahvat-ukrainy-detali-razoblacheniya.
69 "Dubins'komu povidomyly pro pidozru v derzhzradi – SBU," Radio Svoboda, 15:55 2023, https://www.radiosvoboda.org/a/news-dubinskyy-pidozra-derzhzrada-sbu/32682887.html.
70 Galeotti, "Putin's Hydra: Inside Russia's Intelligence Services," 4.
71 "Upravlenie s prigranichnymi vozmozhnostyami."
72 Hosaka, "Welcome to Surkov's Theater," 765.
73 "Spoils of War Failures: FSB General Who Botched Russia's War Preparations Owns Millions of Dollars Worth of Real Estate," The Insider, October 12, 2023, https://theins.info/en/corruption/265802.
74 Hosaka, "Welcome to Surkov's Theater," 762–63; For Glazyev's asset sharing, see Hosaka, "The Kremlin's Active Measures Failed in 2013," 346–47.
75 Hosaka, "Welcome to Surkov's Theater," 764.
76 "Moldova's Turn," Dossier, October 7, 2022, https://fsb.dossier.center/mld-en/.
77 Catherine Belton, "Russia's Security Service Works to Subvert Moldova's Pro-Western Government," The Washington Post, October 28, 2022, https://www.washingtonpost.com/world/2022/10/28/russia-fsb-moldova-manipulation/.
78 "Exclusive: Far-Right German Parliamentary Aide Tasked by Russia with Stopping Leopard Tanks to Ukraine," The Insider, August 4, 2023, https://theins.info/en/politics/264014.
79 Matthias Carlsson, "Kremls Bluff: Falsk Plan Ska Utnyttja Svenska Miljövänner Och Forskare," Expressen, April 27, 2023, https://www.expressen.se/nyheter/kremls-bluff-falsk-plan-ska-utnyttja-svenska-miljovanner-och-forskare/.

Bibliography

Bellingcat. (2019, November 23). *The dreadful eight: GRU's unit 29155 and the 2015 poisoning of Emilian Gebrev*. https://www.bellingcat.com/news/uk-and-europe/2019/11/23/the-dreadful-eight-grus-unit-29155-and-the-2015-poisoning-of-emilian-gebrev/

Belton, C. (2022, October 28). Russia's security service works to subvert Moldova's pro-Western Government. *The Washington Post*. https://www.washingtonpost.com/world/2022/10/28/russia-fsb-moldova-manipulation/

Boiko, V. (2020, October 22). Bakanov rozplatyvsya z Sivkovychem [Bakanov paid off with Sivkovich]. *Naspravdi.Today*. https://naspravdi.today/uk/2020/10/22/bakanov-rozplatyvsia-z-sivkovychem/

Carlsson, M. (2023, April 27). Kremls Bluff: Falsk Plan Ska Utnyttja Svenska Miljövänner Och Forskare. *Expressen*. https://www.expressen.se/nyheter/kremls-bluff-falsk-plan-ska-utnyttja-svenska-miljovanner-och-forskare/

Chebrikov, V. (1983, July 4). Prikaz predsedatelya Komiteta gosudarstvennoi bezopasnosti SSSR № 00140 o vvedenii v deistvie Polozheniya ob agenturnom apparate i doverennykh litsakh organov gosudarstvennoi bezopasnosti SSSR [Order of the Chairman of the State Security Committee of the USSR No. 00140 on the introduction into force of the Regulations on the agent apparatus and proxies of the state security bodies of the USSR]. *Komitet gosudarstvennoi bezopasnosti SSSR*. https://s.kgb.arhivi.lv/public/psrs_vdk_pavele_1983_00140/psrs_vdk_pavele_1983_00140.pdf

Derzhavne byuro rozsliduvan '. (2023, November 13). Kolyshniy nachal ' nyk holovnoho upravlinnya SBU v AR Krym Oleh Kulinich postane pered sudom (VIDEO) [Oleg Kulinich, the former head of the main department of the SBU in the Autonomous Republic of Crimea, will appear in court (VIDEO)]. https://dbr.gov.ua/news/kolishnij-nachalnik-golovnogo-upravlinnya-sbu-v-ar-krim-oleg-kulinich-postane-pered-sudom

Dossier. (2020, September 2). Kak Kreml' vmeshivaetsya vo vnutrennyuyu politiku sosednikh stran. Chast' tret'ya: general Chernov, kuratory i filbifily [How the Kremlin interferes in the internal politics of neighboring countries. Part Three: General Chernov, curators and philbiphiles]. https://dossier.center/chernov/

Dossier. (2021, February 24). Chto takoe i zachem nuzhno Rossotrudnichestvo [What is Rossotrudnichestvo and why is it needed? https://dossier.center/rossotr/

Dossier. (2022, October 7). Moldova's turn. https://fsb.dossier.center/mld-en/

Dziak, J. J. (1980). Soviet intelligence and security services in the 1980's: The Paramilitary dimension. In *Intelligence requirements for the 1980's: Counter-intelligence*. National Strategy Information Center, Inc.

Eckel, M. (2020, February 26). In first interview since departure, Russia's former 'gray cardinal' questions existence of Ukraine. *Radio Free Europe/Radio Liberty*. https://www.rferl.org/a/in-first-interview-since-departure-russia-s-former-gray-cardinal-questions-existence-of-ukraine/30456301.html

Fedorov, Y. (2016). *Hybrid War à La Russe*. Center for Army, Conversion and Disarmament Studies.

Galeotti, M. (2016, May 11). *Putin's hydra: Inside Russia's intelligence services*. The European Council on Foreign Relations. https://ecfr.eu/publication/putins_hydra_inside_russias_intelligence_services/

Godson, R. (1981). Covert action: An introduction. In *Intelligence requirements for the 1980's: Covert action*. National Strategy Information Center, Inc.

Hosaka, S. (2018). The Kremlin's active measures failed in 2013: That's when Russia remembered its last resort – Crimea. *Demokratizatsiya: The Journal of Post-Soviet Democratization*, *26*(3), 321–364.

Hosaka, S. (2019a). Welcome to Surkov's theater: Russian political technology in the Donbas war. *Nationalities Papers*, *47*(5), 750–773.

Hosaka, S. (2019b, July 3). Putin the 'peacemaker'? – Russian reflexive control during the 2014 August invasion of Ukraine. *Journal of Slavic Military Studies*, *32*(3), 324–346.

Hosaka, S. (2022). Repeating history: Soviet offensive counterintelligence active measures. *International Journal of Intelligence and CounterIntelligence*, *35*(3), 429–458.

Hosaka, S. (2023). Cold war active measures. In R. Arcos, I. Chiru, & C. Ivan (Eds.), *Routledge handbook of disinformation and national security*. Routledge.

The Insider. (2023a, August 4). *Exclusive: Far-right German parliamentary aide tasked by Russia with stopping leopard tanks to Ukraine.* https://theins.ru/en/politics/264014

The Insider. (2023b, October 12). *Spoils of war failures: FSB general who Botched Russia's war preparations owns millions of dollars worth of real estate.* https://theins.info/en/politics/264014

Kamenev, M. (2023, November 2). "Narodzhenyi v SRSR". U Kyevi pochaly sudyty Oleha Kulinicha – kolyshn ' oho nachal ' nyka kryms ' koho hlavku SBU, yakoho zvynuvatyly u derzhavniy zradi ["Born in the USSR." The trial of Oleg Kulinich, the former head of the Crimean SBU headquarters, who was accused of treason, has begun in Kyiv]. *Graty.* https://graty.me/narodzhenij-v-srsr-u-ki%d1%94vi-pochali-suditi-olega-kulinicha-kolishnogo-nachalnika-krimskogo-glavku-sbu-yakogo-zvinuvatili-u-derzhavnij-zradi-ta-shpigunstvi/

Knight, A. (1990). *The KGB: Police and politics in the soviet union* (Rev. ed.). Unwin Hyman.

Kolpakidi, A. I., & Prokhorov, D. (2001). *Imperiya GRU: Ocherki istorii rossiiskoi voennoi razvedki* (2 Vols.). OLMA-Press.

Kommersant. (2018, October 2). *Upravlenie s prigranichnymi vozmozhnostyami* [Management with cross-border capabilities]. https://www.kommersant.ru/doc/3759024

Krutov, M. (2017, November 10). Oskolki 'Russkogo mira' [Fragments of the 'Russian World']. *Radio Svoboda.* https://www.svoboda.org/a/28846639.html

Kux, D. (1985, July 4). Soviet active measures and disinformation: Overview and assessment. *The US Army War College Quarterly: Parameters*, *15*(1).

LIGA. (2022a, July 25). *'FSB i GRU konkurirovali v Ukraine'. Kak proval spetssluzhb stoil Rossii 40 000 zhiznei* ["FSB and GRU competed in Ukraine". How the failure of the special services cost Russia 40,000 lives]. https://www.liga.net/politics/articles/fsb-i-gru-konkurirovali-v-ukraine-kak-proval-spetsslujb-stoil-rossii-40-000-jizney

LIGA. (2022b, June 24). *GRU vydelyalo desyatki millionov dollarov Derkachu na zakhvat Ukrainy: detali razoblacheniya* [The GRU allocated tens of millions of dollars to Derkach to seize Ukraine: Details of the exposure]. https://www.liga.net/politics/articles/gru-vydelyalo-desyatki-millionov-dollarov-derkachu-na-zahvat-ukrainy-detali-razoblacheniya

LIGA. (2022c, July 5). *SBU podozrevaet v rabote na FSB pomoshchnika nardepa iz Slugi naroda – video doprosa* [SBU suspects assistant to people's deputy from Servant of the People of working for the FSB – video of interrogation]. https://news.liga.net/politics/video/sbu-podozrevaet-v-rabote-na-fsb-pomoschnika-nardepa-iz-slugi-naroda-video-doprosa

LIGA. (2022d, June 29). *SBU rasskazala, na kogo Rossiya delala stavku v Ukraine, planiruya vtorzheniye* [The SBU told who Russia relied on in Ukraine when planning the invasion]. https://news.liga.net/politics/news/dehtyarenko-rasskazal-na-kogo-rossiya-delala-stavku-planiruya-vtorjenie

LIGA. (2022e, July 17). *Zelenskiii ob"yasnil otstranenie glav SBU i OGP bol'shim kolichestvom predatelei v organakh* [Zelensky explained the removal of the heads of the SBU and the OGP by a large number of traitors in the authorities]. https://news.liga.net/all/video/zelenskiy-obyasnil-otstranenie-glav-sbu-i-ogp-bolshim-kolichestvom-predateley-v-organah

Matthews, O. (2022). *Overreach: The inside Story of Putin's War against Ukraine*. Mudlark.

Miller, C. (2019, August 8). Exclusive: Enigmatic attorney says he hired D.C. Lobbyists for Ukrainian president. *Radio Free Europe/Radio Liberty.* https://www.rferl.org/a/ukraine-zelenskiy-cohen-lobbyist-washington/30100141.html

Miller, G., & Belton, C. (2022, August 19). FSB errors played crucial role in Russia's failed war plans in Ukraine. *The Washington Post.* https://www.washingtonpost.com/world/interactive/2022/russia-fsb-intelligence-ukraine-war/?tid=ss_tw

Office of the United Nations High Commissioner for Human Rights. (2016). *Report on the human rights situation in Ukraine 16 February to 15 May 2016.* https://www.ohchr.org/sites/default/files/Documents/Countries/UA/Ukraine_14th_HRMMU_Report.pdf

Petr Shuklinov. (2016, June 22). Direktor Voiny [Director of war]. *LIGA.Net*. http://www.liga.net/projects/gricak_interview/

Politie. (2019). *Witness appeal June 2019: Chain of Responsibility in the Russian Federation, 4(8)*. https://www.youtube.com/watch?v=hPGmFJH2ZO8

Povidomlennya pro pidozru Chumakovu I.O. [Notice of suspicion of Chumakov I.O. (2022, July 22). *Derzhavne byuro rozsliduvan'*. https://www.gp.gov.ua/ua/posts/povidomlennya-pro-pidozru-cumakovu-io-original-ta-pereklad

Prav.DA! (2016, October 12). *SBU zakryla delo 'Sotsial'nogo Zaporozh'ya'* [SBU closed the case of Social Zaporozhzhia]. https://pravda.in.ua/news-zaporozhye/sbu-zakrila-delo-sotsialnogo-zaporozhya/

Radio Svoboda. (2022). *Ekszastupnyka holovy HUR pidozryuyut ' u derzhavniy zradi* [The ex-deputy head of the GUR is suspected of treason]. https://www.radiosvoboda.org/a/news-gur-ekszastupnyk-derzhzrada/31975346.html

Radio Svoboda. (2023). *Dubins ' komu povidomyly pro pidozru v derzhzradi – SBU*. https://www.radiosvoboda.org/a/news-dubinskyy-pidozra-derzhzrada-sbu/32682887.html

RBK. (2021, November 17). *Eks-Kurator Ukrainy v Kremle Vozglavil MID Abkhazii* [The ex-Curator of Ukraine in the Kremlin headed the foreign Ministry of Abkhazia]. https://www.rbc.ru/politics/17/11/2021/619525fe9a794763d07eee57?from=from_main_1

Shandra, A., & Seely, R. (2019). The Surkov leaks: The inner workings of Russia's hybrid war in Ukraine (Occasional paper). *The Royal United Services Institute for Defence and Security Studies*. https://rusi.org/sites/default/files/201907_op_surkov_leaks_web_final.pdf

Sherr, J., & Gretskiy, I. (2023). *Why Russia went to war: A three-dimensional perspective*. International Centre for Defence and Security. https://icds.ee/en/why-russia-went-to-war-a-three-dimensional-perspective/

Skorkin, K. (2023, January 20). What the return of Kremlin Ally Medvedchuk means for the war in Ukraine. *Carnegie Endowment for International Peace*. https://carnegieendowment.org/politika/88849

Tucker, M. (2025, February 8). How the Kim Philby of Kyiv Infiltrated Team Zelensky. *The Times*. https://www.thetimes.com/world/russia-ukraine-war/article/kim-philby-in-kyiv-became-zelenskys-aide-then-helped-putin-invade-m2703gl5c

War and sanctions. (2023, November 17). *KAZAKOV Alexander Yurievich – Biography, Dossier, Assets*. https://sanctions.nazk.gov.ua/en/sanction-person/8954/

Watling, J., Danylyuk, O. V., & Reynolds, N. (2023, March 29). Preliminary lessons from Russia's unconventional operations during the Russo-Ukrainian War, February 2022–February 2023. *The Royal United Services Institute*. https://rusi.org/explore-our-research/publications/special-resources/preliminary-lessons-russias-unconventional-operations-during-russo-ukrainian-war-february-2022

Watling, J., & Reynolds, N. (2022, February 15). The plot to destroy Ukraine. *The Royal United Services Institute*. https://rusi.org/explore-our-research/publications/special-resources/plot-destroy-ukraine

Wilson, A. (2014). *Ukraine crisis: What it means for the west*. Yale University Press.

ZN.ua. (2020, April 14). *SBU razoblachila general-mayora, kotoryy rabotal na FSB RF i planiroval ubiystvo Osmayeva* [The SBU exposed a major general who worked for the FSB of the Russian Federation and planned the murder of Osmayev]. https://zn.ua/UKRAINE/sbu-razoblachila-general-mayora-kotoryy-rabotal-na-fsb-rf-i-planiroval-ubiystvo-osmaeva-351140_.html

5 Cyber Crisis Management in Non-Military Warfare

Benjamin Jensen

Introduction

What is a clash of wills in a world connected by trade dependencies, changing popular attitudes toward violence, and entangled conventional and nuclear deterrent capabilities?[1] The question, which animates the larger edited volume, is particularly central to understanding what military officers and foreign policy leaders do for most of their professional lives: shape competition. Non-military interactions seeking a position of advantage are the constant, and war is the anomaly.[2] This axiom is especially on display in everything from economic warfare and disinformation campaigns to the methods of contesting space and the revival of cutting fiber optic cables amidst militarized disputes.[3] It is especially acute in cyber operation's a 21st century form of political warfare and subversion.[4] While all war involves politics, political warfare employs a mix of propaganda, messaging and other instruments of power, to include covert action, to influence an adversary's decisions.[5]

States increasingly use cyber operations to manage crisis interactions. As a form of covert action and communication, cyber operations support tacit bargaining in the shadows.[6] They allow states a means of ambiguous signaling, especially in situations defined by rivalry and an acute risk of escalation.[7] Cyber operations balance espionage and deception with the ability to shape strategic interactions and produce a position of relative advantage despite a weak coercive effect.[8] They are tailor made for escalation management, providing states options short of war to signal during militarized disputes.

This chapter evaluates the prospect of cyber operations as a non-military means of managing contemporary crisis interactions. Through a series of crisis studies, it illustrates how states turn to cyber operations as a means of managing escalation. The chapter proceeds by situating the logic of cyber operations in relation to the theory of coercion and escalation dynamics. From this vantage point, it reviews a set of illustrative cases in which states leveraged cyber interactions during tense standoffs on the precipice of war. The chapter concludes with a policy-oriented discussion on the implications for future crisis interactions and risks associated with holding modern digital connectivity hostage to great power competition in the 21st century.

DOI: 10.4324/9781003616993-5

Situating Non-Military Warfare

As defined earlier in this edited volume, non-military warfare embodies the use of non-kinetic tools of statecraft employed as a form of strategic coercion used to produce a strategic advantage. This implies that there is sufficient escalation risk and pain to change how each side evaluates their options. While modifying the word "warfare" is fraught with peril, the concept of non-military warfare explored in this edited volume offers a bridge between Cold War literature on deterrence and 21st century statecraft that takes place across more diverse networks and involves more instrument of power and manipulation as well as more actors beyond the state.[9] This competitive space sees an almost constant, often indirect, effort by actors to pressure target leaders, governments, and populations short of assuming the risk of employing military force. It harkens back to Quincy Wright's definition of conflict as "an extreme intensification of military activity, psychological tension, legal power, and social integration" albeit short of armed struggle since actors can leverage new means to achieve old ends: the power to compel rivals.[10]

Technological trends associated with connectivity offer new forms of competition. New technological means such as social media accelerate the weaponization of words.[11] Varied groups from diasporas to commercial cybersecurity firms (CSF) find themselves part of a larger web of intrigue.[12] Low-cost mechanisms to spread propaganda, send money, or start a movement amplify discord. As you add more ways and means for influencing the behavior of an adversary, the number and type of actors involved increases. This expansion makes it difficult to define participation in non-military warfare using the same norms and legal regimes used for interstate conflict.[13] Victory and defeat become elusive.[14] New escalation pressures emerge.

Non-military warfare increases the complexity of statecraft and increased complexity produces new uncertainty and the possibility for entirely new forms of inadvertent and accidental escalation risks. For example, state leaders must now consider how third-party intelligence – such as the coverage of Ukraine – can undermine state signaling, possibly leading to public pressure to act outside the range of flexible deterrent and response options under consideration by a national security bureaucracy.[15] Covert communication breaks down.[16] Leaders have to respond to an increasingly transnational public as opposed to directly, and in private, to other states. This dynamic is especially acute with respect to domestic audience costs. In the game theory literature on bargaining, crises, and conflict, there are two ways leaders signal each other during a crisis in an effort to deter an adversary: sinking costs and hand-tying.[17,18,19] Leaders who make public statements incur audience costs.[20,21] As a result, their threats are more credible.

Non-military means of compelling an adversary do not take military responses off the table. Leaders, especially those replying to larger transnational networks in real time, may find themselves prone to miscalculation or pressured to attack.[22] The central question then is what ways and means in modern strategy provide options for leaders constantly under siege in a non-military war to respond to crises short of war? In other words, how to respond during a militarized dispute without triggering a major war.

Crisis, Escalation, and War

Cyber operations do not take place in a vacuum. They are often part of a larger national strategy in which states seek a position of relative advantage against a rival. The question is how the domain affects statecraft and patterns of crisis escalation, and more specifically, what role the digital domain plays in providing non-military crisis offramps.

Since the 1970s, empirical research in international relations has differentiated between international crises, militarized disputes, and war.[23] In the Militarized Interstate Dispute (MID) dataset, these crises involve, "the threat, display or use of military force short of war by one member state is explicitly directed towards the government, official representatives, official forces, property, or territory of another state."[24] The international crisis behavior dataset developed coding schemes and narrative overviews of crises since the end of World War I, with an eye toward understanding patterns in crisis behavior.[25] The dataset defined a foreign policy crisis in relation to who each party in the dispute perceives the likelihood of conflict and threats to their values and interests. There had to be some threat of escalation in the present or future to count as a crisis.

Most studies on escalation historically focus on crises and disputes that occur before war along these lines as a means of better understanding coercion, to include deterrence, and bargaining. Escalation, as a process, helps states determine the balance of capabilities and adversary's level of resolve short of conflict.[26] Most statecraft involves the threat of force as opposed to its actual use, even if limited (i.e., conflict). Therefore, crises involve credible threats or demonstrations that keep violence always on the horizon, even if the episode never reaches conflict. As a form of competitive risk taking, escalation involves learning and probing the understanding of adversary capabilities and resolve.[27] The question is whether or not these interactions can be controlled or is subject to larger social, psychological, and institutional effects that can create self-reinforcing feedback loops and a runaway train to war.[28]

Escalation denotes increasing the level or location of a threat or taking additional coercive activity within a dispute in a manner that crosses a threshold that is significant to at least one party. It can be vertical (i.e., increase in magnitude of the threat) or horizontal (i.e., issue a new threat or take expanded coercive action in a new domain or geographic location). These categories relate to earlier ideas in Herman Kahn's seminal study, On Escalation, where he differentiated between increasing the intensity, widening the area, and compounding escalation as a combination of the two.[29] Escalation can be unilateral or reciprocal and takes different forms depending on the degree of control each party has (i.e., deliberate, inadvertent, accidental).

Generally, discussions about escalation dynamics fall within a broader literature on deterrence and crisis bargaining, and more recently, first-image perspectives on reputation and leadership experience that skew decision-making.[30] While coercion is as old as statecraft, the Cold War made the modern understanding of deterrence and the strategic logic of deterrence defined the Cold War era. Since the dawn of the nuclear age, most work on escalation focuses on the conditions likely to produce dangerous threat spirals, inadvertent escalation or chain reactions that change the character of a foreign policy crisis and lead states down a dangerous path to war. Much of this literature focused on escalation thresholds[31] and the idea of clear demarcations, firebreaks,[32] between different types of non-military (i.e., sub crisis maneuvering) and military activities on the escalation ladder.[33]

Early work on crisis dynamics and escalation combined formal models with empirical tests that treated escalation as a bargaining process.[34] Much of the literature built on foundational theoretical work on deterrence.[35] In early game theory work, escalation was often treated as variants of a threat game.[36] Players, whether individuals or great powers, select between a set of finite choices given imperfect information about what the adversary will do and calibrate their actions accordingly. These moves can stabilize interactions putting a cap on escalation.[37] Actors tend to base their decision on estimates of the other parties cost tolerance, how much pain they will endure or advantages they will lose to achieve a demand. Consistent with literature on issue salience, this cost tolerance is what determines escalation dynamics.[38] States with a high tolerance for the costs of escalation will take risks and endure hardship to compel an adversary to concede during crisis bargaining. The challenge is that cost tolerance is private information, making it difficult to gauge the optimal response during a crisis. States may be bluffing, or they may be willing to assume high costs associated with retaliation.

In studying escalation as part of crisis, bargaining scholars tend to differentiate between structural and decision-theoretic factors that shape how actors respond to perceived threats as well as between accounts that focus on current crisis decision-making or prevailing reputations.[39] The decision to escalate is either a question of underlying structural conditions like the balance of power or related to iterated decision-making under imperfect information linked to reputations and what states assume the other side will do. Traditionally, structural accounts examine the balance of capabilities between actors, using power assessments to determine relative stability in a crisis dyad. Intervening factors can alter how states perceive the balance of capabilities, such as geographic proximity, strategic weapons (i.e., nuclear, ballistic missiles), issue salience, and regime type.

Multiple empirical studies in the 1980s and 1990s found that military power plays a secondary role in bargaining processes. Hence, while military capabilities likely shape decisions about how to respond to a crisis, the decision to escalate is often more dependent on the issue at stake.[40] A favorable military balance does not determine whether or not a state will concede in a crisis.[41] Past studies demonstrated that the difference between observed and actual capabilities alters how rational actors calculate cost and benefit during a crisis interaction.[42] There is a non-monotonic relationship, causing successive interactions to produce new information about capabilities that can alter the probability of escalation.[43] In game theory literature, arms races tend to increase the risk of escalation,[44] a finding confirmed by quantitative analysis.[45] The effects of regime type, advanced economies, continuity, nuclear weapons, rivalry, and rough power parity do not alter the relationship between arms races and escalation risks.[46]

Structural conditions often shape how a state in a crisis perceives the other party and calculates credibility.[47] Consistent with the "current calculus" school of thought, states determine the credibility of a threat, and based on it make decisions about escalation, based on available information in context.[48] For example, regional interests likely loom larger than global interests. States will pay higher costs, assume more pain, and take more risk based on their geographic proximity.[49] In addition to geography, issue salience influences crisis tactics. Higher stakes issues tend to see more escalation and coercion.[50] Key capabilities play a role, with nuclear weapons associated with less escalation involving violence historically[51] or in other studies, increase diplomatic status but do not alter crisis behavior.[52] These findings transfer to non-nuclear strategic weapons, such as ballistic missiles, with empirical studies demonstrating that states are less likely to escalate in a crisis involving a party that possesses ballistic missiles.[53]

Other factors create intangibles likely to skew crisis decision-making. Where states are with respect to a power cycle, changing position relative to other states in the international system, can change risk propensity.[54] Specifically, it can alter the psychology of decision-making, creating new cognitive frames based on the perceived loss of power and status. There is a relationship between increasing uncertainty and risk propensity.[55] Low resolve implies a disposition in which actors will not assume risk in a crisis to achieve their objectives. High resolve alternatively means a state is willing to assume high costs and risks to achieve their objectives. The challenge is that this resolve is private information. There are also structural aspects of the strategic environment that can heighten the security dilemma and the prospect of inadvertent escalation. In strategic environments where actors have a difficult time assessing the military balance and distinguishing between offensive and defense force postures and capabilities, rational calculations tend to be skewed.[56]

Alternatively, decision-centric approaches examine how hypothetical agents process information and make decisions, usually assumed to be utility maximizing given imperfect information.[57] Consistent with bargaining theory, crisis interactions are defined by private information.[58] Crisis bargaining is generally thought of as a communitive process involving threats, bluffs, and discrete escalatory moves.[59] Signaling plays a critical role, as each side seeks to achieve the most leverage without paying high costs. While earlier studies assumed escalation was a mechanism for signaling resolve, game-theoretic treatments illustrated that because of private information about cost tolerance and objectives less committed actors can bluff.[60] In these models, states try to get the other party to concede by making irrevocable commitments that limit the ability to back down[61] or making a "threat that leaves something to chance" – raising the risks of an escalation spiral beyond what both parties want. Beyond the classic early Cold War ideas about the balance of risk in Schelling and balance of terror in Wohlstetter, there was a balance of information that defined bargaining processes.[62]

Seen in this light, the decision to escalate is a deliberate act to alter the projected costs of continuing the crisis interaction or a move to reveal the opponent's resolve. It signals the costs the opponent will have to pay as well as the costs the antagonist is willing to assume to achieve their objective.[63] The escalation moves and larger crisis bargaining continues until one side assesses that the relative costs of future interactions outweigh the anticipated benefits of further escalation and opts to concede.[64]

This iterated learning tends to mean states put a premium on reputations. Past diplomatic behavior, what Schelling treated as the interdependence of commitments, can create reputations that enhance credibility.[65] Alternatively, prior beliefs and current crisis considerations (e.g., balance of power, interests at stake) can alter reputations and state preferences.[66] A reputation for resolve is key to crisis bargaining. Resolve involves "the extent to which a state will risk war to achieve its objectives."[67] For example, past work finds that countries that did not escalate to use force in prior crises are more likely to face challenges in the future.[68] Assumptions about credibility are the key determinant of behavior.[69] Bargaining power is a function of the "perceived comparative resolve" producing incentives to escalate early.[70] In empirical studies, the first party to threaten, display, or use limited force is more likely to prevail.[71]

Despite the importance of reputations, in a crisis, states tend to prioritize new information.[72] There are also reasons linked to cognition and psychology to assume that state leaders will not base their decisions on the past.[73] Other studies find that personal interactions and even emotions alter strategic assessments in ways unaccounted for by game theoretic accounts or past empirical studies of structure factors linked to crisis behavior.[74]

Regardless of whether one takes a structural or decision-theoretic approach, there is a question about whether disruptive capabilities should alter escalation dynamics. In studies from the late Cold War, precision strike capabilities – the manifestation of the Revolution in Military Affairs – destabilized nuclear dyads and increased the risk of inadvertent escalation.[75] Along similar lines, Caitlin Talmadge saw similar risks in the combination of cyber, electronic warfare, and stealth interdiction of Chinese military targets called for by the AirSea Battle concept.[76] Separately, multiple studies see a relationship between the security dilemma and changes in the offense-defense balance.[77] Any new capability that alters the ease of the offense has the potential to destabilize crisis interactions. Other studies contend that disruptive capabilities are either insufficient to alter crises or that technology itself tends to be misunderstood. Technology, in the narrow material sense of the term, is rarely "an independent, primary driver of otherwise avoidable escalation."[78]

Cyber Operations and Crisis Escalation

Many of the mechanisms of non-military warfare theorized in this volume rely on the connectivity of the internet even if just to share information and mobilize target populations. As such, cyber operations and the ability to conduct offensive and defensive operations along physical, logical, and personal layers of cyberspace should take a central role in modern strategy.

Cybersecurity scholars are split on the extent to which cyber operations are revolutionary and alter escalation dynamics. In multiple studies, Borghard and Lonergan find that cyber operations are a weak coercive instrument making them "poor tools of escalation."[79] More specifically, four attributes limit the utility of cyber operations as an instrument of escalation: (1) the likely absence of offensive options at the time of need; (2) uncertain effects in the complex networks that make up cyberspace; (3) inherent political, technical, and intelligence trade-offs in using (and likely losing) cyber tools alongside the spillover risk; and (4) the limited costs cyber operations can impose compared with more directly instruments of power, such as military strikes, severing diplomatic relations, and economic coercion. Cyber operations tend to focus on espionage, deception, and shaping activities.[80] As a result, "cyber strategies can be viewed as ambiguous signals that limit escalation. Escalation in cyberspace is likely a function of norms, past behavior, and resolve rather than capability leveling."[81] For Rebecca Slayton, there is a dyadic offense-defense balance in cyberspace, but it is dependent on human capital and does not extend to the physical world.[82]

Other scholars see unique offense-defense dynamics and increased risk of escalation in cyberspace. For Ben Buchan, cyberspace is escalation prone because the high levels of uncertainty acerbate the security dilemma.[83] For Lucas Kello cyber capabilities expand, "the range of possible harm and outcomes between the concepts of war and peace – with important consequences for national and international security."[84]

Applying escalation theory to cyberspace, cyber escalation is defined as situations where "the target responds with more intense and costly cyber means (cyber escalation within the domain) or through breaching the cyber-kinetic threshold (cross-domain escalation)."[85] According to Libicki, cyber escalation is defined by two attributes, "intensity (striking deeper, lasting longer) or more extensive (striking new targets) [and] . . . attacks can jump from cyberspace to physical space."[86] These attributes parallel the notion of vertical and horizontal escalation.[87]

Yet empirical studies as well as survey experiments show that escalation is not the norm with respect to cyber operations.[88] First, because of the unique character of the cyber domain and the fact that it is one of many instruments of power states use for leverage, it is difficult to isolate whether changes in intensity or cross – domain dynamics are a function of the instrument (i.e., cyber) or the larger strategy and available options open to governments during a crisis. When we do see escalatory options used, they tend to be demonstrated in ongoing conflict situations, such as Israel's kinetic attack on Hamas cyber operators in 2019.[89] There are also clear cases, such as the crisis between the United States and Iran in 2019, where cyber operations are as likely to be used to stabilize a crisis and even facilitate de-escalation by shifting the domain of competition away from military forces to network intrusions.[90] This finding finds further resonance in randomized controlled experiments on national attitudes toward using cyber operations during a crisis.[91]

Cyber operations can work as a form of signaling. As Libicki notes, "escalation can convey, for instance, that lesser activities (for example, cyberespionage) have reached a point at which the pain is tantamount to that of a cyberattack. Russians call this escalate to de-escalate."[92] Fischerkeller and Harknett suggest that "escalation in cyberspace, then, is defined as an increase from the level of agreed competition to conflict."[93] As they note,

> the objective of this coupling [the nexus between deterrence and escalation] is to achieve escalation dominance – the condition in which an adversary's next move must be to accept the status quo or back down, because further

escalation would be counterproductive to interests. Coercive escalation strategies like those developed by Kahn are viable in the strategic space of armed conflict, including cyber, due to the nature (and threat) of war.[94]

Because rival states tend to have repeated crisis interactions, the threat of force is always on the horizon. Non-military warfare – herein active campaigning and competition – becomes the norm and offers a range of options for signaling rivals short of resorting to the costly gamble of military force and risk of a larger war. Back to Schelling, both acute crises and everyday competition become a competition in risk taking. In these interactions, cyber operations offer both an instrument of escalation beneath the threshold of armed conflict and another tool for crisis management. This phase space and characteristic doesn't limit the wartime utility of cyber operations. Rather, it points to their greater utility in non-military warfare and disputes as a shaping and signaling mechanism, setting the conditions for transitioning to military operations and demonstrating the risks of not backing down. Even though cyber operations are weak coercive tools in isolation, the fact that they are deniable makes them the perfect tool of covert communication between rivals, providing political leaders the option of doing something that can be denied, be reversed, and is unlikely to trigger a disproportionate military response.

Since much of the cyber literature and popular wisdom focuses on escalation and how cyber operations offer complements to espionage and military campaigns, this chapter explores the alternative: cases in which cyber operations offered an offramp to an otherwise escalating crisis. Each illustrative case maps a crisis between at least two states and shows how cyber operations offered both a signaling and crisis management tool. Rather than produce dangerous escalation spirals, cyber operations appear to have had a moderating influence that helped create the time and space for leaders to back down and recalibrate their approach to a rival during a crisis.

Crisis Management Cases

Based on the insights in the proceeding section, contemporary crisis cases should show the following. First, cyber operations are likely to be used alongside other instruments of power during a crisis that involves the threat of force consistent with definitions of crises and disputes discussed above. Second, cyber operations offer an offramp during periods of escalating military confrontations. Both dynamics should be apparent in crises. To the extent these dynamics are present and plausible it suggests that cyber operations can, at least in part, offer states a mechanism for responding to a crisis short of escalating to the more costly and risk laden game of combat.

Russia-Turkey 2015

In late fall 2015, Russia and Turkey descended into a MID. The dispute marked the highpoint of a series of military actions each took in support of different factions in the Syrian civil war, which started in 2011. Starting in September 2015,

Russian forces deployed to Syria not only in support of the regime at the request of the regime but also to secure military and energy access to the Mediterranean.[95] The initial deployments focused on airpower, with over thirty aircraft alongside intelligence and logistics flowing into Domna airbase.[96] The aircraft, which grew to include helicopters later, consisted predominantly of fighter/attack platforms, including modernized Su-25M, Su-24M, and Su-34. During the first month of the air campaign, Russian forces focused on attacking opposition groups, including the United States and partner-backed rebel organizations, in support of the Syrian regime despite an official counter-Daesh/ISIL narrative.[97] These attacks included aerial raids in the cities of Homs and Hama that led to significant civilian casualties.[98]

In early October, Russian airstrikes continued, including combined offensives with Hezbollah and Iran-backed militias supporting the Syrian regime.[99] As the fighting escalated, Russian aircraft violated Turkish airspace along the border in the Hatay region despite repeated warnings from the Turkish air force resulting in a formal alliance consultation on October 5, 2015.[100] Both NATO and Turkish officials condemned the incursions as intentional-based MiG-29 aircraft and missile batteries in Syria both locking radar on Turkish F-16s patrolling inside Turkish airspace.[101] The episode was especially threatening to Ankara given past air incursion by Syrian forces and the 2012 downing of a Turkish F-4 over the Mediterranean by a Syrian SAM battery.

Over the course of the fall of 2015, Turkey made repeated claims that Russia violated its airspace.[102] Events took a dramatic turn on November 24, 2015, when Turkish F-16s engaged a Su-24 with air-to-air missiles inside Turkish airspace.[103] Turkish authorities claimed the Russian aircraft were warned ten times they were violating Turkish airspace before the F-16s engaged. Russian authorities denied receiving warnings and, without evidence, claimed the attack was a deliberate ambush orchestrated by the United States and NATO. The situation further escalated after Syrian rebels, reportedly from the village the Russians were bombing, killed one of the two Russian pilots who ejected. The other pilot, along with a Russian naval infantryman, was killed during the recovery mission, likely by the same rebel faction.

The incident marked the first time a NATO member state engaged a Russian plane since the Cold War. Of note, within twenty-four hours of the air exchange, suspected Turkish hackers conducted a distributed denial-of-service (DDoS) attack on the Russian Central Bank website, taking it offline for ten minutes.[104]

The Russian response to the crisis illustrated the contours of contemporary coercive campaigns and the supporting role of cyber operations. Within four days, Russia deployed S-400 capable of ranging large sections of Turkey to forward airbases in Syria while imposing a number of punitive economic measures. First, Russia announces a suspension of visa-free travel and cancels charter flights in a deliberate effort to undermine the Turkish tourist sector. Second, Putin signed a 28 November decree limiting imports from Turkey and the ability of Turkish firms and citizens to operate inside Russia. These initial measures focused on Turkish agricultural exports and, similar to Chinese economic warfare, use a mix of inspections and quasi-legal measures to restrict access instead of sanctions.[105] To expand its non-kinetic options, by the end of the month, Russia deployed electronic warfare capabilities as well.[106]

By December, both sides conducted a series of cyber disruption operations.[107] Hacker groups linked to Turkey defaced the Russian communication minister's social media accounts. Separately, Turkish media claimed Russian hackers targeted Turkish internet service providers responsible for over 300,000 websites.

Russian cyber operations shifted from infrastructure to financial targets by the end of the month with Turkish banks reporting challenges with credit card transactions. These efforts were accompanied by a broad-based information operation, with propaganda and misinformation seeking to skew how media outlets portrayed the event.[108]

While cyber operations escalated over the course of January 2016, there was not a corresponding increase in military activities by either country. On January 4, 2016, the Turkish technology minister announced that over the course of December, 400,000 Turkish websites were subject to DDoS attacks, forcing the government to temporarily cut off foreign traffic to .tr websites.[109] In a subsequent study, CrowdStrike reported Turkish groups responded with cyber operations targeting the Russian banking infrastructure and the Russian Federal Security Service.[110] In response, Russian groups continued probing Turkish critical infrastructure and released the personally identifiable information for nearly 50 million Turkish citizens.[111] The crisis ended by summer 2016 when the Turkish president apologized for downing the Russian jet despite its clear violation of Turkish airspace.

The case illustrates how states can substitute cyber operations for more costly military operations as a form of crisis signaling. Despite starting as a traditional militarized dispute with the downing of a Russian aircraft, as the crisis evolved both states reached for non-military measures to manage escalation risks without ceding the initiative. Officials in Ankara and Moscow both opted to respond with economic and cyber effects while minimizing the role of military forces in an effort to respond to the crisis without triggering dangerous escalation pathways. In this

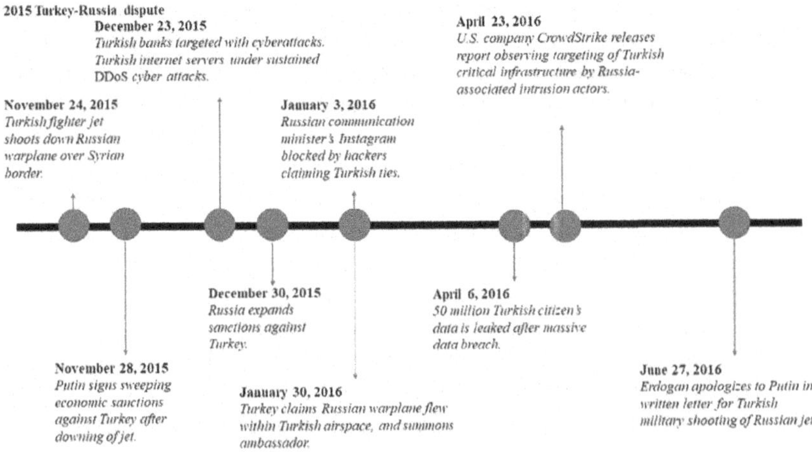

Figure 5.1 A Timeline of the 2015 Russia-Turkey Dispute

manner, the case illustrates the utility of cyber operations as a tool of crisis management. Cyber operations offer states a weak coercive instrument but effective signaling mechanism to respond to rivals.

China-India 2020

In Spring 2020, China and India descended into a militarized dispute over competing territorial claims near Tibet and the Pangong Lake along their border. Since 1947, the countries have had multiple border crisis events, the deadliest of which descended into a short war in 1967. After the winter thaw in 2020, both China and India started to build up military forces and installations along the Line of Actual Control (LAC) in the Ladakh region of the Himalayan Mountain range.[112] In the first week of May, Chinese and Indian forces engaged in a melee after Indian forces accused Chinese troops of illegally crossing the India border.[113]

By June, the situation escalated. On June 15, 2020, in the Galway Valley, a second encounter between Chinese and Indian forces resulted in the death of 20 Indian and four Chinese troops.[114] In response, Indian citizens called for boycotts of Chinese goods.[115] Over the course of June, there were multiple street protests across India.[116] By the end of the month, the Indian government announced a ban on multiple Chinese apps, including TikTok, citing national security concerns.[117]

The Chinese responded to the escalating dispute in the Himalayas through cyber operations. On June 23, 2020, multiple news outlets reported a significant rise in suspected Chinese intrusions, including efforts to hack the Indian banking sector.[118] They also included a long-term effort to hack the Indian power grid uncovered in 2022.[119] These activities included up to 14 trojan horses installed in Maharashtra State Electricity Transmission Company's network.[120]

Chinese and Indian forces continued to deploy forces and expand military infrastructure along the LAC through the summer and into the fall of 2020. On September 8, 2020, the forces clashed again, increasing tensions as soldiers moved from using fists and makeshift weapons to exchanging gunfire.[121] Within a month, Mumbai, India, experienced significant blackouts and disruptions to its power grid leaving millions without power and halting the train system.[122] *The Indian Press* claimed the blackout was a direct result of Chinese cyber operations.[123] Subsequent reporting by *The New York Times* and Recorded Futures attribute the blackouts to Chinese cyber operations.[124] Yet the Indian government refused to attribute the blackouts to Chinese cyber operations, claiming while there were cyberattacks targeting Indian infrastructure the actual blackout was the result of human error.[125] At the same time, there are reports that China conducted a widespread cyber espionage campaign targeting Indian vaccine producers.[126]

Much of the cyber operation on the Chinese side involved the Red Echo Group, ShadowPad, and APT41. Prior to the dispute with India, these groups had been associated with Chinese-linked intrusions across the Asia-Pacific, including major operations in Taiwan and the Philippines.[127] This pattern is consistent with larger studies on China's cyber strategy and their preference for a mix of espionage and signaling activities.[128]

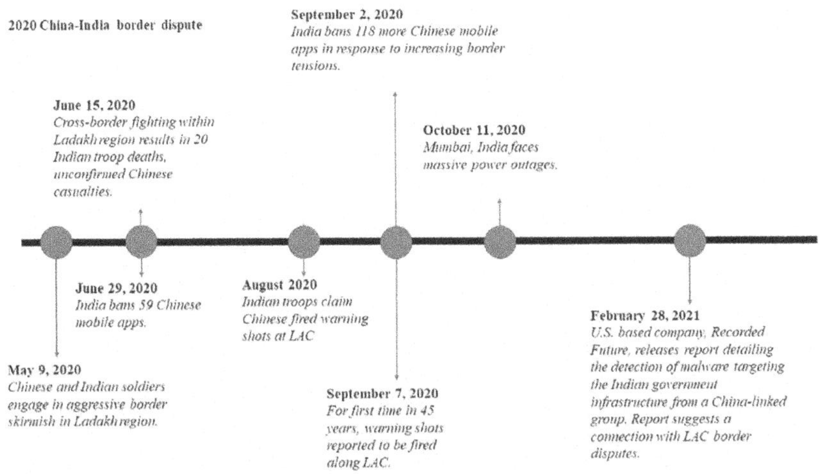

Figure 5.2 A Timeline of the 2020 China-India Dispute

The case illustrates how nuclear powers can substitute cyber effects for higher risk military responses during a crisis. Initiated as a territorial dispute, the crises between China and India after 2020 saw cyber operations employed as a means of signaling and shaping. Deniable cyber operations gave Beijing a mechanism to signal the risk of escalation while also setting conditions for digital sabotage in the event a larger conflict broke out. Like the previous case, the crisis interaction demonstrates the utility of cyber operations as a tool of crisis management.

Of particular interest, this case involved nuclear-armed states with a history of border disputes. In the broader conflict literature, past disputes, rivalry, and ongoing territorial disputes increase the probability of more tense, hostile interactions.[129] Given this dynamic, cyber operations appear to have offered an offramp, with each state employing them to signal the risks of escalation without triggering an event that created a conflict spiral or incentivizing the broader use of military force. In this respect, the case further illustrates the crisis management utility of cyber operations in modern militarized disputes.

Looking Back to Look Ahead: The Next Crisis

The clash of wills in a connected world takes place in and through cyber networks. This clash takes the form of war and non-military warfare, incessant competition, between rival states and increasingly involves whole new categories of actors and attack targets from CSF and Advanced Persistent Threat groups to banks and power plants. The push of new technology and the pull of gray zone competition create a world never at peace and always on the cusp of confrontation. While the cases reviewed here show how cyber operations offer response options in crises short of war, they do not show a world bound for a new era of peace.

Cyber operations between rival states are becoming a norm of modern crisis interactions and a major instrument of non-military warfare. They enable states to compete in the shadows and signal the risks of escalation short of taking non-reversible actions that could trigger a broader conflict. Because the networks in cyberspace are largely private, this new competition will blur the line between state and non-state as it relates to targeting and strategy.[130] This dimension is clear from the illustrative cases above. Banks, power plants, and individual identity are all targets of coercion in non-military warfare and competition along cyber networks. While cyber operations offer a means of managing modern crisis interactions, the extent to which they provide an offramp is not cost free. The cost appears to be shifted from government to civilians.

While the cases show cyber operations used as substitutes and complements during a crisis, there is no guarantee hacking guarantees the peace. As seen in Ukraine, cyber operations can also accompany major combat operations. Cyber tools and malware are instruments and the intentions of those who wield them are what tip the balance between war and peace. Used as a form of non-military warfare, they offer viable mechanisms to shape and signal beneath the threshold of armed conflict. Used in war, they can help shape the battlespace and support intelligence-driven targeting.

What do these observations imply about the future of great power competition in the 21st century? First, the world is likely entering a new phase of competition, likely to see each of our personnel and private lives become part of statecraft. The more states turn to cyber operations as a way and means to wage non-military warfare against their rivals and manage crisis interactions, the more likely services people around the world rely on are to become targets. This character of competition increases the importance of policymakers creating new forms of national strategy to ensure the connectivity we all rely on for future crises and isn't held hostage.[131] Second, just because cyber operations have offered a crisis management tool to date does not mean they will do so in the future. The cases, though illustrative, cannot conclusively prove that it wasn't nuclear weapons that didn't stabilize the disputes. This observation resonates with other work on cyber strategy that finds that it is almost impossible to control for the effects of national wealth and military capabilities when analyzing dyadic cyber interactions.[132]

Notes

1 Charles Cleveland, Benjamin Jensen, Arnel David and Susan Bryant. Military Strategy in the 21st Century: People, Connectivity, and Competition (New York: Cambria, 2018). James M. Acton "Escalation through Entanglement: How the Vulnerability of Command-and-Control Systems Raises the Risks of an Inadvertent Nuclear War," International Security 43, no. 1 (2018): 56–99.

2 See Chapter 1, Introduction; Therese Pettersson and Magnus Öberg, "Organized Violence, 1989–2019," Journal of Peace Research 57, no. 4 (2020): 597–613. Kristi.

3 Nicholas Mulder, The Economic Weapon: The Rise of Sanctions as Tool of Modern War (New Haven: Yale University Press, 2022); S. C. Woolley and Philip Howard, Computational Propaganda Political Parties, Politicians, and Political Manipulation

on Social Media (New York: Oxford University Press, 2019); Philip Howard, Lie Machines How to Save Democracy from Troll Armies, Deceitful Robots, Junk News Operations, and Political Operatives (New Haven: Yale University Press, 2020). Todd Harrison, Kaitlyn Johnson and Makena Young, Defense Against the Dark Arts: Protecting Space Systems from Counterspace Weapons (Washington, DC: Center for Strategic and International Studies, 2021); Thomas Nilsen, "Disruption at One of Two Undersea Cables to Svalbard," Barent Observer, January 9, 2022.

4 Benjamin Jensen, "The Cyber Character of Political Warfare," Brown Journal of International Affairs 24, no. 1 (2017); Lennart Maschmeyer, "The Subversion Trilemma: Why Cyber Operations Fall Short of Expectations," International Security 46, no. 2 (2021): 51–90.

5 Paul Smith, On Political Warfare (Washington, DC: National Defense University Press, 1989).

6 Austin Carson and Keren Yarhi-Milo, "Covert Communication: The Intelligibility and Credibility of Signaling in Secret," Security Studies 26, no. 1 (January 2, 2017): 124–56, https://doi.org/10.1080/09636412.2017.1243921.; Kerhi-Milo.

7 Brandon Valeriano, Benjamin Jensen and Ryan Maness, Cyber Strategy: The Evolving Character of Power and Coercion (New York: Oxford University Press, 2018); Brandon Valeriano and Benjamin Jensen, "De-escalation Pathways and Disruptive Technology: Cyber Operations as Off-Ramps to War," in Scott Shackelford, Frederick Douzet and Chris Ankersen (Eds.), Cyber Peace: Charting a Path Towards a Sustainable, Stable, and Secure Cyberspace (Cambridge University Press, 2022); Brandon Valeriano and Benjamin Jensen, "The Myth of the Cyber Offense: The Case for Restraint," Cato Institute Policy Analysis 862 (2019).

8 Erik Gartzke and Jon R. Lindsay, "Weaving Tangled Webs: Offense, Defense, and Deception in Cyberspace," Security Studies 24, no. 2 (2015): 316–48.

9 Anne-Marie Slaughter, The Chessboard and the Web: Strategies of Connection in a Networked World (New Haven: Yale University Press, 2017).

10 See Chapter 1, Introduction; David F. Rudgers, "The Origins of Covert Action," Journal of Contemporary History 35, no. 2 (2000): 249–62. Quincy Wright, A Study of War (Chicago: University of Chicago, 1964 reprint).

11 Philip Howard, The Digital Origins of Dictatorship and Democracy: Information Technology and Political Islam (Oxford: Oxford University Press, 2010); Samuel Wooley and Philip Howard (Eds.), Computational Propaganda: Political Parties, Politicians, and Political Manipulation on Social Media (Oxford: Oxford University Press, 2018); Philip Howard, Lie Machines: How to Save Democracy from Troll Armies, Deceitful Robots, Junk News Operations, and Political Operatives (Oxford: Oxford University Press, 2018).

12 See Chapter 6, Nina Kollars and Michael Poznansky, The Cybersecurity Firm and the State, See Chapter 9, Audrye Wong, "Diaspora Statecraft and Geopolitical Influence," in A War of Our Time: The Threat and Dynamics of Non-Military Warfare, Forthcoming.

13 See Chapter 3, Jack McDonald, "Who Are Participants in Non-Military Warfare"; See Chapter 7, Justin Canfil, "War by Law: Legal Statecraft with International Aims," in A War of Our Time: The Threat and Dynamics of Non-Military Warfare, Forthcoming.

14 See Chapter 2, Jan Angstrom, "The Meaning of Victory and Defeat in Non-Military Warfare," in A War of Our Time: The Threat and Dynamics of Non-Military Warfare, Forthcoming.

15 A flexible deterrent option is a crisis response option developed as part of contingency planning. On the concept of flexible deterrent options, see U.S. Department of Defense, JP 5–0 Joint Planning Process (Arlington: Chairman of the Joint Chiefs of Staff, 2020), E-1, and COL Matthew Morton, COL Keith Burkepile and COL VeRonica Oswald-Hrutkay (Eds.), Campaigning Planning Handbook, Academic Year 2021 (Carlisle: U.S. Army War College, 2021).

16 Austin Carson and Keren Yarhi-Milo, "Covert Communication: The Intelligibility and Credibility of Signaling in Secret," Security Studies 26, no. 1 (2016): 124–56, https://doi.org/10.1080/09636412.2017.1243921.
17 James D. Fearon, "Rationalist Explanations for War," International Organization 49, no. 3 (1995): 379–414, https://doi.org/10.1017/s0020818300033324.
18 Robert Powell, "Bargaining Theory and International Conflict," Annual Review of Political Science 5, no. 1 (2002): 1–30, https://doi.org/10.1146/annurev.polisci.5.092601.141138.
19 James D. Fearon, "Signaling Foreign Policy Interests," Journal of Conflict Resolution 41, no. 1 (1997): 68–90, https://doi.org/10.1177/0022002797041001004.
20 Jack S. Levy, "Coercive Threats, Audience Costs, and Case Studies," Security Studies 21, no. 3 (2012): 383–90, https://doi.org/10.1080/09636412.2012.706477.
21 James D. Fearon, "Domestic Political Audiences and the Escalation of International Disputes," American Political Science Review 88, no. 3 (1994): 577–92, https://doi.org/10.2307/2944796.
22 Robert Jervis, Richard Ned Lebow and Janice Gross Stein, Psychology and Deterrence (Baltimore: John Hopkins University Press, 1989); Daniel Kahneman, Olivier Sibony and Cass R. Sunstein, Noise: A Flaw in Human Judgment (New York, NY: Little, Brown Spark, 2021).
23 Ole Holsti, Crisis, Escalation and War (Montreal: McQuill University Press, 1972).
24 Daniel M. Jones, Stuart A. Bremer and J. David Singer, "Militarized Interstate Disputes, 1816–1992: Rationale, Coding Rules, and Empirical Patterns," Conflict Management and Peace Science 15 (1996): 163–213, p. 163. To access the dataset, see https://correlatesofwar.org/data-sets/MIDs/mids.
25 Michael Brecher and Jonathan Wilkenfeld, A Study of Crisis (Ann Arbor: University of Michigan Press, 1997). The ICB dataset is still active, see: https://sites.duke.edu/icbdata/.
26 Glenn Snyder and Paul Diesing, Conflict Among Nations: Bargaining, Decision Making, and System Structure in International Crises (Princeton: Princeton University Press, 1977).
27 Herman Kahn, On Escalation: Metaphors and Scenarios (New York: Frederak A. Praeger, 1965), 3.
28 Russell Leng, "Escalation: Competing Perspectives and Empirical Evidence," International Studies Review 6 (2004): 51–64.
29 Kahn, 1965.
30 Kilgour, Marc D. and Frank C. Zagare, "Credibility, Uncertainty and Deterrence," American Journal of Political Science 35 (1991): 305–34; Frank C. Zagare, "NATO, Rational Escalation, and Flexible Response," Journal of Peace Research 29 (1992): 435–54; Harrison R. Wagner, "Deterrence and Bargaining," Journal of Conflict Resolution 26 (1982): 329–58; James D. Morrow, "Capabilities, Uncertainty, and Resolve: A Limited Information Model of Crisis Bargaining," American Journal of Political Science 33; Paul K. Huth and Bruce Russett, "Deterrence Failure and Crisis Escalation," International Studies Quarterly 32 (1988): 29–45. Keren Yarhi-Milo, Knowing the Adversary: Leaders, Intelligence Organizations, and Assessments in International Relation (Princeton: Princeton University Press, 2014); Keren Yarhi-Milo, Who Fights for Reputation? The Psychology of International Conflict (Princeton: Princeton University Press, 2018); Michael Horowitz, Allen Stam and Cali M. Ellis, Why Leaders Fight (New York: Cambridge University Press, 2015); and Joshua Kertzer and Dustin Tingley, "Political Psychology in International Relations: Beyond the Paradigms," Annual Review of Political Psychology 21 (2018): 319–39.
31 Schelling, 1966, 153–68 and 283–86.
32 Sarah Kreps and Jacquelyn Schneider, "Escalation Firebreaks in the Cyber, Conventional, and Nuclear Domains: Moving beyond Effects-Based Logics," Journal of Cybersecurity 5, no. 1 (2019).

33 Kahn On Escalation, 1965.
34 Marc D. Kilgour and Frank C. Zagare, "Credibility, Uncertainty and Deterrence," American Journal of Political Science 35 (1991): 305–34; Frank C. Zagare, "NATO, Rational Escalation, and Flexible Response," Journal of Peace Research 29 (1992): 435–54; Harrison R. Wagner, "Deterrence and Bargaining," Journal of Conflict Resolution 26 (1982): 329–58; James D. Morrow, "Capabilities, Uncertainty, and Resolve: A Limited Information Model of Crisis Bargaining," American Journal of Political Science 33 (1989); Paul K. Huth and Bruce Russett, "Deterrence Failure and Crisis Escalation," International Studies Quarterly 32 (1988): 29–45.
35 Bernard Brodie, Strategy in the Missile Age (Princeton: Princeton University Press, 1959); Herman Kahn, On Escalation: Metaphors and Scenarios (New York: Praeger, 1965); Patrick M. Morgan, Deterrence: A Conceptual Analysis (Beverly Hills: SAGE, 1977); Thomas C. Schelling, Arms and Influence (New Haven: Yale University Press, 1966); Thomas C. Schelling, The Strategy of Conflict (Cambridge: Harvard University Press, 1960); Glenn H. Snyder and Paul Diesing, Conflict Among Nations: Bargaining, Decision Making and System Structure in International Crises (Princeton: Princeton University Press, 1977); Richard Smoke, War: Controlling Escalation (Cambridge: Harvard University Press, 1977); Glenn H. Snyder, Deterrence and Defense: Toward a Theory of National Security (Princeton: Princeton University Press, 1961).
36 Steven J. Brams and D. Marc Kilgour, "Threat Escalation and Crisis Stability: A Game-Theoretic Analysis," American Political Science Review 81 (1987): 833–50; T. Clifton Morgan, "A Spatial Model of Crisis Bargaining," International Studies Quarterly 28 (1984): 408–14.
37 For an empirical test of the threat game, see Patrick James and Frank Harvey, "Threat Escalation and Crisis Stability: Superpower Cases, 1948–1979," Canadian Journal of Political Science 22, no. 3 (September 1989): 523–45, Earlier treatments model this as a de-escalation game based on the prisoners' dilemma. See Steven Brams and Marc Kilgour, "Winding Down If Preemption or Escalation Occurs: A Game-Theoretic Analysis," Journal of Conflict Resolution 31, no. 4 (December 1987): 547–72.
38 On cost tolerance, see Lisa Carlson, "Crisis Escalation: An Empirical Test in the Context of Extended Deterrence," International Interactions 24, no. 3 (1998): 225–52. On issue salience see the Issue Correlates of War Russell Leng and J. David Singer, "Militarized Interstate Crises: The BCOW Typology and Its Application," International Studies Quarterly 32: 155–73.
39 Frank Zagare, "Classical Deterrence Theory: A Critical Assessment," International Interactions 21 (1996): 365–87.
40 Paul K. Huth and Bruce Russett, "What Makes Deterrence Work? Cases from 1900 to 1980," World Politics 36 (1984): 496–526. Paul K. Huth and Bruce Russett, "Deterrence Failure and Crisis Escalation," International Studies Quarterly 32 (1988): 29–45.
41 Todd Sescher, "Militarized Compellent Threats, 1918–2001," 28, no. 4 (2011).
42 Bruce Bueno de Mesquita, James Morrow and Ethan Zorick, "Capabilities, Perception, and Escalation," American Political Science Review 91, no. 1 (1997): 15–27.
43 de Mesquita et al., 1997, 16.
44 Andrew Kydd, "Arms Races and Arms Control: Modeling the Hawk Perspective," American Journal of Political Science 44, no. 2 (2000): 228–44.
45 Wallace, Michael D. "Armaments and Escalation: Two Competing Hypotheses," International Studies Quarterly 26, no. 1 (1982): 37–56. https://doi.org/10.2307/2600598. Paul E. Diehl, "Arms Race and Escalation: A Closer Look," Journal of Peace Research 20, no. 3 (1983): 249–59. Susan Sample, "Arms Races and Dispute Escalation: Resolving the Debate," Journal of Peace Research 34, no. 1 (1997): 7–22. Susan Sample, "The Outcomes of Military Buildups: Minor States vs. Major Powers," Journal of Peace Research 39, no. 6 (2002): 669–91.

46 Douglas Gibbler, Toby Rider and Marc Hutchinson, "Taking Arms Against a Sea of Troubles: Conventional Arms Races During Periods of Rivalry," Journal of Peace Research 42, no. 2 (2005): 131–47.
47 Paul Huth, "Reputations and Deterrence: A Theoretical and Empirical Assessment," Security Studies 7, no. 1 (1997): 72–99.
48 Darryl Press, Calculating Credibility: How Leaders Assess Military Threats (Ithaca: Cornell University Press, 2005).
49 Vesna Danilovic, When the Stakes Are High: Deterrence and Conflict Among Major Powers (Ann Arbor: University of Michigan Press, 2002).
50 Russell Leng, Interstate Crisis Behavior, 1816–1980: Realism vs. Reciprocity (Cambridge: Cambridge University Press, 1993).
51 Victor Asal and Kyle Beardsley, "Proliferation and International Crisis Behavior," Journal of Peace Research 44, no. 2 (2007): 139–55.
52 Erik Gartzke and Dong-Joon Jo, "Bargaining, Nuclear Proliferation, and Interstate Disputes," Journal of Conflict Resolution 53, no. 2 (2009): 209–33.
53 Simon Mettler and Dan Reiter, "Ballistic Missiles and International Conflict," Journal of Conflict Resolution 57, no. 5 (2013): 854–80.
54 Brock Tessman and Steve Chan, "Power Cycles, Risk Propensity, and Great-Power Deterrence," Journal of Conflict Resolution 48, no. 2 (2004): 131–53. Charles F. Doran and Wes Parsons, "War and the Cycle of Relative Power," American Political Science Review 74, no. 4 (1980): 947–65; Charles Doran, Systems in Crisis: New Imperatives of High Politics at Century's End (Cambridge: Cambridge University Press).
55 Paul Huth, D. Scott Bennett and Christopher Gelpi, "System Uncertainty, Risk Propensity, and International Conflict," Journal of Conflict Resolution 36, no. 3 (1992): 478–517. Paul Huth, Christopher Gelpi and D. Scott Bennett, "The Escalation of Great Power Disputes: Testing Rational Deterrence Theory and Structural Realism," American Political Science Review 87, no. 3 (1993): 609–23.
56 Robert Jervis, "Cooperation Under the Security Dilemma," World Politics 30, no. 2 (1978). Charles Glaser, "The Security Dilemma Revisited," World Politics 50, no. 1 (1997): 171–201; Charles Glaser and Chaim Kaufmann, "What Is the Offense-Defense Balance and Can We Measure It?," International Security 22, no. 4 (1998).
57 Stephen Quackenbush, "Deterrence Theory: Where Do We Stand?," Review of International Studies 37, no. 2 (2011): 741–62, 743.
58 Robert Powell, "Bargaining Theory and International Conflict," Annual Review of Political Science 5 (2002): 1–30.
59 James Morrow, "Capabilities, Uncertainty and Resolve: A Limited Information Model of Crisis Bargaining," American Journal of Political Science 33, no. 4 (1989): 941–72.
60 Robert Powell, "Nuclear Brinkmanship with Two-Sided Incomplete Information," American Political Science Review 82 (1988): 155–78.
61 Quackenbush, 2011, 744; Herman Kahn, Thinking About the Unthinkable (New York: New Horizon Press, 1962), Schelling Arms and Influence.
62 James D. Fearon, "Signaling versus the Balance of Power and Interests: An Empirical Test of a Crisis Bargaining Model," Journal of Conflict Resolution 38 (1994): 236–36. James D. Fearon, "Rationalist Explanations for War," International Organization 49, no. 3 (1995): 379–414.
63 Kahn 1965, Schelling 1960, 1966.
64 Carlson, 1998, 230
65 Anne Sartori, Deterrence by Diplomacy (Princeton: Princeton University Press, 1998).
66 Alex Weisiger and Keren Yarhi-Milo, "Revisiting Reputation: How Past Actions Matter in International Politics," International Organization 69 (Spring 2015): 473–95. For work that draws on psychology to question the extent reputations play a role, see Mercer, Reputations and International Politics.
67 Mercer, 1996, 1

68 Joseph Grieco, "Repetitive Military Challenges and Recurrent International Conflicts, 1918–1994," International Studies Quarterly 45, no. 2 (2001): 295–316.
69 Frank C. Zagare and D. Marc Kilgour, Perfect Deterrence (Cambridge: Cambridge University Press, 2000).
70 Snyder and Diesing, 1977, 190
71 Melvin Small and J. David Singer, Resort to Arms: International and Civil Wars, 1816–1980 (Beverly Hills: Sage Publications, 1982); Zeev Maoz, "Resolve, Capabilities, and Outcomes of Interstate Disputes, 1816–1976," Journal of Conflict Resolution 27 (1983): 195–229.
72 Fearon 1994a.
73 Mercer, 1996.
74 Keren Yarhi-Milo, Knowing the Adversary: Leaders, Intelligence and Assessment of Intentions in International Relations (Princeton: Princeton University Press, 2014).
75 Barry R. Posen, Inadvertent Escalation: Conventional War and Nuclear Risks (Ithaca, NY: Cornell University Press, 1991). On the RMA, see, Benjamin M. Jensen, "The Role of Ideas in Defense Planning: Revisiting the Revolution in Military Affairs," Defence Studies 18, no. 3 (2018): 302–17. https://doi.org/10.1080/14702436.2018.1497928.
76 Cailtlin Talmadge, "Would China Go Nuclear? Assessing the Risk of Chinese Nuclear Escalation in a Conventional War with the United States," International Security 41, no. 4 (2017): 50–92.
77 Karen Ruth Adams, "Attack and Conquer? International Anarchy and the Offense-Defense-Deterrence Balance," International Security 28, no. 3 (2004): 45–83; Stanislav Andreski, Military Organizations and Society (Berkeley: University of California Press, 1968). Stephen Biddle, "Rebuilding the Foundations of the Offense-Defense Theory," The Journal of Politics 63, no. 3 (2000): 741–74. James W. Davis Jr., Bernard I. Finel, Stacie E. Goddard, Stephen Van Evera, Charles L. Glaser and Chaim Kaufmann, "Correspondence: Taking Offense at Offense-Defense Theory," International Security 23, no. 3 (1999): 179–206; Robert Gilpin, War and Change in World Politics (Cambridge: Cambridge University Press, 1981). Charles L. Glaser, Rational Theory of International Politics: The Logic of Competition and Cooperation (Princeton: Princeton University Press, 2010), Glaser, "What Is the Offense-Defense Balance and Can We Measure It?," 44–82, Ted Hopf, "Polarity, the Offense-Defense Balance, and War," American Political Science Review 85, no. 2 (1991); Robert Jervis, "Cooperation Under the Security Dilemma," World Politics 30, no. 2 (1978): 167–214; Robert Jervis, Perception and Misperception in International Politics (Princeton: Princeton University Press, 1976). Jack Levy, "The Offense/Defense Balance of Military Technology: A Theoretical and Historical Analysis," International Studies Quarterly 28, no. 2 (1984): 219–38. Lieber, "Grasping the Technological Peace: The Offense-Defense Balance," 71–105; Kier A. Lieber, "Mission Impossible: Measuring the Offense-Defense Balance with Military Net Assessment," Security Studies 20, no. 3 (2011); Kier A. Lieber, War and the Engineers: The Primacy of Politics Over Technology (Ithaca: Cornell University Press, 2005). Sean M. Lynn-Jones, "Offense-Defense Theory and Its Critics," Security Studies 4, no. 4 (1995): 660–91; George H. Quester, Offense and Defense in the International System (New York: Wiley, 1977); Jonathan Shimshoni, "Technology, Military Advantage, and World War I," International Security 15, no. 3 (1991): 187–215; Shiping Tang, "Offense-Defense Theory: Toward a Definitive Understanding," Chinese Journal of International Politics 4, no. 1 (2010): 213–60. Stephen Van Evera, Causes of War (Ithaca: Cornell University Press, 1999); Stephen Van Evera, "Offense, Defense, and Causes of War," International Security 22, no. 4 (1998): 5–43; Quincy Wright, A Study of War (Chicago: Chicago University Press, 1965). For a complete annotated bibliography, see Michael E. Brown, Owen R. Cote Jr., Sean M. Lynn-Jones and Steven E. Miller (Eds.), Offense, Defense, and War (Cambridge: MIT University Press, 2004).

78 Caitlin Talmadge, "Emerging Technology and Intra-War Escalation Risks: Evidence from the Cold War, Implications for Today," Journal of Strategic Studies 42, no. 6 (2019): 864.
79 Borghard and Lonergan, "Cyber Operations as Imperfect Tools of Escalation."
80 Gartzke and Lindsay, "Weaving Tangled Webs"; Lennart Maschmeyer, "The Subversive Trilemma: Why Cyber Operations Fall Short of Expectations," International Security 46, no. 2 (2021): 51–90.
81 Valeriano et al., Cyber Strategy, 146.
82 Rebecca Slayton, "What Is the Cyber Offense-Defense Balance?," International Security 41, no. 3 (Winter 2016/17): 72–109.
83 Ben Buchanan, The Cybersecurity Dilemma: Hacking, Trust, and Fear between Nations (New York: Oxford University Press, 2017).
84 Lucas Kello, "The Meaning of the Cyber Revolution: Perils to Theory and Statecraft," International Security 38, no. 2 (Fall 2013): 7–40.
85 Borghard and Lonergan, "Cyber Operations as Imperfect Tools of Escalation," 124.
86 Martin Libicki, Cyberspace in Peace and War (Annapolis: Naval Institute Press, 2016), 273.
87 Forrest E. Morgan, Karl P. Mueller, Evan S. Medeiros, Kevin L. Pollpeter, Roger Cliff, Dangerous Thresholds: Managing Escalation Risks in the 21st Century (Santa Monica: RAND Corporation, 2008).
88 Sarah Krebs and Jacqueline Schneider, "Escalation Firebreaks in the Cyber, Conventional, and Nuclear Domains: Moving Beyond Effects-Based Logics," Journal of Cybersecurity 5, no. 1 (2019).
89 Erica Borghard and Jacquelyn Schneider, "Israel Responded to a Hamas Cyberattack with an Airstrike. That's Not Such a Big Deal," The Washington Post, May 2019, https://www.washingtonpost.com/politics/2019/05/09/israel-responded-hamas-cyberattack-with-an-airstrike-thats-big-deal/.
90 Brandon Valeriano and Benjamin Jensen, "De-escalation Pathways and Disruptive Technology: Cyber Operations as Off-Ramps to War," in Scott Shackelford, Frederick Douzet and Chris Ankersen (Eds.), Cyber Peace: Charting a Path Towards a Sustainable, Stable, and Secure Cyberspace (Cambridge University Press, 2022).
91 Benjamin Jensen and Brandon Valeriano, "What Do We Know About Cyber Escalation? Observations from Simulations and Surveys," Atlantic Council Report for the Cyber Statecraft Initiative, 2019.
92 Libicki, Cyberspace in Peace and War, 273.
93 Michael Fischerkeller and Richard Harknett, "A Response on Persistent Engagement and Agreed Competition," Lawfare Blog, June 2019, https://www.lawfareblog.com/response-persistent-engagement-and-agreed-competition.
94 Fischerkeller and Harknett, "What Is Agreed Competition in Cyberspace?"
95 Jiri Valenta and Leni Friedman Valenta, "Why Putin Wants Syria," Middle East Quarterly (2016), https://www.meforum.org/5876/why-putin-wants-syria
96 Anton Lavrov, The Russian Air Campaign in Syria: A Preliminary Analysis (Arlington: Center for Naval Analysis, 2018).
97 Hugo Spaulding, Russia's False ISIS Narrative in Syria, Washington: Institute for the Study of War, 2015), https://www.understandingwar.org/sites/default/files/Russia%20False%20Narrative%20in%20Syria_1.pdf
98 Patrick J. McDonnell, "Russia Launches Airstrikes in Syria Amid U.S. Concern About Targets," Los Angeles Times, September 30, 2015.
99 Sam Dagher and Asa Fitch, "Iran Expands Role in Syria in Conjunction with Russia's Airstrikes," The Wall Street Journal, October 2, 2015, Retrieved October 3, 2015. https://www.wsj.com/articles/iran-expands-role-in-syria-in-conjunction-with-russias-airstrikes-1443811030
100 NATO. "Statement by the North Atlantic Council on Incursions into Turkey's Airspace by Russian Aircraft," October 5, 2015, http://www.nato.int/cps/en/natohq/news_123392.htm.

101 "Syria Conflict: Russia Violation of Turkish Airspace 'No Accident.'" BBC News, October 6, 2015, sec. Europe. https://www.bbc.com/news/world-europe-34453739.
102 Julian Barnes, Emre Peker and Gordon Lubold, "Turkey Says Russian Fighter Jet Violated Its Airspace with Syria," Wall Street Journal, October 5, 2015, sec. World. https://www.wsj.com/articles/turkey-says-russian-fighter-jet-violated-its-airspace-with-syria-1444040488
103 "Turkey Downing of Russia Jet 'Stab in the Back' – Putin," BBC News, November 24, 2015, sec. Middle East. https://www.bbc.com/news/world-middle-east-34913173; Thomas Gibbons-Neff, "The Last Time a Russian Jet Was Shot down by a NATO Jet Was in 1952," Washington Post, November 24, 2015, https://www.washingtonpost.com/news/checkpoint/wp/2015/11/24/the-last-time-a-russian-jet-was-shot-down-by-a-nato-jet-was-in-1952/; Neil MacFarquhar and Steven Erlanger, "NATO-Russia Tensions Rise After Turkey Downs Jet," The New York Times, November 24, 2015, sec. World. https://www.nytimes.com/2015/11/25/world/europe/turkey-syria-russia-military-plane.html; "Sorting Out What Russia and Turkey Say Happened in the Sky," The New York Times, November 24, 2015, https://www.nytimes.com/interactive/2015/11/24/world/middleeast/russia-turkey-jet-shoot-down-maps.html
104 HackRead, "Turkish Attackers Shut Down Russian Central Bank Website," November 25, 2015, https://www.hackread.com/turkish-hackers-target-russian-central-bank-site/
105 Edward Fishman, "Even Smarter Sanctions: How to Fight in the Era of Economic Warfare," Foreign Affairs 96, no. 6 (November/December 2017): 102–10.
106 Defense World, "Russia to Wage Electronic Warfare Against Turkey," November 30, 2015, https://www.defenseworld.net/news/14727/Russia_To_Wage_Electronic_Warfare_Against_Turkey
107 "Hackers Block Russian Minister's Cyber Account with Turkish Symbols," Reuters, January 3, 2016, sec. Internet News, https://www.reuters.com/article/us-russia-turkey-minister-cybersecurity-idUSKBN0UH0HJ20160103
108 Katherine Costello, Russia's Use of Media and Information Operations in Turkey: Implications for the United States (RAND Corporation, 2018), https://www.rand.org/pubs/perspectives/PE278.html
109 Cory Bennett, "Turkey: Rift with Russia Inciting Hackers," The Hill, January 4, 2016, https://thehill.com/policy/cybersecurity/264671-turkey-russia-rift-inciting-hackers-official-says
110 Adam Meyers, "Cyber Skirmish: Russia v. Turkey," Crowdstrike, April 13, 2016, https://www.crowdstrike.com/blog/cyber-skirmish-russia-v-turkey/
111 Adam Meyers, "Cyber Skirmish: Russia v. Turkey," Crowdstrike, April 13, 2016, https://www.crowdstrike.com/blog/cyber-skirmish-russia-v-turkey/
112 The Economist, "Military Tensions Mount on the India-China Border," The Economist, May 29, 2020.
113 Reuters, "China Reveals Four Soldiers Killed in June 2020 Border Clash with India," Reuters, February 18, 2021.
114 Jeffrey Gettleman, Hari Kumar and Sameer Yasir, "Worst Clash in Decades on Disputed India-China Border Kills 20 Indian Troops," The New York Times, June 16, 2020, sec. World, https://www.nytimes.com/2020/06/16/world/asia/indian-china-border-clash.html; "India-China Clash: 20 Indian Troops Killed in Ladakh Fighting," BBC News, June 16, 2020, sec. Asia. https://www.bbc.com/news/world-asia-53061476.;
115 Russell Goldman, "India-China Border Dispute: A Conflict Explained," The New York Times, June 17, 2020, sec. World, https://www.nytimes.com/2020/06/17/world/asia/india-china-border-clashes.html; "China Reveals Four Soldiers Killed in June 2020 Border Clash with India," Reuters, February 18, 2021.
116 Hannah Ellis-Petersen, "Indians Call for Boycott of Chinese Goods After Fatal Border Clashes," The Guardian, June 18, 2020, https://www.theguardian.com/world/2020/jun/18/indians-call-for-boycott-of-chinese-goods-after-fatal-border-clashes

117 Public Information Bureau of the Government of India, "Government Bans 59 Mobile Apps Which Are Prejudicial to Sovereignty and Integrity of India, Defence of India, Security of State and Public Order," June 29, 2020, https://pib.gov.in/PressReleseDetail.aspx?PRID=1635206
118 Manu Kaushik, "Notorious Chinese Hackers Attack Indian Entities; Defence Ministry, Jio, Airtel, Cipla, L&T Top Targets," Business Today, June 23, 2020, https://www.businesstoday.in/latest/economy-politics/story/notorious-chinese-hackers-attack-indian-entities-defence-ministry-jio-airtel-cipla-lt-top-targets-262008-2020-06-23
119 Agence France Presse, "India Claims It Foiled Chinese Cyber-attack On Disputed Border," AFP, April 7, 2022; Associated Press, "Chinese Hackers Reportedly Target India's Power Grid," VOA, April 7, 2022.
120 P. J. George, "Explained | Red Echo, ShadowPad, and the Targeting of India's Power Grid," The Hindu, March 6, 2021, sec. Technology, https://www.thehindu.com/sci-tech/technology/red-echo-over-india/article34008299.ece.
121 Jeffrey Gettleman, "Shots Fired Along India-China Border for First Time in Years," The New York Times, September 8, 2020, sec. World, https://www.nytimes.com/2020/09/08/world/asia/india-china-border.html.
122 Rajesh Roy, "Mumbai Blackout Leaves Millions Without Power for Hours," Wall Street Journal, October 12, 2020, sec. World, https://www.wsj.com/articles/mumbai-blackout-leaves-millions-without-power-for-hours-11602507630.
123 Sahil Joshi and Divyesh Singh, "Mega Mumbai Power Outage May Be Result of Cyber Attack, Final Report Awaited," India Today, November 20, 2020, https://www.indiatoday.in/india/story/mumbai-power-outage-malware-attack-1742538-2020-11-20; Abhishek Sharan, "October 12 Blackout Was a Sabotage," Mumbai Mirror, November 20, 2020, https://mumbaimirror.indiatimes.com/mumbai/cover-story/oct-12-blackout-was-a-sabotage/articleshow/79312959.cms.
124 Insikt Group, "China-Linked Group RedEcho Targets the Indian Power Sector Amid Heightened Border Tensions," Recorded Future, February 28, 2021, https://www.recordedfuture.com/redecho-targeting-indian-power-sector/. David E. Sanger and Emily Schmall, "China Appears to Warn India: Push Too Hard and the Lights Could Go Out," The New York Times, February 28, 2021, sec. U.S., https://www.nytimes.com/2021/02/28/us/politics/china-india-hacking-electricity.html.
125 The Times of India, "Mumbai Blackout: Government Denies China's Cyber Campaign against Indian Power Grid," March 1, 2021, https://timesofindia.indiatimes.com/india/government-denies-any-impact-of-chinas-cyber-campaign-against-indian-power-grid/articleshow/81277602.cms; The Times of India, " 'It Was Human Error': Cyberattacks Took Place But Didn't Cause Mumbai Power Outage, Says Govt," March 2, 2021, https://timesofindia.indiatimes.com/india/2020-mumbai-power-outage-caused-by-human-error-not-cyber-attack-union-power-minister/articleshow/81292545.cms
126 Krishna N. Das, "Chinese Hackers Target Indian Vaccine Makers SII, Bharat Biotech, Says Security Firm," Reuters, March 1, 2021, sec. China, https://www.reuters.com/world/china/chinese-hackers-target-indian-vaccine-makers-sii-bharat-biotech-says-security-2021-03-01/.
127 Tonya Riley, "Suspected Chinese Hackers Target Telecoms, Research in Taiwan, Recorded Future Says," CyberScoop (blog), July 8, 2021, https://cyberscoop.com/chinese-apt-hacking-taiwan-semiconductor/.
128 Valeriano et al., Cyber Strategy.
129 John A. Vasquez, The War Puzzle (Cambridge: Cambridge University Press, 1993); John A. Vasquez, The War Puzzle Revisited (Cambridge: Cambridge University Press, 2009); John A. Vasquez and Marie T. Henehan, Territory, War, and Peace (New York: Routledge, 2010); Monica Toft, "Territory and War," Journal of Peace Research 51, no. 2 (2014): 185–98.
130 M. Montgomery, B. Jensen, E. D. Borghard, J. Costello, V. Cornfeld, C. Simpson and B. Valeriano, Cyberspace Solarium Commission Report (Washington, DC, 2020),

https://www.solarium.gov/report; Brandon Valeriano and Benjamin Jensen, "Building a National Cyber Strategy: The Process and Implications of the Cyberspace Solarium Commission," in T. Jančárková, L. Lindström, G. Visky and P. Zotz (Eds.), Going Viral (Report 2021 13th International Conference on Cyber Conflict 2021 © NATO CCDCOE Publications, Tallinn)

131 Montgomery et al., Cyberspace Solarium Report.

132 Valeriano et al., Cyber Strategy.

Bibliography

Acton, J. M. (2018). Escalation through entanglement: How the vulnerability of command-and-control systems raises the risks of an inadvertent nuclear war. *International Security*, *43*(1), 56–99. https://doi.org/10.1162/isec_a_00320

Adams, K. R. (2003). Attack and conquer? International anarchy and the offense-defense-deterrence balance. *International Security*, *28*(3), 45–83.

Andreski, S. (1971). *Military organization and society*. University of California Press.

Asal, V., & Beardsley, K. (2007). Proliferation and international crisis behavior. *Journal of Peace Research*, *44*(2), 139–155.

Bennett. (n.d.). *Turkey: Rift with Russia inciting hackers*. Retrieved January 11, 2025, from https://thehill.com/policy/cybersecurity/264671-turkey-russia-rift-inciting-hackers-official-says/

Biddle, S. (2001). Rebuilding the foundations of offense-defense theory. *The Journal of Politics*, *63*(3), 741–774.

Borghard, E. D., & Lonergan, S. W. (2019). Cyber operations as imperfect tools of escalation. *Strategic Studies Quarterly*, *13*(3), 122–145.

Borghard, E. D., & Schneider, J. (2019, May). Israel responded to a Hamas cyberattack with an airstrike. That's not such a big deal. *The Washington Post*. https://www.washingtonpost.com/politics/2019/05/09/israel-responded-hamas-cyberattack-with-an-airstrike-thats-big-deal/

Brams, S. J., & Kilgour, D. M. (1987a). Threat escalation and crisis stability: A game-theoretic analysis. *The American Political Science Review*, *81*(3), 833–850. https://doi.org/10.2307/1962679

Brams, S. J., & Kilgour, D. M. (1987b). Winding down if preemption or escalation occurs: A game-theoretic analysis. *The Journal of Conflict Resolution*, *31*(4), 547–572.

Brandon, V., Jensen, B. M., & Maness, B. C. (2018). *Cyber strategy: The evolving character of power and coercion*. Oxford University Press.

Brecher, M., & Wilkenfeld, J. (1997). *A study of crisis*. University of Michigan Press. https://doi.org/10.3998/mpub.14982

Brodie, B. (2007). *Strategy in the missile age* (New RAND ed.). RAND Corp.

Brown, M. E. (Ed.). (2004). *Offense, defense, and war*. MIT Press.

Buchanan, B. (2017). *The cybersecurity dilemma* (Vol. 1). Oxford University Press. https://doi.org/10.1093/acprof:oso/9780190665012.001.0001

Carlson, L. J. (1998). Crisis escalation: An empirical test in the context of extended deterrence. *International Interactions*, *24*(3), 225–253. https://doi.org/10.1080/03050629808434930

Carson, A., & Yarhi-Milo, K. (2017). Covert communication: The intelligibility and credibility of signaling in secret. *Security Studies*, *26*(1), 124–156. https://doi.org/10.1080/09636412.2017.1243921

China reveals four soldiers killed in June 2020 border clash with India. (2021, February 19). *Reuters*. https://www.reuters.com/article/world/china-reveals-four-soldiers-killed-in-june-2020-border-clash-with-india-idUSKBN2AJ05A/

Cleveland, C., Jensen, B. M., David, A., & Bryant, S. (2018). *Military strategy for the 21st century: People, connectivity, and influence*. Cambria Press.

Costello, K. (2018). *Russia's use of media and information operations in Turkey: Implications for the United States*. RAND Corporation. https://www.rand.org/pubs/perspectives/PE278.html

Danilovic, V. (2002). *When the stakes are high: Deterrence and conflict among major powers*. University of Michigan Press. https://doi.org/10.3998/mpub.16953
Das, K. N. (2021, March 1). Chinese hackers target Indian vaccine makers SII, Bharat Biotech, says security firm. *Reuters*. https://www.reuters.com/world/china/chinese-hackers-target-indian-vaccine-makers-sii-bharat-biotech-says-security-2021-03-01/
Davis, J. W., Finel, B. I., Goddard, S. E., Evera, S. V., Glaser, C. L., & Kaufmann, C. (1998). Taking offense at offense-defense theory. *International Security*, *23*(3), 179. https://doi.org/10.2307/2539342
De Mesquita, B. B., Morrow, J. D., & Zorick, E. R. (1997). Capabilities, perception, and escalation. *American Political Science Review*, *91*(1), 15–27. https://doi.org/10.2307/2952256
Diehl, P. F. (1983). Arms races and escalation: A closer look. *Journal of Peace Research*, *20*(3), 205–212.
Doran, C. F. (2008). *Systems in crisis: New imperatives of high politics at century's end*. Cambridge University Press.
Doran, C. F., & Parsons, W. (1980). War and the cycle of relative power. *The American Political Science Review*, *74*(4), 947–965. https://doi.org/10.2307/1954315
Ellis-Petersen, H. (2020, June 18). Indians call for boycott of Chinese goods after fatal border clashes. *The Guardian*. https://www.theguardian.com/world/2020/jun/18/indians-call-for-boycott-of-chinese-goods-after-fatal-border-clashes
Fearon, J. D. (1994a). Domestic political audiences and the escalation of international disputes. *American Political Science Review*, *88*(3), 577–592. https://doi.org/10.2307/2944796
Fearon, J. D. (1994b). Signaling versus the balance of power and interests: An empirical test of a crisis bargaining model. *The Journal of Conflict Resolution*, 38(2), 236–269.
Fearon, J. D. (1995). Rationalist explanations for war. *International Organization*, *49*(3), 379–414. https://doi.org/10.1017/S0020818300033324
Fearon, J. D. (1997). Signaling foreign policy interests: Tying hands versus sinking costs. *Journal of Conflict Resolution*, *41*(1), 68–90. https://doi.org/10.1177/0022002797041001004
Fischerkeller, M., & Harknett, R. (2019, June). A response on persistent engagement and agreed competition. *Lawfare Blog*.
Fischerkeller, M., & Harknett, R. (n.d.). What is agreed competition in cyberspace? *Lawfare Blog*. Retrieved January 11, 2025, from https://www.lawfaremedia.org/article/what-agreed-competition-cyberspace
Fishman, E. (2017, October 16). Even smarter sanctions: How to fight in the era of economic warfare. *Foreign Affairs*, *96*(6), 102–110.
Fitch, A., & Dagher, S. (2015, October 2). Iran expands role in Syria in conjunction with Russia's airstrikes. *Wall Street Journal*. http://www.wsj.com/articles/iran-expands-role-in-syria-in-conjunction-with-russias-airstrikes-1443811030
Gartzke, E., & Jo, D.-J. (2009). Bargaining, nuclear proliferation, and interstate disputes. *The Journal of Conflict Resolution*, *53*(2), 209–233.
Gartzke, E., & Lindsay, J. R. (2015). Weaving tangled webs: Offense, defense, and deception in cyberspace. *Security Studies*, *24*(2), 316–348. https://doi.org/10.1080/09636412.2015.1038188
George, P. J. (2021, March 6). Explained | Red echo, ShadowPad, and the targeting of India's power grid. *The Hindu*. https://www.thehindu.com/sci-tech/technology/red-echo-over-india/article34008299.ece
Gettleman, J. (2020, September 8). Shots fired along India-China border for first time in years. *The New York Times*. https://www.nytimes.com/2020/09/08/world/asia/india-china-border.html
Gettleman, J., Kumar, H., & Yasir, S. (2020, June 16). Worst clash in decades on disputed India-China border kills 20 Indian troops. *The New York Times*. https://www.nytimes.com/2020/06/16/world/asia/indian-china-border-clash.html
Gibbons-Neff, T. (2015, November 24). The last time a Russian jet was shot down by a NATO jet was in 1952. *Washington Post*. https://www.washingtonpost.com/news/checkpoint/wp/2015/11/24/the-last-time-a-russian-jet-was-shot-down-by-a-nato-jet-was-in-1952/

Gibler, D. M., Rider, T. J., & Hutchison, M. L. (2005). Taking arms against a sea of troubles: Conventional arms races during periods of rivalry. *Journal of Peace Research*, *42*(2), 131–147.

Gilpin, R. (1991). *War and change in world politics*. Cambridge University Press.

Glaser, C. L. (1997). The security dilemma revisited. *World Politics*, *50*(1), 171–201.

Glaser, C. L. (2010). *Rational theory of international politics: The logic of competition and cooperation*. Princeton University Press.

Glaser, C. L., & Kaufmann, C. (1998). What is the offense-defense balance and can we measure it? *International Security*, *22*(4), 44. https://doi.org/10.2307/2539240

Goldman, R. (2020, June 17). India-China border dispute: A conflict explained. *The New York Times*. https://www.nytimes.com/2020/06/17/world/asia/india-china-border-clashes.html

Government Bans 59 mobile apps which are prejudicial to sovereignty and integrity of India, defence of India, security of state and public order. (2020, June 29). *Public Information Bureau of the Government of India*. https://pib.gov.in/pib.gov.in/Pressreleaseshare.aspx?PRID=1635206

Grieco, J. M. (2001). Repetitive military challenges and recurrent international conflicts, 1918–1994. *International Studies Quarterly*, *45*(2), 295–316.

Hackers block Russian minister's cyber account with Turkish symbols | Reuters. (2016, January 3). *Reuters*. https://www.reuters.com/article/technology/hackers-block-russian-minister-s-cyber-account-with-turkish-symbols-idUSL8N14N0DR/

Harrison, T. (with Johnson, K., & Young, M.). (2021). *Defense against the dark arts in space: Protecting space systems from counterspace weapons* (1st ed). Center for Strategic & International Studies.

Holsti, O. R. (1972). *Crisis escalation war*. MQUP.

Hopf, T. (1991). Polarity, the offense defense balance, and war. *The American Political Science Review*, *85*(2), 475–493. https://doi.org/10.2307/1963170

Horowitz, M., Stam, A. C., & Ellis, C. M. (2015). *Why leaders fight*. Cambridge University Press.

Howard, P. N. (2010). *The digital origins of dictatorship and democracy: Information technology and political Islam*. Oxford University Press. https://doi.org/10.1093/acprof:oso/9780199736416.001.0001

Howard, P. N. (2020). *Lie machines: How to save democracy from troll armies, deceitful robots, junk news operations, and political operatives*. Yale University Press.

Huth, P. K. (1997). Reputations and deterrence: A theoretical and empirical assessment. *Security Studies*, *7*(1), 72–99. https://doi.org/10.1080/09636419708429334

Huth, P., Bennett, D. S., & Gelpi, C. (1992). System uncertainty, risk propensity, and international conflict among the great powers. *The Journal of Conflict Resolution*, *36*(3), 478–517.

Huth, P., Gelpi, C., & Bennett, D. S. (1993). The escalation of great power militarized disputes: Testing rational deterrence theory and structural realism. *American Political Science Review*, *87*(3), 609–623. https://doi.org/10.2307/2938739

Huth, P., & Russett, B. (1984). What makes deterrence work? Cases from 1900 to 1980. *World Politics*, *36*(4), 496–526. https://doi.org/10.2307/2010184

Huth, P., & Russett, B. (1988). Deterrence failure and crisis escalation. *International Studies Quarterly*, *32*(1), 29. https://doi.org/10.2307/2600411

India-China clash: 20 Indian troops killed in Ladakh fighting. (2020, June 16). https://www.bbc.com/news/world-asia-53061476

"It was human error": Cyberattacks took place but didn't cause Mumbai power outage, says govt. (2021, March 2). *The Times of India*. https://timesofindia.indiatimes.com/india/2020-mumbai-power-outage-caused-by-human-error-not-cyber-attack-union-power-minister/articleshow/81292545.cms

James, P., & Harvey, F. (1989). Threat escalation and crisis stability: Superpower cases, 1948–1979. *Canadian Journal of Political Science / Revue Canadienne de Science Politique*, *22*(3), 523–545.

Jensen, B. (2017). The cyber character of political warfare. *The Brown Journal of World Affairs*, *24*(1), 159–172.

Jensen, B. M. (2018). The role of ideas in defense planning: Revisiting the revolution in military affairs. *Defence Studies*, *18*(3), 302–317. https://doi.org/10.1080/14702436.2018.1497928

Jensen, B., & Valeriano, B. (n.d.). *What do we know about cyber escalation?* Observations from Simulations and Surveys [Atlantic Council Report for the Cyber Statecraft Initiative]. https://www.atlanticcouncil.org/wp-content/uploads/2019/11/What_do_we_know_about_cyber_escalation_.pdf

Jervis, R. (1976). *Perception and misperception in international politics*. Princeton University Press. https://doi.org/10.2307/j.ctvc77bx3

Jervis, R. (1978). Cooperation under the security dilemma. *World Politics*, *30*(2), 167–214. https://doi.org/10.2307/2009958

Jervis, R., Lebow, R. N., Stein, J. G., Morgan, P. M., & Snyder, J. L. (1989). *Psychology and deterrence*. The John Hopkins University Press.

Jones, D. M., Bremer, S. A., & Singer, J. D. (1996). Militarized interstate disputes, 1816–1992: Rationale, coding rules, and empirical patterns. *Conflict Management and Peace Science*, *15*(2), 163–213.

Joshi, S., & Singh, D. (2020, November 20). Mega Mumbai power outage may be result of cyber attack, final report awaited. *India Today*. https://www.indiatoday.in/india/story/mumbai-power-outage-malware-attack-1742538-2020-11-20

Kahn, H. (1962). *Thinking about the unthinkable*. Horizon Press.

Kahn, H. (1965). *On escalation: Metaphors and scenarios*. Transaction Publishers.

Kahneman, D., Sibony, O., & Sunstein, C. R. (2022). *Noise: A flaw in human judgment* (First Little, Brown Spark paperback edition). Little, Brown Spark.

Kaushik. (2020, June 23). Notorious Chinese hackers attack Indian entities; Defence Ministry, Jio, Airtel, Cipla, L&T top targets. *Business Today*. https://www.businesstoday.in/latest/economy-politics/story/notorious-chinese-hackers-attack-indian-entities-defence-ministry-jio-airtel-cipla-lt-top-targets-262008-2020-06-23

Kello, L. (2013). The meaning of the cyber revolution: Perils to theory and statecraft. *International Security*, 38(2), 7–40.

Kertzer, J. D., & Tingley, D. (2018). Political psychology in international relations: Beyond the paradigms. *Annual Review of Political Science*, *21*(1), 319–339. https://doi.org/10.1146/annurev-polisci-041916-020042

Kilgour, D. M., & Zagare, F. C. (1991). Credibility, uncertainty, and deterrence. *American Journal of Political Science*, *35*(2), 305. https://doi.org/10.2307/2111365

Kreps, S., & Schneider, J. (2019). Escalation firebreaks in the cyber, conventional, and nuclear domains: Moving beyond effects-based logics. *Journal of Cybersecurity*, *5*(1), tyz007. https://doi.org/10.1093/cybsec/tyz007

Kydd, A. (2000). Arms races and arms control: Modeling the hawk perspective. *American Journal of Political Science*, *44*(2), 228. https://doi.org/10.2307/2669307

Lavrov, A. (2018). *The Russian air campaign in Syria*. Centre for Analysis of Strategies and Technologies.

Leng, R. (1994). Interstate crisis behaviour, 1816–1980: Realism versus reciprocity. *International Affairs*, *70*(3), 534–534. https://doi.org/10.2307/2623728

Leng, R. J. (2004). Escalation: Competing perspectives and empirical evidence. *International Studies Review*, *6*(4), 51–64. https://doi.org/10.1111/j.1521-9488.2004.00449.x

Leng, R. J., & Singer, J. D. (1988). Militarized interstate crises: The BCOW typology and its applications. *International Studies Quarterly*, *32*(2), 155. https://doi.org/10.2307/2600625

Levy, J. S. (1984). The offensive/defensive balance of military technology: A theoretical and historical analysis. *International Studies Quarterly*, *28*(2), 219. https://doi.org/10.2307/2600696

Levy, J. S. (2012). Coercive threats, audience costs, and case studies. *Security Studies*, *21*(3), 383–390. https://doi.org/10.1080/09636412.2012.706477

Libicki, M. C. (2016). *Cyberspace in peace and war*. Naval Institute Press.
Lieber, K. A. (2011). Mission impossible: Measuring the offense-defense balance with military net assessment. *Security Studies*, *20*(3), 451–459. https://doi.org/10.1080/09636412.2011.599193
Lieber, K. A. (2018). *War and the engineers: The primacy of politics over technology*. Cornell University Press. https://doi.org/10.7591/9781501724466
Lynn-Jones, S. M. (1995). Offense-defense theory and its critics. *Security Studies*, *4*(4), 660–691. https://doi.org/10.1080/09636419509347600
MacFarquhar, N., & Erlanger, S. (2015, November 24). NATO-Russia tensions rise after Turkey downs jet. *The New York Times*. https://www.nytimes.com/2015/11/25/world/europe/turkey-syria-russia-military-plane.html
Maoz, Z. (1983). Resolve, capabilities, and the outcomes of interstate disputes, 1816–1976. *The Journal of Conflict Resolution*, *27*(2), 195–229.
Maschmeyer, L. (2021). The subversive trilemma: Why cyber operations fall short of expectations. *International Security*, *46*(2), 51–90. https://doi.org/10.1162/isec_a_00418
McDonnell, P. (2015, September 30). Russia launches airstrikes in Syria amid U.S. concern about targets. *Los Angeles Times*. https://www.latimes.com/world/europe/la-fg-kremlin-oks-troops-20150930-story.html
Mercer, J. (1996). *Reputation and international politics*. Cornell University Press.
Mettler, S. A., & Reiter, D. (2013). Ballistic missiles and international conflict. *The Journal of Conflict Resolution*, *57*(5), 854–880.
Meyers, A. (n.d.). Cyber Skirmish: Russia v. Turkey. *CrowdStrike.com*. Retrieved January 11, 2025, from https://www.crowdstrike.com/en-us/blog/cyber-skirmish-russia-v-turkey/
Military tensions mount on the India-China border. (n.d.). *The Economist*. Retrieved January 11, 2025, from https://www.economist.com/asia/2020/05/29/military-tensions-mount-on-the-india-china-border
Montgomery, M., Jensen, B., Borghard, E., Costello, J., Cornfeld, V., Simpson, C., & Valeriano, C. (2020). *Cyberspace solarium commission report*. Cyberspace Solarium Commission
Morgan, F. E., Mueller, K. P., Medeiros, E. S., Pollpeter, K. L., & Cliff, R. (2008). *Dangerous thresholds: Managing escalation in the 21st century*. RAND Corporation. https://www.rand.org/pubs/monographs/MG614.html
Morgan, P. M. (1977). *Deterrence: A conceptual analysis*. SAGE.
Morgan, T. C. (1984). A spatial model of crisis bargaining. *International Studies Quarterly*, *28*(4), 407. https://doi.org/10.2307/2600563
Morrow, J. D. (1989). Capabilities, uncertainty, and resolve: A limited information model of crisis bargaining. *American Journal of Political Science*, *33*(4), 941. https://doi.org/10.2307/2111116
Morton, M. D., Burkepile, K., & Oswald-Hrutkay, V. (n.d.). *Campaign planning handbook: Academic year 2021*. Retrieved January 11, 2025, from https://apps.dtic.mil/sti/citations/AD1133069
Mulder, N. (2022). *The economic weapon: The rise of sanctions as a tool of modern war*. Yale University Press.
Mumbai blackout: Government denies China's cyber campaign against Indian power grid. (2021, March 1). *The Times of India*. https://timesofindia.indiatimes.com/india/government-denies-any-impact-of-chinas-cyber-campaign-against-indian-power-grid/articleshow/81277602.cms
NATO. (n.d.). Statement by the North Atlantic Council on incursions into Turkey's airspace by Russian aircraft. *NATO*. Retrieved January 11, 2025, from https://www.nato.int/cps/en/natohq/news_123392.htm
The New York Times. (2015, November 24). Sorting out what Russia and Turkey say happened in the sky. *The New York Times*. https://www.nytimes.com/interactive/2015/11/24/world/middleeast/russia-turkey-jet-shoot-down-maps.html, https://www.nytimes.com/interactive/2015/11/24/world/middleeast/russia-turkey-jet-shoot-down-maps.html

Peker, E., Lubold, G., & Barnes, J. E. (2015, October 5). Turkey says Russian fighter jet violated its airspace with Syria. *Wall Street Journal*. http://www.wsj.com/articles/turkey-says-russian-fighter-jet-violated-its-airspace-with-syria-1444040488

Pettersson, T., & Öberg, M. (2020). Organized violence, 1989–2019. *Journal of Peace Research*, *57*(4), 597–613. https://doi.org/10.1177/0022343320934986

Powell, R. (1988). Nuclear brinkmanship with two-sided incomplete information. *American Political Science Review*, *82*(1), 155–178. https://doi.org/10.2307/1958063

Powell, R. (2002). Bargaining theory and international conflict. *Annual Review of Political Science*, *5*(1), 1–30. https://doi.org/10.1146/annurev.polisci.5.092601.141138

Press, D. G. (2005). *Calculating credibility: How leaders assess evaluate military threats*. Cornell University Press.

Presse, A.-A. F. (n.d.). *India claims it foiled Chinese cyber-attack on disputed border*. Retrieved January 11, 2025, from https://www.barrons.com/news/india-claims-it-foiled-chinese-cyber-attack-on-disputed-border-01649336107

Quackenbush, S. L. (2011). Deterrence theory: Where do we stand? *Review of International Studies*, *37*(2), 741–762. https://doi.org/10.1017/S0260210510000896

Quester, G. H. (2003). *Offense and defense in the international system*. Transaction Publishers.

RecordedFuture. (n.d.). China-linked group RedEcho targets the Indian power sector amid heightened border tensions. Retrieved January 11, 2025, from https://www.recordedfuture.com/research/redecho-targeting-indian-power-sector

Riley, T. (2021, July 8). Suspected Chinese hackers target telecoms, research in Taiwan, recorded future says. *CyberScoop*. https://cyberscoop.com/chinese-apt-hacking-taiwan-semiconductor/

Roy, R. (2020, October 12). Mumbai blackout leaves millions without power for hours. *Wall Street Journal*. https://www.wsj.com/articles/mumbai-blackout-leaves-millions-without-power-for-hours-11602507630

Rudgers, D. F. (2000). The origins of covert action. *Journal of Contemporary History*, *35*(2), 249–262. https://doi.org/10.1177/002200940003500206

Russia to wage electronic warfare against Turkey. (2015, November 30). *Defense World*.

Sample, S. G. (1997). Arms races and dispute escalation: Resolving the debate. *Journal of Peace Research*, *34*(1), 7–22.

Sample, S. G. (2002). The outcomes of military buildups: Minor states vs. major powers. *Journal of Peace Research*, *39*(6), 669–691.

Sanger, D. E., & Schmall, E. (2021, February 28). China appears to warn India: Push too hard and the lights could go out. *The New York Times*. https://www.nytimes.com/2021/02/28/us/politics/china-india-hacking-electricity.html

Sartori, A. E. (2013). *Deterrence by diplomacy*. Princeton University Press.

Schelling, T. C. (1960). *The strategy of conflict: With a new preface*. Harvard University Press.

Schelling, T. C. (1966). *Arms and influence*. Yale University Press. https://doi.org/10.2307/j.ctt5vm52s

Sechser, T. S. (2011). Militarized compellent threats, 1918–2001. *Conflict Management and Peace Science*, *28*(4), 377–401.

Sharan, A. (n.d.). October 12 blackout was a sabotage. *Mumbai Mirror*. Retrieved January 11, 2025, from https://mumbaimirror.indiatimes.com/mumbai/cover-story/oct-12-blackout-was-a-sabotage/articleshow/79312959.cms

Shimshoni, J. (1990). Technology, military advantage, and World War I: A case for military entrepreneurship. *International Security*, *15*(3), 187. https://doi.org/10.2307/2538911

Shiping, T. (2010). Offence-defence theory: Towards a definitive understanding. *The Chinese Journal of International Politics*, *3*(2), 213–260. https://doi.org/10.1093/cjip/poq004

Slaughter, A.-M. (2017). *The chessboard and the web: Strategies of connection in a networked world*. Yale University Press.

Slayton, R. (2016). What is the cyber offense-defense balance? Conceptions, causes, and assessment. *International Security*, *41*(3), 72–109.

Small, M., & Singer, J. D. (1982). *Resort to arms: International and civil wars, 1816–1980* (2nd ed.). SAGE.

Smith, P. A. (1989). *On political war*. National Defense University Press.

Smoke, R. (1977). *War: Controlling escalation*. Harvard University Press.

Snyder, G. H. (1975). *Deterrence and defense: Toward a theory of national security*. Greenwood Press.

Snyder, G. H., & Diesing, P. (2015). *Conflict among nations: Bargaining, decision making, and system structure in international crises*. Princeton University Press. https://doi.org/10.1515/9781400871186

Spaulding, H. (2015). *Russia's false ISIS narrative in Syria*. Institute for the Study of War.

Syria conflict: Russia violation of Turkish airspace "no accident." (n.d.). *BBC News*. Retrieved January 11, 2025, from https://www.bbc.com/news/world-europe-34453739

Talmadge, C. (2017). Would China go nuclear? Assessing the risk of Chinese nuclear escalation in a conventional war with the United States. *International Security*, *41*(4), 50–92. https://doi.org/10.1162/ISEC_a_00274

Talmadge, C. (2019). Emerging technology and intra-war escalation risks: Evidence from the cold war, implications for today. *Journal of Strategic Studies*, *42*(6), 864–887. https://doi.org/10.1080/01402390.2019.1631811

Tessman, B. F., & Chan, S. (2004). Power cycles, risk propensity, and great-power deterrence. *The Journal of Conflict Resolution*, *48*(2), 131–153.

Toft, M. D. (2014). Territory and war. *Journal of Peace Research*, *51*(2), 185–198.

Turkey downing of Russia jet 'stab in the back' – Putin. (2015, November 24). *BBC News*. https://www.bbc.com/news/world-middle-east-34913173

Turkish attackers shut down Russian Central Bank website. (2015, November 25). https://hackread.com/turkish-hackers-target-russian-central-bank-site/

U.S. Department of Defense. (2020). *JP 5-0 joint planning process*. Chairman of the Joint Chiefs of Staff.

Valenta, J., & Valenta, L. F. (2016, March 1). Why Putin wants Syria. *Middle East Forum*. https://www.meforum.org/middle-east-quarterly/why-putin-wants-syria

Valeriano, B., & Jensen, B. (2019, January 15). The myth of the cyber offense: The case for restraint. *Cato Institute*. https://www.cato.org/policy-analysis/myth-cyber-offense-case-restraint

Valeriano, B., & Jensen, B. (2021). *Building a National Cyber Strategy: The process and implications of the cyberspace solarium commission report*. 2021 13th International Conference on Cyber Conflict (CyCon), 189–214. https://doi.org/10.23919/CyCon51939.2021.9467806

Valeriano, B., & Jensen, B. (2022). De-escalation pathways and disruptive technology: Cyber operations as off-ramps to war. In S. J. Shackelford, F. Douzet, & C. Ankersen (Eds.), *Cyber peace: Charting a path toward a sustainable, stable, and secure cyberspace* (pp. 64–93). Cambridge University Press.

Van Evera, S. (1998). Offense, defense, and the causes of war. *International Security*, *22*(4), 5–43. https://doi.org/10.1162/isec.22.4.5

Van Evera, S. (2013). *Causes of war: Power and the roots of conflict*. Cornell University Press.

Vasquez, J. A. (1993). *The war puzzle* (1st ed.). Cambridge University Press. https://doi.org/10.1017/CBO9780511583483

Vasquez, J. A. (2009). *The war puzzle revisited*. Cambridge University Press. https://doi.org/10.1017/CBO9780511627224

Vasquez, J. A., & Henehan, M. T. (2010). *Territory, war, and peace*. Routledge.

Wagner, R. H. (1982). Deterrence and bargaining. *Journal of Conflict Resolution*, *26*(2), 329–358. https://doi.org/10.1177/0022002782026002007

Wallace, Michael D. (1982). Armaments and escalation: Two competing hypotheses. *International Studies Quarterly 26*(1), 37–56. https://doi.org/10.2307/2600598.

Weisiger, A., & Yarhi-Milo, K. (2015). Revisiting reputation: How past actions matter in international politics. *International Organization, 69*(2), 473–495. https://doi.org/10.1017/S0020818314000393

Wooley, S. C., & Howard, P. N. (Eds.). (2018). *Computational propaganda* (Vol. 1). Oxford University Press. https://doi.org/10.1093/oso/9780190931407.001.0001

Wright, Q. (1964). *A study of war: Vol. reprint*. The University of Chicago.

Yarhi-Milo, K. (2014). *Knowing the adversary: Leaders, intelligence, and assessment of intentions in international relations*. Princeton University Press.

Yarhi-Milo, K. (2018). *Who fights for reputation: The psychology of leaders in international conflict*. Princeton University Press.

Zagare, F. C. (1992). NATO, rational escalation and flexible response. *Journal of Peace Research, 29*(4), 435–454.

Zagare, F. C. (1996). Classical deterrence theory: A critical assessment. *International Interactions, 21*(4), 365–387. https://doi.org/10.1080/03050629608434873

Zagare, F. C., & Kilgour, D. M. (2000). *Perfect deterrence* (1st ed.). Cambridge University Press. https://doi.org/10.1017/CBO9780511491788

6 The Cybersecurity Firm and the State

Disclosure, Escalation, and Non-Military Warfare

*Nina Kollars and Michael Poznansky**

While cyber operations will surely play a role in any future kinetic conflict between nation states and their militaries – the war in Ukraine being a prime example – it is a domain for competition and conflict in its own right.[1] It is an arena of non-military warfare in which actors of various stripes exploit information and communications technology for their own ends with increasing sophistication and intensity while the victims seek to defend and mount resistance against such intrusions.

For their part, states compete to steal one another's information while protecting their own, leveraging that information for advantage while trying to deny the same to rivals.[2] Terrorist groups, criminal actors, and a range of other non-state entities exploit cyberspace to extort companies and individuals, disrupt systems, and influence publics.[3] There is also the hybrid state/non-state teaming, a form of proxy conflict,[4] with well-established organized crime syndicates and criminal gangs opportunistically building upon all of the prior behaviors. This is far from a complete list of the phenomena of malicious behaviors that go on.

Understanding how states and non-state actors use cyberspace for strategic advantage, economic gain, and the like is critical. The role of malicious non-state actors as key players in this domain is particularly important to wrestle with. But there is more to the story than myriad attackers and new ways of attacking. There are also new defenders. This chapter highlights one of the most important players in the daily competition and non-military conflict playing out in cyberspace, albeit one that has received relatively scant scholarly attention: cybersecurity firms (CSFs).[5] The main goal is to highlight some of the ways these firms affect the contours of non-military warfare on numerous levels.

The discussion begins by situating this chapter within the context of non-military warfare and highlighting the challenges facing states when trying to address cyber intrusions that occur within their own borders, particularly when the victims are non-state actors, and how this has created the conditions for CSFs to step in and fill the void. Attention then turns to how states and CSFs often have misaligned incentives, as well as the implications of that misalignment. We illustrate these dynamics with two important cases of digital espionage: Russia's hack of the widely used SolarWinds platform and China's hack into Microsoft Exchange. The chapter concludes by discussing the opportunities and challenges facing states when it comes to better integration with CSFs and its consequences for non-military warfare.

DOI: 10.4324/9781003616993-6

Ownership, Control, and the Limits of Government

This volume defines non-military warfare as "non-kinetic tools of statecraft employed as means of strategic coercion and effect employed with an intensity and purpose such that the target perceives them as significant escalations and/or commensurable to warfare" (see Chapter 1). Any account of cyberspace, a major arena of non-military warfare, must consider the unique role of CSFs in providing protection for the networks and devices through which, and upon which, cyber conflict takes place. And therefore, CSFs are inadvertently agents in the defense of and potentially agents of non-military warfare itself.

The concept of non-military warfare captures much of what goes on in the international system today, from how host nations leverage diaspora communities to achieve strategic advantage to the ways in which states wield the law as a weapon to undermine rivals (see Chapters 7 and 9). Because of this wide scope, however, ascertaining who the relevant participants are or what constitutes victory and defeat can be challenging (see Chapters 2 and 3). In many cases, participants in non-military warfare are not state-affiliated military personnel wearing official uniforms but lawyers, immigrants, private companies, and so forth. And because competition is almost always ongoing, traditional concepts of winning and losing are less relevant than for discrete episodes of interstate war or crisis bargaining (see Chapter 5).

Cyberspace, a key locus of non-military warfare in the 21st century and the primary focus of this chapter, offers a unique window into the complexities inherent in non-military warfare. Some of the most fruitful debates in recent years have focused on whether cyber operations are escalatory,[6] whether they can be useful tools of signaling or are better at shaping the strategic environment,[7] and whether they are best thought of as an intelligence contest or a military problem.[8] Jensen's chapter in this volume wades into the center of these debates, arguing that rather than being simply a complement to spying and military operations or a potential source of escalation, cyber operations typically provide states an offramp during moments of escalating military confrontations (see Chapter 5).

This chapter argues that to fully understand the dynamics of non-military warfare in cyberspace, it is necessary to come to terms with who primarily controls it. Contrary to what one might expect given the focus of much of the existing literature, it is not states. Rather, cyberspace is predominantly owned – and protected – by the private sector.[9] When states attack one another in cyberspace, they overwhelmingly target each other's private sectors. The conflict space is often not government systems, although the intended effects, particularly when state-sponsored, are usually meant to have political or military ramifications.[10]

When North Korea decided to strike out at Western insults to their leader, it was Sony pictures that became the target.[11] Obama vowed to "respond proportionately."[12] Russian efforts to influence the U.S. election in 2016 and 2020 included a notable focus on manipulating social media platforms such as Facebook and Twitter.[13] U.S. intelligence estimates explicitly note that the Russian government refrained from tampering with the voting systems themselves.[14] Chinese efforts to steal intellectual property and eavesdrop on U.S. government and military planning occur overwhelmingly on infrastructure not owned by military or government entities.[15] In 2014, Iran carried out a cyberattack against Sands Casino in Las Vegas in an apparent act of retaliation against Sheldon Adelson, a mega-donor in Republican politics who at the time advocated for bombing Iran.[16]

While many of these intrusions and attacks occur in the private sector, governments have not sat idly by. In the United States, there is a patchwork of institutions to deal with cybersecurity threats of various kinds. Some entities, like the Federal Bureau of Investigation or the Department of Homeland Security (DHA), which includes the Cybersecurity and Infrastructure Security Agency (CISA), are internally focused. Others, like U.S. Cyber Command and the National Security Agency, are externally focused. Together, these institutions work to identify, warn, mitigate, and address threats to a range of entities beyond government and military networks.

But there are several reasons why governments are limited in what they can do, or are willing to do, when it comes to defending individuals and companies against malicious cyber activity. These are particularly salient for democracies, a point we take up in the conclusion. For starters, there is a public-private divide, both in terms of physical infrastructure and legal and normative considerations. In many cases, governments may be prohibited from accessing commercial systems or mandating policies in ways that limit their ability to defend against and gain insight into vulnerabilities outside their networks. Moreover, commercial firms and individuals simply may not want governments directly protecting them, particularly if it means providing access to sensitive data and intellectual property.[17]

Another constraint is the sheer size of the attack surface. No government has the resources to fully protect against the dizzying array of intrusions that occur daily, often against systems that are not well protected. As JD Work puts it, "a single prominent financial sector institution may be responsible for more endpoints than the entirety of the Department of Defense network."[18]

Birth of a Marketplace

An unanticipated side-effect of states' inability to provide security to the predominantly non-military structures of cyberspace has been the concomitant rise of CSFs to fill the void.[19] As such, these firms play a crucial role in the dynamics of non-military warfare in cyberspace. CSFs come in many shapes and sizes and perform lots of security-related functions for clients. Some are large, like CrowdStrike or Mandiant, with massive portfolios. Others are more niche entities, like Dragos Inc., that specialize in areas like industrial control systems or operational technology. Still others are security teams set within major information technology and software companies like Apple, Google, and Microsoft. CSFs are the frontline defenders in most cases in cyberspace.

In hindsight, the simultaneous evolution of private sector solutions to fill a massive gap in security should have been just as obvious as the birth of Cyber Command. For CSFs to provide better security, they needed direct access to their clients' private networks, data storage systems, and all the points of access between them. The more sensors and data CSFs collected on a client's system, the better they could observe attempts at intrusion and exfiltration of data. The more clients a CSF has, the broader their sampling of the behaviors unfolding in cyberspace. With access to more and more systems, CSFs then needed to increase their workforce size and sophistication. Over time, the leading firms developed better tools to identify, diagnose, and mitigate damage from attacks and breaches more quickly. Those leading firms are now the global standard in cybersecurity.

Since the vast majority of cyberspace is not owned and operated by states or their militaries, the sheer number of contracts available, and the competitive incentives to stay ahead of other firms, means that CSFs continually use their experience in the day-to-day management of attacks and mitigation to improve their data gathering and analysis capabilities.[20] CSF expertise develops, therefore, in lockstep with new attack trends – the rise of ransomware, for instance – and exploitation tools.

Across the privately held data systems and structures, it is therefore CSFs, and not really states' intelligence agencies or military forces, that have a front-row seat to attacks.[21] The result is that defense, attack detection, response, and mitigation begin with CSFs. While states may be included in all these elements, the first to know and respond is usually a CSF. The key point is that CSFs are challenging states on each of these fronts, that is defending targets that have direct implications for national security, whether by virtue of their contributions to economic prosperity or the functioning of civil society, and holding sensitive intelligence.

If CSFs and states were fully simpatico, we would have a situation in which the former were stepping in to pick up the slack that the latter was unable to fully address but in ways that were usually harmonious with state interests. That is far from the case. A large part of what makes this relationship so interesting is that states and CSFs operate according to different incentives. The potential lack of alignment, and the resulting limits on the degree of information sharing and access that state-level institutions are granted, is not well understood.

Firm and State Misalignments: A Tension

What are the prevailing incentives for CSFs and how do they differ from states and militaries? Let us begin with governments. At the risk of oversimplification, they care first and foremost about national security. Governments typically want to promote the national interest and economic prosperity of their own country as well as their allies and undermine rivals.

Toward this end, states make calculations that look different from those CSFs would make. Consider, for example, states' incentives when it comes to what Erik Lin-Greenberg and Theo Milonopoulos call the "disclosure decision," or "the government's choice to publicly release or acknowledge information that was initially secret and concealed from one or more audiences" in the cyber context.[22]

Many factors influence whether states will disclose secret information.[23] The desire to protect sources and methods may lead states to conceal certain information. Relatedly, revelation of intelligence may lead to calls for escalation from hawkish publics or undermine international rules and norms, thereby pushing leaders to keep it secret.[24] Disclosure may also make states look weak or induce widespread panic. But there may also be cases when states want to disclose secrets. Revealing hostile state activity may embarrass rivals and shore up domestic and international support for countermeasures.[25] It can also demonstrate a state's bona fides when it comes to detection and response.

With respect to mitigating and addressing vulnerabilities at home, states also have different priorities than CSFs. For example, if they had their way, policy-makers would probably push companies to limit certain forms of multinational cooperation, restrict use of certain supply chains that may have a high risk of vulnerabilities, and require expensive defensive measures that may be insensitive to risk-reward considerations. In the case of ransomware, states may want companies to avoid paying ransoms to discourage would-be extortionists from continuing to leverage this capability even if the immediate costs of doing so are larger than simply paying up. In general, national security concerns often take precedence over other considerations.[26]

CSFs operate according to a different logic. First and foremost, they generate profit by establishing a reputation for effectiveness as well as discretion. CSFs are technology-heavy ventures working with a scarce labor market heavily backed in many cases by venture capital markets that have continually increased year after year.[27] Firms gain new clients when they demonstrate effective technical savvy in defense and detection. These demonstrations appear in the form of white papers and public disclosure on social media and news outlets. One of the potential pathologies of this racing to announce big breaches is CSFs' hyperfocus on high-dollar, high-return security issues.[28] Nevertheless, public disclosure is part of the performative element of CSFs building a reputation for being the best. Being the first to announce a major attack or malware in the wild demonstrates that a firm has the technical know-how, tools, and workforce to protect clients.

In contrast to the performative disclosure incentive, firms retain clients by assuring them they can be discreet about breaches and vulnerabilities. CSFs have special access to the information systems they are contracted to protect. Depending on the kinds of security services, software, and hardware they provide, firms have first-hand insight into the systemic weaknesses of their clients' systems. Knowledge of a specific client's vulnerabilities is sacred data bound by the contract's non-disclosure agreement. This type of data is tantalizing for intelligence agencies, especially when they are high-value clients. In addition to specific knowledge of a particular client, large globally footprinted CSFs have insights across multiple clients. This means they have aggregated data. Between specific knowledge, aggregated data, and across networked working groups, CSFs can produce finished intelligence that can then be made available to clients or marketed to draw in new clients.[29]

Thus, firms exercise things like disclosure decisions with this balance in mind – selective and timely public release of technical analysis, coupled with careful adherence to non-disclosure agreements (NDAs), and protection of back-channel and working groups that help develop intelligence ("friend-DAs"). And, because of the sensitivities of their level of access to other people's crown vulnerability jewels, lawyers are almost always in the room advising at each step of the way on the disclosure decision, whether to the public, their clients, or governments.

Overall, the disclosure of breaches and other sorts of malicious activity are beneficial for CSFs. The incentive to disclose, whether selectively or publicly, is strong since it improves a firm's reputation not just for expertise but also as being a responsible defender situated in a broader network of firms who often work together to understand the bigger picture.

Two caveats limit disclosure however: direct attribution to states and specific details of attack. Many CSFs operate according to a norm of non-attribution to states as attackers, preferring to establish a reputation more akin to firefighters putting out a blaze but avoiding the messy second step of assigning blame. Non-attribution is not simply a matter of technical difficulty. It is also a matter of the emerging, albeit informal, separation of roles that govern CSF behavior.

Increasingly, though, there are other CSFs that lean closer to the edge of state attribution while still leaving room for state-on-state politics. Consider that in the development of profiles of an Advanced Persistent Threat (APT) – defined as "a sophisticated, sustained cyberattack in which an intruder establishes an undetected presence in a network in order to steal sensitive data over a prolonged period of time"[30] – CSFs like CrowdStrike use hedging language to avoid a specific pin.[31] Fancy Bear is a "Russia-based" attacker. Wicked Panda is a "China-based" attacker.

The prevarication is a subtle one, but CSFs have historically been reluctant to name governments or government leaders in tracking malicious activity because CSFs work to preserve their autonomy from the ebb of flow of state politics. Not only does blaming a country's cyber military forces necessarily vault the firm into global politics, but it can also put the firm directly in the line of fire. Blaming a regime politicizes a firm's reputation, and that is a choice that a board and CEO must make in planning their marketing strategy. This norm may be shifting.[32]

Ultimately, the delicate balance of doing the work and preserving business space for profit cannot be overstated. CSFs attempt to hold the line in demonstrating elite expertise in knowing the state of breaches and attacks while working to ensure that they are understood as the cyber medics, the first responders, and the cyber firefighters. CSFs in the United States, in particular, want to be known as the cyber troubleshooter, not the cyber shooter. That is the job of agencies like U.S. Cyber Command. The autonomy of CSFs in other countries, particularly authoritarian states, is sometimes an open question, a point we take up in the conclusion. The case of Kaspersky and their potential ties to Russia is just one example. U.S. officials warned that Kaspersky, whose headquarters is in Moscow, "could be coerced into providing or helping establish remote access into their customers' computers by Russian law enforcement or intelligence agencies."[33]

Implications of State-CSF Misalignment

Given the unique position of access that CSFs have when it comes to providing security in cyberspace, and especially the different incentives that drive their actions relative to states, several interesting implications follow. These affect how states experience cyberattacks and the broader dynamics of non-military competition in cyberspace. First, there is slack or lag in escalation. Second, there is the degradation of disclosure decisions.

Escalation Lag

Whether we want them to or not, and whether they want to or not, CSFs find themselves working amidst, despite, and in aid to states. The special access CSFs have to telemetry and the networks of their vast client base means they will naturally stumble across significant state activities, as the case studies below on SolarWinds and Microsoft Exchange reveals. The different incentives, scope, and alignments CSFs have with state political agendas naturally affect the awareness governments have about cyberattacks. One outcome of this misalignment, and the special position of CSFs, is to introduce lag in potential state-on-state escalation.

The existing literature has consistently argued that cyberattacks among states are not escalatory.[34] There are numerous possibilities. Scholarship has focused on the empirical evidence of non-response,[35] as well as the perceptions of appropriate response by citizens in understanding cyberattacks as conceptually distinct from kinetic attacks.[36]

An additional or potentially competing factor in this non-escalatory dynamic is the unique position CSFs play in orienting others to activity in the systems they protect. States may not have immediate access to cyberattack data in real time. The entities that do have that data, and that are equipped to conduct the research to determine the scope of the attack and its implications, are the CSFs.

To be clear, there are many reasons CSFs might hesitate to notify states when they detect what could be state-based cyberattacks. Attacks are plentiful. The vast majority are not state attackers. The difficulty of attribution,[37] and complex compilations of attackers and opportunistic actors achieving other goals alongside them, creates confusion and therefore uncertainty about whether and when state authorities should be notified of potentially malicious behavior that could be attributed to adversary states.

An initial detection of a compromised system may or may not be a malicious attacker, and even if it is, the discovery of the malign activity is almost always delayed by weeks if not months. Additionally, the process of researching an attack's scope and scale takes time, usually involving the formulation of ad hoc working groups and backchannel discussions among trusted and known researchers. Regardless of the time delay to discovery and determining the scope and scale of an attack, CSFs have their own internal process of business decision that include consulting with lawyers about their obligations regarding who to notify, when, and with what degree of detail. Finally, those internal processes guide the pace at which research is conducted internally, and who is invited to work on the problem from the broader community.

Thus, regardless of whether states want to respond immediately to cyberattacks and regardless of whether publics would support timely retaliation leading to escalation, the pace at which CSFs conduct their research, publish their findings, work through their business process, and disclose will naturally create its own escalation lag. Figuring out which direction this cuts–whether the additional time helps to defuse the impact of a cyberattack, or whether it generates greater surprise and therefore pressure to respond–is a ripe area for future work.

Disclosure Decision Degradation

Another implication of misaligned incentives between states and CSFs is that the latter may publicly disclose the existence of a cyber intrusion from a hostile actor, one that bears on state networks and interests, in cases where governments might have either preferred to keep the operation secret or to disclose later. This has several potential effects.

First, premature disclosure (from the state's view) may put the government in an uncomfortable position, pushing leaders to respond to incidents more quickly–and more publicly–than they otherwise would have. By publicizing information independently of government calculations, CSFs remove policymakers' ability to choose when or if to disclose an intrusion. As noted above, there may be cases when governments wish to keep the existence of an intrusion quiet in the hopes of watching and learning more about an adversary, mitigating escalation, or preserving international norms. A lack of control of disclosure decisions may generate domestic pressure or reputational imperatives to respond, as noted above.

In addition to premature disclosure of malicious behavior of rivals, states may also find that CSFs operating within their borders may expose their own operations. As JD Work writes, "many private sector entities do not distinguish between offensive cyber operations conducted by democratic states under legitimate oversight mechanisms from unrestrained actions by authoritarian regimes."[38] What this means in practice is that even when states go through the proper channels for carrying out covert cyber operations, CSFs that discover them may go public just as they would have for malicious attacks conducted by autocracies.

To be sure, states have faced this sort of problem for decades. During the Cold War and beyond, journalists routinely reported the details of covert operations that states would have preferred to keep secret. But the sophistication of CSFs and the ease with which they can piece together and disseminate forensic information related to cyber intrusions renders this problem particularly salient in the digital age.

Finally, there may be cases when governments are happy to let private actors call out adversaries for malicious behavior.[39] By allowing a non-state actor to reveal an intrusion, the government avoids the diplomatic implications of publicly calling out a rival while still netting some of the benefits. They can credibly accuse a foreign rival of meddling without divulging their own sources and methods.[40] Because CSF disclosures often entail painstaking forensics, they have credibility that states can leverage.[41]

Case Illustrations

We offer up two case studies, both of which occurred during 2020–2021, that illustrate these phenomena to varying degrees and the complicated nature of the role of CSFs on the global stage. The first case is the SolarWinds attack which is currently attributed to Russian state hackers seeking to gain access to the U.S. government systems. The second case is the Microsoft Exchange hack which has been attributed to Chinese state efforts to exfiltrate communications data.

Together, they demonstrate several things. First, they showcase the centrality of CSFs in the detection of the attacks, development of early working groups to discern a response plan, the dilemmas of disclosure both publicly and to states, and a fundamental lag in state response to both state-based attacks on U.S. systems. While we are not privy to the backchannel discussions within the working groups and between the CSF and the U.S. Federal officials, it is possible to establish a reasonable record of timelines and discussions through CSF white papers, official statements on social media by both parties, as well as public discussions by leading cybersecurity practitioners.

More broadly, the two cases show the important role the private sector plays, wittingly or unwittingly, in national security and non-military warfare. Whether these were purely cases of espionage or were the beginning of something more disruptive, CSFs are on the front lines of the fierce contest between nation states jockeying for influence against one another in the digital domain.

SolarWinds

On December 8, 2020, what we now know as the SolarWinds/SUNBURST attack, was first reported by FireEye in a public announcement that their own system had been compromised and that their Red Team toolkit – an assembly of different software programs and scripts intended to test their clients' cybersecurity defense posture – had been stolen by Russian hackers.[42] The theft was worrying because red team kits, particularly one as powerful as the industry leader FireEye, should be closely guarded since they are essentially hacking toolkits that can be as easily used for nefarious purposes as for good.[43]

FireEye posted links to a public GitHub repository to help responders protect themselves against malicious use of the toolkit.[44] FireEye further embarked upon its own investigation of how its systems had been breached, a timeline most were certain would reach far further back than December.

Three to four days later, roughly December 11 and 12, researchers in FireEye discovered that the attackers had used a backdoor through a software service they used to manage their internal systems and that the backdoor was established because of a software update provided by the service provider. The name of the software company providing those services was SolarWinds. The attack was deemed a software "supply chain" attack, meaning that attackers had inserted malicious code directly into SolarWinds' software update, and when the update was pushed out to firms, their clients downloaded the malware without knowing. FireEye named this malware SUNBURST.

While it is unclear precisely which Federal agencies were involved in emergency response, the announcement of the supply chain attack via SolarWinds undoubtedly set off alarms all across the U.S. national security enterprise since SolarWinds was a major service provider to many of its agencies. These included the Department of Energy, Department of State, as well as DHS itself.

On December 13, SolarWinds generated a public notice to its customers via Twitter to "upgrade immediately to Orion Platform version 2020.2.1 HF 1 to address a security vulnerability."[45] Meanwhile, the company also submitted an SEC form 8-K, which is a notice to the Securities and Exchange Commission that there has been an unanticipated event for the company that shareholders should know about. On that same day, CISA – the organization tasked with protecting federal civilian government networks[46] – issued an emergency notification that requires federal agencies to disable SolarWinds connections.[47]

In the weeks following these announcements a dizzying number of public notices, patches, internal assessments, and journalistic investigations dominated the news. In early January, a joint statement by U.S. intelligence agencies indicated that

> an advanced persistent threat actor, likely Russian in origin, is responsible for most or all of the recently discovered, ongoing cyber compromises of both government and non-governmental networks. At this time, we believe this was, and continues to be, an intelligence gathering effort.[48]

Finally, on February 19, 2021, the newly inaugurated Biden administration's U.S. National Security Advisor Jake Sullivan announced that

> [w]e are in the process now of working through, with the intelligence community and national security team, a series of steps to respond to SolarWinds, including steps that will hold who we believe is responsible for this and accountable, and you will be hearing about this in short order.

Sullivan added, "We're not talking about months from now, but weeks from now, that the United States will be prepared to take the first steps in response to solar winds."[49]

In the aftermath of the first discovery of SUNBURST, FireEye and other research working groups discerned that the date of initial attack appears to have been in September of 2019. While we are likely far from ever knowing the scope of the compromised systems inside the Federal government, if the numbers are anything akin to that of the nongovernmental systems, the attack was widespread and could have been dangerous for private and government systems alike. Since the attackers managed to compromise SolarWinds' software update processes, the malware that the system pushed could have done far worse than lurk and steal files. It is perhaps fortunate that the attackers were so audacious as to compromise FireEye's systems. Had their own internal processes not discovered the theft, it is unclear how much or whether we would know of SolarWinds more broadly.

Microsoft Exchange

On the heels of the U.S. government's official response to the SolarWinds attack, a separate systemic compromise was beginning to unravel involving Microsoft Exchange. As in the prior attack, the discovery of a vulnerability and its active exploitation occurred outside government awareness, though is undoubtedly being leveraged within them over time. The other element this story highlights is the difficulty of managing resolutions to vulnerabilities without actively pointing to the issue itself. Unlike SolarWinds, the active exploitation of the vulnerability in the Microsoft Exchange case escalated massively after the report and software patches were issued, adding additional pressure to the racing dynamic of public reporting.

The story starts somewhat quietly, though not unlike many vulnerability reports submitted by researchers to software firms. Microsoft's email and calendar collaboration system, known as Microsoft Exchange Server, had a vulnerability being actively exploited that a security researcher known on Twitter as Orange Tsai discovered and tweeted about tauntingly into cyberspace on January 5, 2021: "Just report a pre-auth RCE chain to the vendor. This might be the most serious RCE I have ever reported! Hope there is no bug collision or duplicate."[50]

Orange Tsai is a researcher for DEVCORE, a Taiwanese cybersecurity and penetration testing firm. Orange Tsai is indicating that they have discovered malicious behavior in a system and have documented it to the vendor. Since there is prestige and money that can come from reporting vulnerabilities and active exploitation of them, Tsai was likely hoping that no one else had discovered it and they would get credit for it. Orange Tsai did not disclose the firm's name or the specific vulnerability publicly and appears to have been operating under existing legal and normative disclosure reporting frameworks.

But DEVCORE was not the only CSF tracking the problem. The very next day, a U.S.-based cybersecurity firm, Volexity, noticed attacks on their client's systems that appeared to be leveraging something in Microsoft Exchange and began research to determine the sources of the vulnerability. Volexity then filed its report to Microsoft on February 2. Additionally, a Danish firm, Dubex, noted attacks on their clients around January 18 and filed its report to Microsoft roughly ten days later, on January 27.[51]

Microsoft's security teams received and escalated the varying reports, began the painstaking process of resolving the vulnerabilities in its Exchange software and issued an exceptional emergency public plea to patch on March 2, 2021.[52] The emergency patch update request is notable since Microsoft, among many other global firms, normally issues automatic updates once a month on Tuesdays, known globally as "patch Tuesday." When a patch occurs out of cycle, defenders and attackers take notice because it means that the threat is so high as to be worth pushing immediately.

In Microsoft's public announcement of the details on March 2, they also named the Chinese state-sponsored APT group Hafnium.[53] Hafnium primarily targets U.S. industrial and government defense contracting systems operating, ironically, on U.S. soil through leased servers. Shortly after Microsoft named Hafnium, several security researchers supported and pushed research about Hafnium and the Microsoft Exchange exploits. Also, notably during this time, Jake Sullivan and former CISA Director Chris Krebs, both tweeted and provided public statements regarding the necessity to patch all public and private systems, although they provided little additional statement regarding Hafnium or a government response.

By March 5, additional criminal actors had figured out how to exploit the Exchange vulnerability and piled on to attack any remaining unpatched systems by installing ransomware in compromised systems to monetize the weakness. Of the remaining systems to be patched, there were thousands, and cybersecurity responders globally canceled their weekend plans to stem the damage. But it would be months, not weeks, before the U.S. Federal government issued a statement about who was responsible and promised a response to the blatant and widespread attack. Estimates indicate that more than 30,000 organizations in the United States were compromised.[54]

On July 19, 2021, the U.S. government, in conjunction with allies in the European Union, the United Kingdom, and NATO, formally accused China of being behind the Exchange attack.[55] The accusation condemned China's activities but stopped short of pressing for retribution whether cyber or otherwise. U.S. Secretary of State, Antony Blinken, added a public statement that called upon partners and allies to provide support to build defense capacity, "investigate and impose consequences on malicious cyber actors, and participate in international conversations on cyber policy."[56]

Delay and Decisions

Considering our prior suppositions about the role of CSFs and the challenge they pose to states, we can observe some evidence worth considering toward our proposed phenomena.

Both cases notably document long and continuous processes of research with uncertain but broad consequences. The Microsoft Exchange attack reveals a not-uncommon, very long chain of queries across cybersecurity researchers globally. Multiple parties noted the activity on their client's systems months in advance of what is later identified as Chinese APT actors. Microsoft received these multiple reports of activity and began working on a solution to the vulnerability along its normal timeframe but, ultimately, escalated the issue to mid-cycle patching outside its normal patch Tuesday routine.

What this tells us about CSFs (as well as Microsoft's own cybersecurity teams) is that the research cycle following discovery of malicious activity is complex and does not necessarily immediately reveal itself to be attributable. Instead, research teams work to resolve one of hundreds, if not thousands, of issues yearly; this is just one among them. The lag appears to be one of problem-solving that is granted higher and higher priority as the scope and scale of the attack is revealed over time. It is only later in the research process that it becomes likely that the U.S. government is notified of the active exploitation of the vulnerability. Prior to this, before scope and scale are established, Microsoft considered it pertinent to stay within their normal update cycle.

Similarly, in terms of the SolarWinds attack, while we cannot know when and how frequently FireEye worked directly with Federal agencies prior to the date of their public announcement, the evidence appears to suggest that the notification likely occurred around the same time as their public release of the theft of their tools. Because the repository for mitigating attacks using the toolkit had already been built, we are likely safe in assuming that FireEye had been working to discern the scope of the theft as well as to generate the mitigation solutions well before its public announcement.

Also interesting in these two case studies is that the talent and expertise to detect, verify, and resolve these issues were firmly outside government capabilities. While FireEye had every reason to be concerned about its clients' potential compromise through their toolkit, the discovery of the much wider compromise of SolarWinds' systems was also initiated by their initial detection of theft within their own systems.

It is also worth noting that CSFs appear to have additional authority in verifying and sometimes even falsifying government statements about what is really going on. FireEye's initial statement generically points to a "highly sophisticated state-sponsored adversary."[57] But, by December 13, that adversary was identified as being a Russian APT group. This is even though the U.S. government had not yet issued its own statement about who they considered to be at fault. Simultaneously, the Russian foreign ministry was already roundly accusing CSFs and the United States of spreading false accusations intended to defame the Kremlin.[58] It is also worth noting that, during the December timeframe, President Trump attempted to suggest that China was behind the supply chain attack.[59] That claim went unsubstantiated by FireEye and SolarWinds, and in many outlets, roundly unrecognized by experts conducting the private sector investigation and was thus soon dropped from public debate. The U.S. Federal Government waited an additional month to make its own declarations and to launch a response to the massive Russian espionage campaign.[60]

In both the Exchange and SolarWinds attacks, the decision of when and how to notify the government is a C-suite decision that emerges from extensive conversations between the technical working groups observing the attack, lawyers working through the thorny legal issues, and CSF leadership. While we do not address, specifically, the potential role that former employment for U.S. military and intelligence agencies may play as an influence in disclosure and notification, we consider it a potential key contributor to the discussion. Former military and intelligence employees, at a minimum, have familiarity with their former employers' processes and organizational structures. Those insights – whether negative or positive – likely affect with whom, how, and when disclosure and collaboration occur between CSFs and states.

Finally, in both cases, the official recognition of the need for a domestic security response – patching, updates, and fixes – occurred within days of public disclosure by the CSFs, if not the very same day. In contrast, the official international response to state-based attacks was significantly delayed by weeks, if not months.

The delay, we should point out, need not necessarily be an indicator of ineptitude or ineffectiveness. What it is an indicator of, however, is a new calculus that states must take into consideration in this new landscape of non-military warfare, wherein governments may well choose to allow CSFs to bear the burden of truth telling to reduce the likelihood of forcing a government's hand into taking actions that it would rather not pursue. There is ample room here for new theorizing about state strategic behavior given these new defense actors in cyberspace. The aim of this chapter has merely been to showcase some of these new dynamics around which states must develop new behavior.

Where to From Here?

One of the most important areas to watch in the coming years is how and whether states may seek to assert greater control over CSFs. While all states are subject to the dynamics of the battlespace and intelligence outlined above to varying degrees, not all states have the same relationship with CSFs. Simply because one state may outlaw some firms' disclosure of vulnerabilities does not mean that all firms will be similarly affected. Instead, it will be a matter of the relationship between CSFs, the firm attacked, and the degree that these are integrated.

One strategy for dealing with CSF access to frontline data is for states to implement a top-down integration with those firms, thereby reducing the degree of misalignment between cybersecurity provision and state awareness of the overarching trends and shenanigans. This is particularly feasible in autocracies able to exert such control.[61]

In this vein, the Chinese government appears to be moving in this direction as it increases its grip on its domestic CSFs and their knowledge of emerging techniques. A 2021 law, for instance, "requires Chinese researchers to report zero-day bugs to the government first instead of vendors."[62] Moreover, as mentioned earlier, there have been questions about Kaspersky's relationship to the Kremlin. In 2017, the U.S. government passed bipartisan legislation banning the use of their services on government networks given these concerns.[63]

In contrast, the degree of CSF integration with the U.S. Federal government is less formalized. CSFs recognize that federal level notification and working with agencies such as DHS CISA is an option, yet there is not yet a full legal mandate requiring total alignment with U.S. government political agendas. Partly, this is due to the limits placed on governments in democracies, particularly those with market economies, which may face legal restrictions on what they can and cannot mandate.

Between these two extremes are other models of implied but not official integration. A prime example of this is Israel's preeminence in the CSF space, most of whose premier technical talents have served in the Israel Defense Forces, and many of whom were part of the elite signals intelligence group, the 8200.[64]

The upside of tightly aligned integration cannot be undersold. It reduces the likelihood of disclosure issues, and theoretically gives states earlier decision space on the attacks and trends being managed by CSFs. This, then, would appear to provide autocratic states with some advantage. Those governments with the power and discretion to force integration provides decision-makers with greater flexibility and control of the happenings in cyberspace.

But the story is not all bad for democracies. The downside to forced integration is changing the landscape of working groups and problem-solving entities. Part of what makes the horizontal problem-solving and threat discovery process useful is the practice of observing existing traffic light protocols of information sharing throughout the CSF ecosystem. Firms too tightly aligned to government entities increase the likelihood of interference by states and law enforcement before CSFs have had an opportunity to assess and orient to the threat.

Moreover, CSFs with greater autonomy have greater insight and greater talent pool attraction. In addition to this careful management of reputation for technical expertise and discretion, autonomous CSFs are likely also aware of the systemic advantages of being in countries where they can act independently, to include legal protections that (at least currently) permit the firms to operate without direct government control over data, access to a highly educated talent pool, and a marketplace that is saturated in hardware and software that needs to be protected.

Interestingly, governments like the United States may be attempting to square the circle with legislative solutions. The proposed Cyber Incident Notification Act of 2021, for instance,

> would require federal government agencies, federal contractors, and critical infrastructure operators to notify the Department of Homeland Security's (DHS) Cybersecurity and Infrastructure Security Agency (CISA) when a breach is detected so that the U.S. government can mobilize to protect critical industries across the country.[65]

Legislation such as this is an attempt to wrest some control back from CSFs when it comes to what is happening in the private sector in cyberspace without direct integration.[66] Whether it works remains to be seen.

Logically, the demands for faster and more information sharing throughout the cybersecurity ecosystem make sense.[67] But calls for more and better information sharing have been around in academic and policy circles for more than three decades now. The reality of moving from detection to determining the scope, scale, and source of an attack is far murkier than state leaders would prefer to admit. Detection of an anomaly is not the same thing as a useful diagnosis of the problem. It remains unclear when, to whom, and how often, notification will be useful. It is also unclear just how many of such notifications CISA can handle at any point in time without overwhelming its own small staff and systems. Legal requirements such as the one above mean well in requiring notification within 24 hours of discovery, but that leaves far too much to be defined to be useful in real terms for CSFs.

It is clear, nonetheless, that the significance of CSFs is real in the cyber domain. As they have different incentive structures, capabilities, and legal frameworks, they are making a notable impact on non-military warfare. Understanding modern cyber conflict is impossible without a strong grasp of the anatomy of CSFs, which is something this chapter seeks to contribute.

Notes

* The views expressed here are the authors' own and do not represent those of the U.S. Naval War College, the Department of the Navy, or the Department of Defense.

1 The actual impact cyber operations have had on the war is a subject of intense debate. See, for example, David Cattler and Daniel Black, "The Myth of the Missing Cyber

War: Russia's Hacking Succeeded in Ukraine–And Poses a Threat Elsewhere, Too," Foreign Affairs (2022).

2 Joshua Rovner, "What Is an Intelligence Contest?," Texas National Security Review 3, no. 4 (2020): 114–20.

3 Audrey Kurth Cronin, Power to the People: How Open Technological Innovation Is Arming Tomorrow's Terrorists (New York: Oxford University Press, 2020).

4 Erica Dr. Lonergan, "Cyber Proxies in the Ukraine Conflict: Implications for International Norms," CFR Net Politics (2022), https://www.cfr.org/blog/cyber-proxies-ukraine-conflict-implications-international-norms.

5 For important exceptions, see Amitai Etzioni, "The Private Sector: A Reluctant Partner in Cybersecurity," Georgetown Journal of International Affairs 15 (2014): 69–78; Lennart Maschmeyer, Ronald J. Deibert and Jon R. Lindsay, "A Tale of Two Cybers – How Threat Reporting by Cybersecurity Firms Systematically Underrepresents Threats to Civil Society," Journal of Information Technology & Politics 18, no. 1 (2020): 1–20; Sasha Romanosky and Benjamin Boudreaux, "Private-Sector Attribution of Cyber Incidents: Benefits and Risks to the U.S. Government," International Journal of Intelligence and Counterintelligence 34, no. 3 (2021): 463–93; Florian Skopik, Giuseppe Settanni and Roman Fiedler, "A Problem Shared Is a Problem Halved: A Survey on the Dimensions of Collective Cyber Defense Through Security Information Sharing," Computers & Society 60 (2016): 154–76; J. D. Work, "Private Actors and the Intelligence Contest in Cyberspace," in Robert Chesney and Max Smeets (Eds.), Deter, Disrupt, or Deceive: Assessing Cyber Conflict as an Intelligence Contest (Washington, DC: Georgetown University Press, 2022).

6 Erica D. Lonergan and Shawn W. Lonergan, "Cyber Operations, Accommodative Signaling, and the De-Escalation of International Crises," Security Studies 31, no. 1 (2022): 32–64; Brandon Valeriano, Benjamin Jensen and Ryan C. Maness, Cyber Strategy: The Evolving Character of Power and Coercion (New York: Oxford University Press, 2018).

7 Ben Buchanan, The Hacker and the State: Cyber Attacks and the New Normal of Geopolitics (Cambridge, MA and London: Harvard University Press, 2020); Michael P. Fischerkeller, Emily O. Goldman and Richard J. Harknett, Cyber Persistence Theory, Redefining National Security in Cyberspace (New York: Oxford University Press, 2022).

8 Jon R. Lindsay, "Cyber Conflict vs. Cyber Command: Hidden Dangers in the American Military Solution to a Large-Scale Intelligence Problem," Intelligence and National Security 36, no. 2 (2021): 260–78; Rovner, "What Is an Intelligence Contest?"

9 Jason Healey and Erik B. Korn, "Defense Support to the Private Sector," Cyber Defense Review, 2019, 232; Work, "Private Actors and the Intelligence Contest in Cyberspace."

10 Etzioni, "The Private Sector: A Reluctant Partner in Cybersecurity," 75.

11 Christopher Whyte, "Ending Cyber Coercion: Computer Network Attack, Exploitation and the Case of North Korea," Comparative Strategy 35, no. 2 (2016): 93–102.

12 BBC, "Sony Hack: Obama Vows Response as FBI Blames North Korea," BBC, December 19, 2014, https://www.bbc.com/news/world-us-canada-30555997.

13 Thomas Rid, Active Measures: The Secret History of Disinformation and Political Warfare (New York: Farrar, Strauss and Giroux, 2020).

14 Office of the Director of National Intelligence, "Background to 'Assessing Russian Activities and Intentions in Recent US Elections': The Analytic Process and Cyber Incident Attribution," Intelligence Community Assessment, January 6, 2017: iii, https://www.dni.gov/files/documents/ICA_2017_01.pdf.

15 Adam Segal, "The Code Not Taken: China, the United States, and the Future of Cyber Espionage," Bulletin of the Atomic Scientists 69, no. 5 (2013): 38–45.

16 Jose Pagliery, "Iran Hacked an American Casino, U.S. Says," CNN, February 27, 2015, https://money.cnn.com/2015/02/27/technology/security/iran-hack-casino/index.html.

17 Etzioni, "The Private Sector: A Reluctant Partner in Cybersecurity," 70.

18 Work, "Private Actors and the Intelligence Contest in Cyberspace," 230.
19 Healey and Korn, "Defense Support to the Private Sector," 235.
20 Romanosky and Boudreaux, "Private-Sector Attribution of Cyber Incidents: Benefits and Risks to the U.S. Government," 478.
21 On the outsourcing of intelligence in a range of contexts, see Damien Van Puyvelde, Outsourcing US Intelligence: Contractors and Government Accountability (Edinburgh: Edinburgh University Press, 2021).
22 Erik Lin-Greenberg and Theo Milonopoulos, "Private Eyes in the Sky: Emerging Technology and the Political Consequences of Eroding Government Secrecy," Journal of Conflict Resolution 65, no. 6 (2021): 1070.
23 Allison Carnegie and Austin Carson, "The Disclosure Dilemma: Nuclear Intelligence and International Organizations," American Journal of Political Science 63, no. 2 (2019): 269–85; Allison Carnegie and Austin Carson, "The Spotlight's Harsh Glare: Rethinking Publicity and International Order," International Organization 72, no. 3 (2018): 627–57.
24 Austin Carson, Secret Wars: Covert Conflict in International Politics (Princeton: Princeton University Press, 2018).
25 Jeff D. Colgan and Nicholas L. Miller, "Rival Hierarchies and the Origins of Nuclear Technology Sharing," International Studies Quarterly 63, no. 2 (2019): 310–21. See also https://history.state.gov/milestones/1953-1960/u2-incident.
26 Work, "Private Actors and the Intelligence Contest in Cyberspace," 228–30.
27 Carly Page, "Cybersecurity VC Funding Surges to a Record $11.5B in 2021," Tech Crunch, August 25, 2021, https://techcrunch.com/2021/08/25/cybersecurity-vc-funding-surges-to-a-record-11-5b-in-2021/.
28 Maschmeyer et al., "A Tale of Two Cybers – How Threat Reporting by Cybersecurity Firms Systematically Underrepresents Threats to Civil Society."
29 Interestingly, CSFs often use the language of threat intelligence, mirroring functions that have long been the purview of state-based organizations. See https://www.mandiant.com/advantage/threat-intelligence.
30 Crowdstrike, "Advanced Persistent Threat (APT)," Crowdstrike, June 15, 2022, https://www.crowdstrike.com/cybersecurity-101/advanced-persistent-threat-apt/.
31 Romanosky and Boudreaux, "Private-Sector Attribution of Cyber Incidents: Benefits and Risks to the U.S. Government," 468.
32 William G. Rich, "The US Leans on Private Firms to Expose Foreign Hackers," Wired, November 29, 2018, https://www.wired.com/story/private-firms-do-government-dirty-work/. To take a recent example, a Mandiant report from 2021 specifically tied a group known as UNC1151 to the Belarusian government. See Gabriella Roncone, Alden Wahlstrom, Alice Revelli, David Mainor, Sam Riddell and Ben Read, "UNC1151 Assessed with High Confidence to Have Links to Belarus, Ghostwriter Campaign Aligned with Belarusian Government Interests," Mandiant Research Team (2021), https://www.mandiant.com/resources/blog/unc1151-linked-to-belarus-government.
33 Christopher Bing, "U.S. Warned Firms About Russia's Kaspersky Software Day After Invasion," Reuters, March 31, 2022, https://www.reuters.com/technology/exclusive-us-warned-firms-about-russias-kaspersky-software-day-after-invasion-2022-03-31/.
34 Michael Fischerkeller and Richard J. Harknett, "Persistent Engagement, Agreed Competition, and Cyberspace Interaction Dynamics and Escalation," The Cyber Defense Review (2019): 267–87.
35 Valeriano et al., Cyber Strategy: The Evolving Character of Power and Coercion.
36 Sarah E. Kreps and Jacquelyn Schneider, "Escalation Firebreaks in the Cyber, Conventional, and Nuclear Domains: Moving Beyond Effects-Based Logics," Journal of Cybersecurity 5, no. 1 (2019): 1–11.
37 Thomas Rid and Ben Buchanan, "Attributing Cyber Attacks," Journal of Strategic Studies 38, no. 1–2 (2015): 4–37.
38 Work, "Private Actors and the Intelligence Contest in Cyberspace," 242.

39 Work, 245.
40 Myriam Dunn Cavelty and Victor Maruer, "Postmodern Intelligence: Strategic Warning in an Age of Reflexive Intelligence," Security Dialogue 40, no. 2 (2009): 123–44; Len Scott, "Sources and Methods in the Study of Intelligence: A British View," Intelligence and National Security 22, no. 2 (2007): 185–205.
41 Romanosky and Boudreaux, "Private-Sector Attribution of Cyber Incidents: Benefits and Risks to the U.S. Government," 464.
42 FireEye, "Highly Evasive Attacker Leverages SolarWinds Supply Chain to Compromise Multiple Global Victims With SUNBURST Backdoor," Mandiant Threat Research Blog (2020), https://www.mandiant.com/resources/blog/evasive-attacker-leverages-solarwinds-supply-chain-compromises-with-sunburst-backdoor. See also https://www.kiuwan.com/solarwinds-hack-timeline/.
43 This relates to debates about "ethical hacking," or penetration testing, in which experts intentionally seek out vulnerabilities in a system with the intention of fixing them. The problem, as this case shows, is that the same methods used for these purposes can be exploited by bad actors.
44 https://github.com/mandiant/sunburst_countermeasures.
45 https://twitter.com/solarwinds/status/1338325699300651018?lang=en.
46 https://www.cisa.gov/about-cisa/.
47 https://www.cisa.gov/supply-chain-compromise.
48 CISA, "Joint Statement by the Federal Bureau of Investigation (FBI), the Cybersecurity and Infrastructure Security Agency (CISA), the Office of the Director of National Intelligence (ODNI), and the National Security Agency (NSA)," (2021), https://www.cisa.gov/news/2021/01/05/joint-statement-federal-bureau-investigation-fbi-cybersecurity-and-infrastructure. See also Brian Carlson, "The Microsoft Exchange Server Hack: A Timeline," CSO (2021), https://www.csoonline.com/article/3616699/the-microsoft-exchange-server-hack-a-timeline.html.
49 Nicole Gaouette, "White House Says It Will Hold Those Responsible for SolarWinds Hack Accountable Within Weeks," CNN, February 19, 2021, https://www.cnn.com/2021/02/19/politics/sullivan-solarwinds-khashoggi/index.html.
50 https://twitter.com/orange_8361/status/1346401788811825153. RCE is remote code execution.
51 Brian Krebs, "A Basic Timeline of the Exchange Mass-Hack," KrebsonSecurity, March 8, 2021, https://krebsonsecurity.com/2021/03/a-basic-timeline-of-the-exchange-mass-hack/.
52 Brian Krebs, "At Least 30,000 U.S. Organizations Newly Hacked Via Holes in Microsoft's Email Software," KrebsonSecurity, March 5, 2021, https://krebsonsecurity.com/2021/03/at-least-30000-u-s-organizations-newly-hacked-via-holes-in-microsofts-email-software/
53 Anthony Eich, "Chinese APT Hafnium Attacking Microsoft Exchange Servers," UHWO Cybersecurity, March 25, 2021, https://westoahu.hawaii.edu/cyber/uncategorized/chinese-apt-hafnium-attacking-microsoft-exchange-servers/
54 Krebs, "A Basic Timeline of the Exchange Mass-Hack."
55 Zolan Kanno-Youngs and David E. Sanger, "U.S. Accuses China of Hacking Microsoft," New York Times, August 26, 2021, https://www.nytimes.com/2021/07/19/us/politics/microsoft-hacking-china-biden.html.
56 Antony J. Blinken, "Responding to the PRC's Destabilizing and Irresponsible Behavior in Cyberspace," U.S. Department of State Press State, Office of the Spokesperson, July 19, 2021, https://www.state.gov/responding-to-the-prcs-destabilizing-and-irresponsible-behavior-in-cyberspace/.
57 Fireeye, "Unauthorized Access of FireEye Red Team Tools," Mandiant, December 8, 2020, https://www.mandiant.com/resources/unauthorized-access-of-fireeye-red-team-tools.
58 David Brennan, "Russia Denies 'Baseless' SolarWinds Claims as Trump Administration Divided on Hack," Newsweek, December 21, 2020, https://www.newsweek.com/russia-denies-baseless-solarwinds-claims-trump-admin-divided-hack-1556316

59 James Crowley, "Trump Downplays SolarWinds Hack, Says May Be Tied to Election Loss and China," Newsweek, December 19, 2020, https://www.newsweek.com/trump-tweets-solarwinds-hack-voter-fraud-election-loss-1556165.
60 Sara Morrison, "Biden Makes Good on His Promise to Punish Russia for the Massive SolarWinds Hack," Vox, April 15, 2021, https://www.vox.com/recode/22385555/biden-solarwinds-hack-russia-sanctions
61 Other chapters in this volume, such as Wong's discussion of diaspora communities in Chapter 9, similarly point out that autocracies wield levers of non-military warfare in ways that differ from democracies.
62 Tonya Riley, "Chinese Researchers Accuse NSA of Being Behind a Powerful Exploit," CyberScoop, February 23, 2022, https://www.cyberscoop.com/chinese-researchers-nsa-exploit-pangu-lab/.
63 Dustin Volz, "Trump Signs Into Law U.S. Government Ban on Kaspersky Lab Software," Reuters, December 12, 2017, https://www.reuters.com/article/us-usa-cyber-kaspersky/trump-signs-into-law-u-s-government-ban-on-kaspersky-lab-software-idUSKBN1E62V4.
64 Gil Press, "6 Reasons Israel Became a Cybersecurity Powerhouse Leading the $82 Billion Industry," July 18, 2017, https://www.forbes.com/sites/gilpress/2017/07/18/6-reasons-israel-became-a-cybersecurity-powerhouse-leading-the-82-billion-industry/?sh=782b05fd420a.
65 Mark R. Warner, "Following SolarWinds & Colonial Hacks, Leading National Security Senators Introduce Bipartisan Cyber Reporting Bill," Press Releases, July 21, 2021, https://www.warner.senate.gov/public/index.cfm/2021/7/following-solarwinds-colonial-hacks-leading-national-security-senators-introduce-bipartisan-cyber-reporting-bill.
66 Governments have previously faced pushback when floating the idea of restricting the intelligence functions of private actors like CSFs. See J. D. Work, "Private Actors and the Intelligence Contest in Cyberspace," 246–47.
67 Stefan Laube and Rainer Bohme, "Strategic Aspects of Cyber Risk Information Sharing," ACM Computing Surveys (CSUR) 50, no. 5 (2017): 1–36; Oscar Serrano, Luc Dandurand and Sarah Brown, "On the Design of a Cyber Security Data Sharing System," Proceedings of the 2014 ACM Workshop on Information Sharing & Collaborative Security, 2014, 61–69; Skopik et al., "A Problem Shared Is a Problem Halved: A Survey on the Dimensions of Collective Cyber Defense Through Security Information Sharing."

Bibliography

BBC. (2014, December 19). Sony Hack: Obama vows response as FBI blames North Korea. *BBC*. https://www.bbc.com/news/world-us-canada-30555997

Bing, C. (2022, March 31). U.S. warmed firms about Russia's Kaspersky software day after invasion. *Reuters*. https://www.reuters.com/technology/exclusive-us-warned-firms-about-russias-kaspersky-software-day-after-invasion-2022-03-31/

Blinken, A. J. (2021, July 19). Responding to the PRC's destabilizing and irresponsible behavior in cyberspace. *U.S. Department of State Press State, Office of the Spokesperson*. https://www.state.gov/responding-to-the-prcs-destabilizing-and-irresponsible-behavior-in-cyberspace/

Brennan, D. (2020, December 21). Russia Denies 'baseless' SolarWinds claims as Trump administration divided on hack. *Newsweek*. https://www.newsweek.com/russia-denies-baseless-solarwinds-claims-trump-admin-divided-hack-1556316

Buchanan, B. (2020). *The hacker and the state: Cyber attacks and the new normal of geopolitics*. Harvard University Press.

Carison, B. (2021). The Microsoft exchange server hack: A timeline. *CSO*. https://www.csoonline.com/article/3616699/the-microsoft-exchange-server-hack-a-timeline.html

Carnegie, A., & Carson, A. (2010). The disclosure dilemma: Nuclear intelligence and international organizations. *American Journal of Political Science*, *63*(2).

Carnegie, A., & Carson, A. (2018). The spotlight's harsh glare: Rethinking publicity and international order. *International Organization*, *72*(3), 627–657.

Carson, A. (2018). *Secret wars: Covert conflict in international politics*. Princeton University Press.

Cattier, D., & Black, D. (2022). The myth of the missing cyber war: Russia's hacking succeeded in Ukraine-and poses a threat elsewhere, too. *Foreign Affairs*.

CISA. (2021). *Joint statement by the Federal Bureau of Investigation (FBI), the Cybersecurity and Infrastructure Security Agency (CISA), the Office of the Director of National Intelligence (ODNI), and the National Security Agency (NSA)*. https://www.cisa.gov/news/2021/01/05/joint-statement-federal-bureau-investigation-fbi-cybersecurity-and-infrastructure

CISA. (2022, February 15). FBI, CISA, and NSA cyber alert: Russian state-sponsored cyber actors gain network access by exploiting default multifactor authentication protocols and "PrintNightmare" vulnerability. *FBI*. https://www.cisa.gov/news-events/cybersecurity-advisories/aa22-074a

Colgan, J. D., & Miller, N. L. (2019). Rival hierarchies and the origins of nuclear technology sharing. *International Studies Quarterly*, *53*(2), 310–321.

Crowdstrike. (2022, June 15). Advanced persistent threat (APT). *Crowdstrike*. https://www.crowdstrike.com/cybersecurity-101/advanced-persistent-threat-apt/

Crowley, J. (2020, December 19). Trump downplays SolarWinds hack, says may be tied to election loss and China. *Newsweek*. https://www.newsweek.com/trump-tweets-solarwinds-hack-voter-fraud-election-loss-1556165

Dunn Cavelty, M., & Mauer, V. (2009). Postmodern intelligence: Strategic warning in an age of reflexive intelligence. *Security Dialogue*, *40*(2), 123–144.

Eich, J. M. (2022, April 7). USCYBERCOM disrupted Russian military hacking capabilities. *SC Media*. https://westoahu.hawaii.edu/cyber/uncategorized/chinese-apt-hafnium-attacking-microsoft-exchange-servers/

Etzioni, A. (2014). *The new normal: Finding a balance between individual rights and the need for security*. Transaction Publishers.

FireEye. (2016, June 16). APT28: A window into Russia's cyber espionage operations? *FireEye*. https://cyber-peace.org/wp-content/uploads/2018/11/rpt-apt28.pdf

FireEye. (2020a, December 8). FireEye: Unauthorized access of FireEye red team tools. *Mandiant*. https://www.mandiant.com/resources/unauthorized-access-of-fireeye-red-team-tools

FireEye. (2020b). Highly evasive attacker leverages SolarWinds supply chain to compromise multiple global victims with SUNBURST backdoor. *Mandiant Threat Research Blog*. https://www.mandiant.com/resources/blog/evasive-attacker-leverages-solarwinds-supply-chain-compromises-with-sunburst-backdoor. See also https://www.kiuwan.com/solarwinds-hack-timeline/

Fischerkeller, M., Goldman, E. O., & Harknett, R. J. (2022). *Cyber persistence theory, redefining national security in cyberspace*. Oxford University Press.

Fischerkeller, M., & Harknett, R. J. (2019). Persistent engagement, agreed competition, and cyberspace interaction dynamics and escalation. *The Cyber Defense Review*, 267–287.

Gaouette, N. (2021, February 19). White house says it will hold those responsible for SolarWinds hack accountable within weeks. *CNN*. https://edition.cnn.com/2021/02/19/politics/sullivan-solarwinds-khashoggi/index.html

Healey, J., & Korn, E. B. (2019). Defense support to the private sector. *Cyber Defense Review, Special Edition* 2019, 227–242.

Kanno-Youngs, Z., & Sanger, D. E. (2021, August 26). U.S. accuses China of hacking Microsoft. *New York Times*. https://www.nytimes.com/2021/07/19/us/politics/microsoft-hacking-china-biden.html

Krebs, B. (2021a, March 5). At least 30,000 U.S. organizations newly hacked via holes in Microsoft's email software. *KrebsonSecurity*. https://krebsonsecurity.com/2021/03/a-basic-timeline-of-the-exchange-mass-hack/

Krebs, B. (2021b, March 8). A basic timeline of the exchange mass-hack. *KrebsonSecurity*. https://krebsonsecurity.com/2021/03/at-least-30000-u-s-organizations-newly-hacked-via-holes-in-microsofts-email-software/

Kreps, S. E., & Schneider, J. (2019). Escalation firebreaks in the cyber, conventional, and nuclear domains: Moving beyond effects-based logics. *Journal of Cybersecurity*, *5*(1), 1–11.

Kurth Cronin, A. (2020). *Power to the people: How open technological innovation is arming tomorrow's terrorists*. Oxford University Press.

Laube, S., & Bohme, R. (2017). Strategic aspects of cyber risk information sharing. *ACM Computing Surveys (CSUR)*, *50*(5), 1–36.

Lindsay, J. R. (2021). Cyber conflict vs. cyber command: Hidden dangers in the American military solution to a large-scale intelligence problem. *Intelligence and National Security*, *36*(2).

Lin-Greenberg, E., & Milonopoulos, T. (2021). Private eyes in the sky: Emerging technology and the political consequences of eroding government secrecy. *Journal of Conflict Resolution*, *65*(6).

Lonergan, E. (2022). Cyber proxies in the Ukraine conflict: Implications for international norms. *CFR Net Politics*. https://www.cfr.org/blog/cyber-proxies-ukraine-conflict-implications-international-norms

Lonergan, E., & Lonergan, S. W. (2022). Cyber operations, accommodative signaling, and the de-escalation of international crises. *Security Studies*, *31*(1).

Maschmeyer, L., Deibert, R. J., & Lindsay, J. R. (2020). A tale of two cybers – how threat reporting by cybersecurity firms systematically underrepresents threats to civil society. *Journal of Information Technology & Politics*, *18*(1).

Morrison, S. (2021, April 15). Biden makes good on his promise to punish Russia for the massive SolarWinds hack. *Vox*. https://www.vox.com/recode/22385555/biden-solarwinds-hack-russia-sanctions

Office of the Director of National Intelligence. (2017, January 6). *Background to 'assessing Russian activities and intentions in recent US elections': The analytic process and cyber incident attribution* (Intelligence Community Assessment). https://www.dni.gov/files/documents/ICA_2017_01.pdf

Page, C. (2021, August 25). Cybersecurity VC funding surges to a record $11.5B in 2021. *Tech Crunch*. https://techcrunch.com/2021/08/25/cybersecurity-vc-funding-surges-to-a-record-11-5b-in-2021/

Pagliery, J. (2015, February 27). Iran hacked an American Casino, U.S. says. *CNN*. https://money.cnn.com/2015/02/27/technology/security/iran-hack-casino/index.html

Press, G. (2017, July 18). 6 reasons Israel became a cybersecurity powerhouse leading the $82 billion industry. *Forbes*. https://www.forbes.com/sites/gilpress/2017/07/18/6-reasons-israel-became-a-cybersecurity-powerhouse-leading-the-82-billion-industry/?sh=782b05fd420a

Rich, W. G. (2018, November 29). The US leans on private firms to expose foreign hackers. *Wired*. https://www.wired.com/story/private-firms-do-government-dirty-work/

Rid, T. (2020). *Active measures: The secret history of disinformation and political warfare*. Farrar, Strauss and Giroux.

Rid, T., & Buchanan, B. (2015). Attributing cyber attacks. *Journal of Strategic Studies*, *38*(1–2), 4–37.

Riley, T. (2022, February 23). Chinese researchers accuse NSA of being behind a powerful exploit. *CyberScoop*. https://www.cyberscoop.com/chinese-researchers-nsa-exploit-pangu-lab/

Romanosky, S., & Boudreaux, B. (2021). Private-sector attribution of cyber incidents: Benefits and risks to the U.S. Government. *International Journal of Intelligence and Counterintelligence*, *34*(3).

Rovner, J. (2020). What is an intelligence contest? *Texas National Security Review*, *3*(4).

Scott, L. (2007). Sources and methods in the study of intelligence: A British view. *Intelligence and National Security, 22*(2), 185–205.
Segal, A. (2013). The code not taken: China, the United States, and the future of cyber espionage. *Bulletin of the Atomic Scientists, 69*(5), 38–45.
Serrano, O., Dandurand, L., & Brown, S. (2014). *On the design of a cyber security data sharing system*. Proceedings of the 2014 ACM Workshop on Information Sharing & Collaborative Security, 61–69.
Skopik, F., Settanni, G., & Fiedler, R. (2016). A problem shared is a problem halved: A survey on the dimensions of collective cyber defense through security information sharing. *Computers & Society, 60*.
Valeriano, B., Jensen, B., & Maness, R. C. (2018). *Cyber strategy: The evolving character of power and coercion*. Oxford University Press.
Van Puyvelde, D. (2021). *Outsourcing US intelligence: Contractors and government accountability*. Edinburgh University Press.
Volz, D. (2017, December 12). Trump signs into law U.S. Government ban on Kaspersky lab software. *Reuters*. https://www.reuters.com/article/us-usa-cyber-kaspersky/trump-signs-into-law-u-s-government-ban-on-kaspersky-lab-software-idUSKBN1E62V4
Warner, M. (2021, July 21). Following SolarWinds & colonial hacks, leading national security senators introduce bipartisan cyber reporting bill. *Press Releases*. https://www.warner.senate.gov/public/index.cfm/2021/7/following-solarwinds-colonial-hacks-leading-national-security-senators-introduce-bipartisan-cyber-reporting-bill
Whyte, C. (2016). Ending cyber coercion: Computer network attack, exploitation and the case of North Korea. *Comparative Strategy, 35*(2), 93–102.
Work, J. D. (2022). Private actors and the intelligence contest in cyberspace. In *Deter, disrupt, or deceive: Assessing cyber conflict as an intelligence contest*. Georgetown University Press.

7 War by Law

Legal Statecraft With International Aims

*Justin Key Canfil**

"Any country? Anyone voted for Russian membership?" Ukraine's Ambassador to the United Nations (UN), Sergiy Kyslytsya, rocked on his heels as he looked around the hall. The silence was deafening. "Anyone? Should I put my glasses [on] if my vision fails me?"

Having been invaded only a few days prior, Ukraine was mounting a diplomatic counteroffensive against its aggressor, Russia, at an emergency meeting convened by the UN General Assembly. Russia had inherited a Permanent 5 (P5) seat on the UN Security Council after the Soviet Union disbanded in 1991, a position that gave it veto power over any UN action to halt the war. P5 powers can also thwart attempts by the General Assembly to expel them from the UN.[1]

But Kyslytsya's point was altogether different. Rather than calling on UN members to vote to condemn Russia or expel it from the UN, Kyslytsya was questioning whether Russia had ever had a legal right to its P5 seat at all, a premise that, until that very moment, not a single one of the UN's 193 member states had thought to question.

> "Please raise your hand," Kyslytsya continued, if your country voted in the formal session of the General Assembly in reply to the letter by President Yeltsin dated December 24 1991 when he told the United Nations that Russia would like to be the continuator state of the demised Soviet Union. . . . Anyone?

He paused for effect, before delivering the next blow: "I wonder if the Russian presence in the United Nations has ever been legitimate!"[2]

Kyslytsya's argument was provocative – even sardonic – and of course politically motivated. But it was compelling because it contained kernel of truth. In fact, the P5 list, enshrined in Chapter V, Article 23(1) of the UN Charter, enumerates only the following five permanent members: the Republic of China, France, the Union of Soviet Socialist Republics [Soviet Union], the United Kingdom of Great Britain and Northern Ireland, and the United States of America.[3] When the Soviet Union dissolved, other former Soviet and the Warsaw Pact countries had to formally reapply for membership. Yet Russia had evidently never done so.[4] Russia had simply assumed the Soviet Union's seat without debate. Kyslytsya of course did not mention that members of the former Soviet Union had agreed among

DOI: 10.4324/9781003616993-7

themselves that Russia should be given the seat, but he was not wrong that other UN members had never technically consented to it.

Although Russia continues to exercise its de facto P5 powers on the Security Council, the speech spurred an active debate about whether Russia is indeed a de jure P5 member.[5] Amid the furor, UN members voted two months later to suspend Russia from the Human Rights Council.[6] It was not the only occasion in which Ukraine used legal sleight-of-hand to advance its interests during the war; an article in Forbes, for example, asserted that Ukraine continues to wage "lawfare" against Russia in other forms, for example by "weaponizing corporations."[7] One state, the People's Republic of China (PRC), has even recognized lawfare as part of its strategic doctrine.[8] Cynics have pointed to this as a reason why China has increased its influence in international regimes, and some proponents of U.S. hegemony have even called on the U.S. government to adopt its own lawfare doctrine in response.[9]

If this volume explores varieties of "nonmilitary warfare," this chapter concentrates on the concept of lawfare. Although dominant accounts of lawfare have offered many definitions (outlined later in this chapter), a common thread is that lawfare involves weaponization of the legal system to achieve politically motivated aims. Through this lens, lawfare includes baseless allegations of noncompliance, frivolous litigation before international courts, intimidation of judges, and other attempted subversions of the legal process.

The term "lawfare" has gained new currency in American domestic politics since 2020. At his criminal sentencing for falsification of business records on January 10, 2025, U.S. President Donald Trump insisted to the court that the Justice Department had treated him "very, very unfairly. . . . They call it lawfare."[10] A year prior, the Colorado Supreme Court had ruled that President Trump's criminal record disqualified him from being named on the state election ballot. In response, Trump's defenders complained that it was part of a campaign of "Democratic lawfare"[11] and advocate a fire-fights-fire approach. Within six months, Republicans were on the offensive, celebrating that "Trump is finally – and rightly – winning the 'lawfare' battle."[12]

Notwithstanding increasing use of the term to describe political fights over domestic matters in domestic courts, classically the concept has referred to weaponization of international law. This chapter will focus on the latter. But, to the extent skeptics of international law have always viewed it as a tool of powerful states, how, really, does "lawfare" differ from day-to-day international legal practice?

I argue in this chapter that, if law is a battleground, legitimacy is the high ground. Legal practitioners have a professional code of ethics, and even in authoritarian countries, legal training is often infused with norms from imported legal-educational models. Not every lawyer is an angel, but even the stereotypical "hired gun" must convince the "court" (figurative or literal) that the defendant has (or has not) broken the law.[13] In order to be effective, these arguments must be perceived as credible to legal audiences – especially, when political interests diverge.[14] In the global arena, international audiences are ultimately one's judge and jury.[15] As such, they decide what is persuasive or not. Preposterous arguments tend to invite blowback.[16] This would explain why legal disputes are not unbounded.[17] Even adversarial states with almost no common interests must still fight legal battles on common ground.[18]

To the extent that it carries a connotation of unfair advantages, this chapter contends that "lawfare" is not a very useful concept. It is true that baseless allegations or frivolous litigation can sometimes pay – when the truth is hard to verify, for example, or when defenders forfeit or concede, often out of impatience, material weakness, or fear of bad press. But lawfare should be carefully distinguished from information warfare. Legal practice is about mutual interpretation of common-knowledge legal rules given the information political actors have. Thus, information warfare can shape the legal battleground but is ancillary to legal statecraft, itself. Since credibility is currency, "lawfare" is most effective at advancing state interests when it aligns with conventional legal strategies.

Within this framework, the chapter advances a typology whereby international legal statecraft can be systematically categorized. It distinguishes between two types of international political aims (offensive and defensive) and the types of law that states can seek to instrumentalize (domestic and international). It concludes by hypothesizing that an overemphasis on "lawfare" – as concept and practice – may undermine states' ability to instrumentalize international law for any purpose at all as cynicism increasingly permeates the terrain. "Flooding the zone" with incredible or illegitimate allegations can risk damaging one's own reputation, but countering it in-kind may eventually reduce the overall faith in the system.[19] Future researchers are invited to explore deeper questions about the effectiveness of legal practices perceived by other states as legally unsound, variation in the sources and prevalence of these sorts of practices in international relations, and their long-run effects on international regimes.

Legal Statecraft Is Not "Lawfare"

Despite the concept's growing popularity, thinkers still have not converged on an agreed-upon definition for lawfare. For example, the United States (U.S.) Department of Defense has defined China's "legal warfare" doctrine as "[the use of] international and domestic law to claim the legal high ground or assert Chinese interests."[20] Yet this type of activity hardly seems unique to China.[21] Indeed, replace "Chinese" with "American," and it could easily describe the way the United States (or any other sovereign actor) has always practiced law in its foreign relations.[22]

We might be tempted to categorize the Ukrainian diplomatic counteroffensive, or attempted Russian justifications, as a form of "lawfare." Consistent with older definitions of lawfare, it hinges specifically on a wartime context. More recent commentary has expanded the concept of lawfare beyond the domain of armed force. Other types might include the extraterritorial application of domestic laws; vexatious litigation; forum-shopping, the creation and exploitation of legal loopholes; or of conflicts between legal obligations. What all of these definitions have in common is the perceived "weaponization" or otherwise bad-faith instrumentalization of legal tools against an adversary.

As such, these definitions rely on knowledge of intent. But lawyers would be the first to acknowledge how notoriously difficult intent is to observe or prove. Who has the authority to classify use cases into either bad-faith "lawfare" or innovative (but good-faith) legal strategy? Absent a way to demonstrate intent, "lawfare" cannot easily be separated from the everyday behavior of lawyers in the foreign policy establishment. While perhaps illustrative, definitions that center on intent offer little empirical or explanatory traction. The effectiveness of legal strategies depends not on whether one's intent is in good bad faith, but whether they are credible enough to be convincing to key audiences.

Instead of "lawfare," which has traditionally connoted the use of law to shape armed conflicts, this chapter focuses on international legal statecraft. International legal statecraft describes a range of ways states use law and legal tools to advance their foreign policy interests.

The chapter explains that the use of law in foreign relations can be typologized along two basic dimensions. First, states can have offensive or defensive purpose in how they leverage legal tools. Second, those legal tools work by leveraging sources – statutes, treaties, court rulings, and so on – that can be either domestic or international. The way legal statecraft translates into legal operations and tactics depends on how sources and purpose combine. The use of law to advance national interests is most effective when states choose the optimal combination for a given issue or contest. Unlike traditional invocations of "lawfare," the international legal statecraft framework is agnostic about whether states use law to advance their interests in good or bad faith. What matters more is whether they succeed.

How Law Works

The first thing students of international relations learn is that the world is "anarchical." By this, we mean only that there is no supreme authority to pass binding laws or resolve disputes between countries. And yet countries have agreed to order themselves under a "law of nations" – a complex system of legal norms. Countries that have consented to those norms are considered bound to obey them. In the most minimalist sense, laws work by defining a focal point around which self-interested actors – here, sovereign states – can coordinate their behavior around expectations of "right" and "wrong."[23] The community of participants look to the focal point when evaluating the behavior of other participants. Participants are incentivized to adhere to the focal point because behaviors that deviate in harmful ways, when observed, can be collectively sanctioned.[24]

Whether international law is "law" depends on who you ask. For some observers, "right" and "wrong" take on a normative angle. Under certain doctrines, especially in certain areas like human rights law or the laws of war, obligations may arguably precede consent. Others are more skeptical. In the 19th century, the jurist Sir Thomas Erskine Holland described international law as the "vanishing point of jurisprudence."[25] But even relativists agree that international law moderates political freedom of action via formalized, increasingly codified sets of community expectations.[26]

Earlier literature saw international law as a hard constraint on states' naked ambitions. Legal scholar Louis Henkin famously described how "almost all nations observe almost all principles of international law and almost all of their obligations almost all of the time."[27] Abram and Antonia Chayes argued that, in part because of the penalties of noncompliance, involvement with international law signals an intent to comply even when compliance falls short.[28] Others have gone further, arguing that experience with international institutions can even shift the preferences of scofflaws toward compliance.[29] Skeptics replied that high rates of compliance are misleading because states only sign on to laws they are willing to faithfully observe in the first place.[30] The rift only deepened between empiricists in the early 2000s,[31] giving way to an argument that attempts to measure compliance might be fruitless.[32]

While scholars have struggled to evaluate compliance, this does not mean that states cannot decide what it means for themselves. But because states are political entities with political aims, there is a tendency in more recent scholarship to suppose that international legal practice must be politics all the way down.

Scholars have argued that, because so much of international law is decidedly ambiguous, compliance, while hard to measure, must be related to the way legal institutions are designed – and specifically, the precision with which the rules are written.[33] This is because the operating system of law is natural language, and language is a coarse tool, rife with indeterminacies.[34] These indeterminacies are a double-edged blade: they can allow the law to be flexibly adapted to new situations and to grow and evolve to suit a changing world. But they can also be exploited for political purposes.

Under this framework, rather than simply taking on obligations, states can seek to "do things" with international law.[35] Most importantly, states can exploit loopholes and ambiguities in legal language – much of it poorly codified, some of it not codified at all – to violate the spirit of international law while maintaining the appearance of being technically compliant with its letter.[36] This tension between obligations and opportunities characterizes the international legal landscape, and especially capable legal actors presumably stand the most to gain.

Realpolitik and Law

The idea that law can be weaponized for political purposes is nothing new. In 1609, a lawyer named Hugo Grotius (Huig de Groot) was hired by the Dutch East India Company to develop a theory under which Holland could justify going to war against the Portuguese for excluding it from the Indian Ocean.[37] Grotius set to work developing a theory that the sea, unlike land territory, belonged to no one and everyone. His treatise, "Mare Liberum," which is still cited today, became the bedrock for the modern principle of freedom of navigation.

As international legal regimes become denser and more complex, the number of constraints on states has risen.[38] But so have the opportunities to bend or leverage those constraints.[39] In addition, domestic laws have been used to effect influence in foreign relations since at least the 19th century. For example, U.S. federal agents adopted a technique of "extraordinary rendition," whereby foreign nationals captured on foreign soil were smuggled back to stand trial in U.S. courts.[40] The practice of applying domestic laws extraterritorially increased significantly in the later 20th century and especially after 2001.[41] Today, for example, non-U.S. nationals can sue other non-U.S. nationals in U.S. courts for crimes committed abroad involving torture. The U.S. DoJ has indicted scores of foreign cyber criminals since 2014, some for crimes committed against other states.[42] The Bush administration infamously used "black" detention sites on foreign soil so it could interrogate suspects in the War on Terror without affording them basic U.S. Constitutional protections.

Other countries have unsurprisingly followed suit. Justly or not, U.S. officials have long-expressed concern that the International Criminal Court would be used by other countries to pursue vendettas against American leaders and military servicemembers.[43] The passage of a harsh new National Security Law in Hong Kong has given rise to fears that critics of the Chinese government could be arrested between airport connections.[44] And in response to a Russian law criminalizing "fake news" about the war, a number of Western media outlets left the country to avoid exposing their journalists to potential 15-year prison terms.[45]

Although some legal scholars and analysts have embraced the term "lawfare," opinions diverge as to its meaning.[46] Orde Kittrie's eponymous book contains the most comprehensive review to date of the literature on this topic.[47] In particular, Kittrie credits U.S. Air Force Major General Charles Dunlap with coining the term in English vernacular. Gen. Dunlap first defined lawfare as "using – or misusing – law as a substitute for traditional military means to achieve an operational objective."[48] In a 2011 article, he elaborated that "lawfare . . . focuses principally on circumstances where law can create the same or similar effects as those ordinarily sought from conventional warmaking approaches"[49] and later explained that it "can have effects utterly indistinguishable from those produced by their kinetic analogs."[50] All of these variations imply two commonalities: substitutability and a narrow focus on the use of force.

Extending Dunlap's framework, Kittrie distinguishes two types of lawfare, in particular. The first, what he calls "instrumental lawfare," involves "the use of legal tools to achieve the same or similar effects as those traditionally sought from conventional kinetic military action." Another interpretation might be the use of law to achieve political aims where better-suited than traditional warfare for this purpose. As examples, he lists pushing for new international rules, reinterpreting old ones, or soliciting advisory opinions from international courts in ways "designed to disadvantage an adversary."[51] He also mentions the initiation of "intrusive and protracted investigations" under international legal bodies, vote-buying in international institutions, instigating international criminal prosecutions of foreign nationals, and various ways of maliciously invoking universal jurisdiction, national laws, and civil litigation to influence foreign targets.

The second, "compliance-leverage disparity lawfare," does not involve using law directly but rather using someone else's legal commitment against them. Compliance-leverage disparity refers to the idea that Party A might exploit Party

B's disproportionate respect for legal norms. An example Kittrie cites is the employment of human shields, which forces an adversary to make an uncomfortable choice between looking bad versus not acting.

Kittrie agrees with predecessors like Dunlap that lawfare means leveraging the law to "create the same or similar effects as those ordinarily sought from conventional warmaking approaches,"[52] and that lawfare is aimed at achieving military ends through legal-procedural – instead of purely kinetic – means. Unlike Dunlap, however, Kittrie sees lawfare as complementary to armed force, rather than a true substitute.[53] This opens up a potentially wider range of possibilities. As examples, Kittrie cites a range of activities he considers to be lawfare – from economic sanctions, to tacit noncompliance with arms control agreements like the Nonproliferation Treaty, to subversive nongovernmental organization (NGO) work, and to Al Qaeda instructing its operatives to falsely lodge torture complaints if detained.

What might question to what extent these use cases qualify under the "law" part of Kittrie's "lawfare" concept, or how they differ from straightforward violations. False complaints, for example, seem less like a legal innovation than a public relations gesture. Given sufficient evidence and a neutral judge, the consistency of any of these measures is falsifiable in court. Kittrie's point is that, because there is rarely a judge in international relations, international law is the wild west. Consequently, the threat of lawfare is everywhere: foreign courts like the ICC claim jurisdiction over U.S. nationals; U.S. adversaries court the Global South to form institutional voting blocs based on "popularity contests," and terrorists exploit safe havens. After surveying examples in which U.S. adversaries putatively used respect for the rule of law against it, Kittrie concludes that "law is becoming an increasingly powerful and prevalent weapon of war" – and one in which the United States has fallen behind in developing.[54]

The PRC was contemplating this concept before Gen. Dunlap put pen to paper. Though Jiang Zemin was far from the first Chinese leader to recognize the importance of international legal capability, and certainly not the last, he purportedly "advised a group of Chinese international law experts" in 1996 "that China must be adept at using international law as a weapon."[55] The People's Liberation Army (PLA) set to work considering how law could be harnessed for China's international statecraft. Two PLA colonels published a book titled *Unrestricted Warfare* in 1999 that referenced "legal warfare" （法律战. This concept of legal warfare, defined principally as "activities conducted by using the law as a weapon," was later adopted as one of "three [nonkinetic] warfares" （三种战法） in the PLA's 2003 Political Work Regulations (alongside public opinion warfare and psychological warfare).[56]

Without a single definition, the concept of lawfare is slippery. Dunlap's concept seems to be scoped on the use of law within armed conflict, rather than in lieu of it. The PLA's concept of legal warfare is apparently much broader in scope than Dunlap's. Rather than functioning only to substitute for armed force, legal warfare would appear to encompass the use of any and all legal levers aimed at advancing Chinese strategic goals. Meanwhile, Kittrie's definition international armed conflict, but his extensive list of examples – which span from NGO work to the extraterritorial application of domestic laws by criminal courts – would seem to place his idea of the concept nearer to the PLA's than Dunlap's. Absent consensus about how oriented lawfare actually is around military action, it is uncertain how germane the concept is for a volume on non-military warfare.

One shortcoming of these schematics is that law can be a shield for its adherents, not only a sword. Not every legal strategy is offensive in the sense that it is intended to cause damage to an opponent's standing. Sometimes law is invoked defensively, to protect its wielder against something an opponent has tried to or could potentially do with it. In any definition of the use of law as a mode of non-military warfare, the role of defense should be considered. Another is that the concept of "lawfare" based on ill intent almost presupposes a kind of exceptionalism. As observers, we make judgments about how genuine other people's acts are everyday. We grow suspicious when we see behaviors that don't feel right, especially when done by people we already find suspicious. But we are sometimes wrong, just as our adversaries are about us. Through the eyes of the most narcissistic states, lawfare is everything every other state does.

Even if nongovernmental, many proponents implicitly view the concept of lawfare through the eyes of their government. Kittrie's premise, for instance, is that the U.S. is hamstrung against adversaries who have little respect for the law and are weaponizing it to their advantage. In Kittrie's view, the same states that egregiously violate international law sit in judgment over others. Other states can use law "to augment their own power and constrain that of the U.S." Kittrie concludes that the U.S. should take the gloves off too. That is not to say proponents like Kittrie are necessarily wrong. But a real case could also be made that, for the U.S., which wields an extraordinary amount of international legal power,[57] the gloves have been off for a long time.

One can easily imagine similar conversations happening within foreign governments in response to perceived instances where lawfare was used against them. According to Kittrie, the Chinese PLA writers who first advanced this doctrine did so by analyzing one hundred prior cases – a tacit acknowledgement that, in the eyes of the PLA, "lawfare" was in common practice well before China adopted it. In the U.S., as in China, there is probably a tendency to view anything an adversary does as lawfare, and anything one's own country does as legitimate.

An optimistic take is that bad-faith interpretations of the law do not actually convey much of an advantage because the community can see through disingenuity, so lawfare probably doesn't work as well as skeptics presume.[58] More generally, definitions of lawfare that hinge on the intent of an adversary may not be analytically meaningful for scholars. Rather than using unobservable quantities like intent to classify state-directed legal strategies into uses and abuses, an alternative would be to assume all state-directed legal strategy is oriented around advancing national goals, examine the types of strategy and what they have in common, and ask what makes some strategies successful while others fail.

Typologizing Legal Statecraft

Proving intent is difficult, especially in the anarchical system, on questions of war and peace, where evidentiary standards are ambiguous, and judges are hard to come by. And yet it is important to note that even rivalrous states don't always disagree on everything. To both sides, some violations are unambiguous,[59] which is why guilty parties work to obfuscate the fact of the violation or their involvement. It must therefore be the case that a nontrivial amount of legal activity in international affairs must be in good faith. And the reason is that states are aware that it is the community that decides how far they've diverged from a normative focal point. Effective lawfare is somewhere in the zone between bad intent and community acceptance.

Absent a way to classify behaviors neatly into good-faith legal practice and lawfare, a more tractable operationalization would be to conceive of all focused uses of law in the international arena by the foreign-facing organs of a government as an extension of statecraft. This section proposes a schematic for understanding the varieties of legal statecraft that states employ.

To visualize the full range of legal statecraft tactics, consider two dimensions. The first dimension, Measures, refers to whether statecraft is advanced using instruments that are rooted in or empowered by domestic (unilateral and nonconsensual on the part of the target) or international (mutual, collective, and ostensibly consensual) legal sources. Domestic legal instruments are often leveraged to exact unilateral pressure on a target state, which cannot – and presumably would not – consent to the use of such an instrument. International legal instruments, conversely, draw on community standards and other collective obligations to which multiple states – and ideally the target state itself – have ostensibly expressed mutual consent.

The second dimension, Political Aims, refers to whether the goal is to protect or expand the sending state's freedom of action over international matters – in the war analogy, to hold or take political ground. Like terrain, freedom of action over contested matters in international relations is often zero-sum: by adopting an undefended position or embarking on an indefensible course, states may draw fire from other states in the community, particularly adversaries.

Defensive measures, such as rhetorical justifications or legal pretexts, shield the sending state or its allies from allegations of noncompliance with status quo expectations and likewise put pressure on target states who attempt to do so. Conversely, offensive measures increase the sending state's freedom of action in international relations. In particular, offensive measures can be used to advance the sending state's political, military, or foreign policy goals beyond the status quo, including by causing direct harm to target states.

Together, these two dimensions can be combined to create a 2 × 2 framework through which we can understand international legal statecraft. For instance, the U.S.'s Alien Tort Statute, which greatly relaxes requirements for "standing" in U.S. domestic courts against torture-related suits, or domestically authorized practices, such as extraordinary rendition to bring foreign nationals to justice in U.S. courts,

Table 7.1 A framework for understanding lawfare

Measures			
		Domestic Law	*International Law*
Political Aims	Defensive	Sluggish or limited implementation of nominal international obligations[60]	"Evasion" or pretext to justify behavior,[61] espouse permissive features in design stage,[62] rhetorical adaptation[63]
	Offensive	Extraterritorial application of domestic laws,[64] export controls,[65] indictment of foreign agents[66]	Allegations of international violations, frivolous suits,[67] espouse constraining features in design stage[68]

are both examples of offensive-domestic tactics – as is Hong Kong SAR's new National Security Law, which has already been used to expand transnational prosecutorial discretion. Similarly, the Bush administration's widely lambasted justification for waterboarding tactics – which it argued could not strictly be considered "torture" – might be an example of evasion, a defensive-international tactic.

While my rough categorization in this 2×2 is far from systematic, scholars have already explored how states employ a variety of measures in each of its cells (see footnoted citations for examples). Future research could be considered in light of the framework, this chapter advances.

Tactical Efficacy

If Table 7.1 describes the range of measures, what determines their efficacy? Essentially, law is about how general purpose rules (which are encoded in words) map onto individual cases (fact constellations). Governments can engage in legal statecraft at either end of this functional relationship: they can question the meaning of the words, themselves, or they can dispute the veracity of factual accounts.

Whether governments prefer to engage in a legal or factual battlespace depends in large part on the forum in which the engagement takes place. Fora is analogous to the terrain of each battlespace in a Clausewitzian sense. Fora vary widely in their membership, constitution, and procedure.[69] The choice of forum can also affect the mapping between words and facts at both ends. The pertinent rules in some fora are more complicated or ambiguous than others, making it not only easier for attackers to initiate disputes but also easier for defenders to riposte.[70] At the factual end, some fora are better at reducing uncertainty around the facts if they provide for information collection and synthesis.[71]

However, many fora, especially in highly institutionalized settings, may bias rule interpretation or fact-finding away from one (or multiple) disputants.[72] Fact-finding bodies or dispute settlement agencies can privilege or refuse to consider certain evidence, raise or lower the evidentiary standards altogether, or affect who ultimately gets to determine the meaning of ambiguous or conflicting rules.[73] Some fora are even independent enough to pursue their own agenda, irrespective of what the litigants want.[74]

Because choice of forum is so important for outcomes, it is no wonder that, in choosing a forum where disputes will take place, attackers seek the high ground.[75] However, defenders may be able to use conflicts of law or overlapping institutions to regain the initiative.[76] Each party will prefer the forum in which its arguments will have maximum effect. The attractiveness of a forum depends on how much it reduces the "fog of law" around rules and facts, and whether increased certainty about rule meaning and factual circumstances disproportionately favors Party A over Party B.[77]

Absent a higher sovereign, international law is an opt-in system. Participation in a legal institution not only offers states a chance to shape the topography of each battlespace but also exposes states to potential attack on its contours. States can protect their flanks by not joining certain institutions. Even when not parties, these states may still be able to carry out rhetorical attacks on states that are parties to those institutions from the outside.[78] However, by forgoing participation, the marginal benefit of attacks is reduced for outsiders because their opponents have opportunities both before and after those attacks to fortify the institutional terrain in a relatively uncontested manner.

Because foreign countries have little say over the national laws of their opponents, use of domestic law to affect international relations sidesteps many of these problems. States can unilaterally perfect the terrain of these laws and use them as a cudgel against external actors. This may explain why attempts at extraterritorial application have surged in recent years. However, extraterritoriality presents downsides of its own. Jurisdiction is the most obvious limitation.[79] In the absence of extradition agreements, for example, states may find it difficult to impose costs on individual targets. For example, although the U.S. Department of Justice (DoJ) has publicly indicted dozens of foreign hackers under U.S. law in the last decade, only a fraction have ever been brought to the United States to stand trial in U.S. courts.[80] The rise of multinational corporations and international nongovernmental organizations also implies that jurisdiction is a two-way street, and that target countries may find it easier to retaliate.

Governments rely extensively on expert legal advisors in the conduct of legal statecraft.[81] Legal advisors are critical both for shaping the terrain in advance of potential disputes (by designing and negotiating fora shape and content) and in navigating actual disputes at the tactical, operational, and occasionally strategic levels. Yet, among international relations scholars, although the importance of expert advisors in general is widely understood,[82] legal advisors remain surprisingly understudied as agents of political change,[83] especially in a cross-national context.[84]

The relationship between good legal advice and foreign policy outcomes is poorly understood, but we can imagine reliance on better lawyers generally translates into more effective statecraft.[85] This idea has certainly motivated China's renewed interest in investing in its international legal capacity, as well as calls in the West to combat it.[86] However, because lawyers tend to optimize over credibility, political leaders may not always be happy with the advice their legal advisors give. Meanwhile, attempts to constrain one's legal agents may backfire by biasing recommendations away from the most credible options and toward what the relevant audiences might find unreasonable.[87]

Legitimacy as the High Ground

Proponents of lawfare have advocated a conscious turn to legal instrumentalism.[88] Cornell Law School Dean Roger C. Cramton famously taught that "the lawyer's task, in an instrumental approach to law, is to facilitate and manipulate legal processes to advance the interest of his client."[89] Similarly, Brian Tamanaha writes of "law as an instrument of power to advance their personal interests or the interests or policies of the individuals or groups they support" and argued that "law is widely viewed as an empty vessel to be filled as desired, and to be manipulated, invoked, and utilized in the furtherance of ends."[90]

Orde Kittrie extensively cites Tamanaha in his book, *Lawfare*, before concluding that because instrumentalism is so pervasive, the United States should either negotiate agreements limiting international lawfare or its executive branch lawyers should learn to take off their gloves as well.[91] Of course, this presupposes that the United States weren't already playing hardball to begin with, for example by formulating legal justifications for the Strategic Defense Initiative, Nicaragua, Guantanamo Bay, the unwilling and unable doctrine, drone warfare, and warrantless wiretaps – or as the chief architect of post-1945 global institutions, most of which continue to benefit U.S. interests.

Instead of embracing instrumentalism, however, Tamanaha's work laments its perceived rise. Nor is instrumentalism as dominant as cynics suppose. Retired Stanford Law professor Robert W. Gordon, the former president of the American Society for Legal History, argues forcefully that a return to the "managerialist ideal" is underway. Under this ideal, lawyers recognize that they must "act with integrity" and consider the basic good in addition to "helping their clients attain their performance goals."[92] And while Kittrie quotes Gordon as having advanced a contrasting "advocacy ideal" that emphasizes ruthless practice in supporting fashion,[93] Gordon writes separately that he believes the advocacy ideal is "dangerously misguided" and "welcomes the revival of . . . lawyer-statesman ideals."[94] And even if most government lawyers saw their role as advancing the client's interests at all costs (rather than the Constitution or international legal rules to which their country has committed itself), which probably is not the case, empirically these lawyers feel constrained from advancing losing propositions.[95]

Instrumentalism is a dangerous idea because the centripetal force of law is its legitimacy. Legitimacy is why cooperation between distrustful adversaries doesn't fly apart when harmful facts come to light – why powerful violators ever accept responsibility. States – even U.S. adversaries – recognize legitimacy's importance: as the famed legal scholar Thomas Franck once wrote, legitimacy – or "the capacity of a rule to pull those to whom it is addressed toward consensual compliance" – "is rather consistently upheld in the rhetoric of all states and the behavior of most."[96] Real lawfare is waging war on the idea that law means nothing when not backed by force – that states can bend it in any direction they wish, at any time.[97] In fact, the credibility of international legal statecraft is precisely what makes it effective.

For states concerned with the threat of lawfare, to respond in-kind would be to hamstring themselves in serious ways. Instead, states should respond in ways where legal credibility is maintained. Legal statecraft is effective when it persuades, or when other countries cannot mount persuasive opposition. Ineffective legal statecraft can undermine itself; absent a higher judge, the community decides what right and wrong look like. And if everyone is practicing ineffective legal statecraft, the law loses its meaning. Absent a world police force to compel states to participate and comply with rules they view as meaningless, the lawfare battlespace itself would evaporate, and states would be left with the threat of traditional warfare, only.

The most legally capable states have the greatest advantages when they uphold the legitimacy of the law. As Orde Kittrie identifies, "the United States is a far more law-oriented society, with a much higher percentage of its best minds going into the legal field and creatively using law to achieve their objectives than is, say, the PRC." In order for these "best minds" to have any effect, the United States must continue to hold the law in basic respect while practicing international legal statecraft, even if its adversaries do not.

Countries that advance interpretations or proposals that the community views as reasonable might get what they want because their opponents fight an uphill battle in pushing back. The law of nations is more than simply the law of the jungle, so preposterous legal arguments only undermine those who espouse them.

Future Research Directions

This chapter has considered the various types of legal statecraft that international actors – and in particular states, which have unique standing under international law – use to accomplish their objectives short of war, including in war and non-military warfare. Sovereignty is a double-edged power in that it empowers states to skirt international obligations to which they do not consent, while acting as a lightning rod for criticism when states appear to skirt obligations to which they have consented.

While significant work has been done to uncover the many tactical instruments states employ, future research could empirically test the scope, limits of, and conditions under which these instruments are effective. Cross-national variation in states' propensity to leverage certain instruments, as well as how regime type, legal tradition, institutional membership, or other variables influence their respective sensitivity to these measures.[98] More research also remains to be done on the theoretical micro foundations of these patterns – in particular, the relationship between the lawyers governments hire to carry out this work and their political principles.[99] And, while there is no substitute for careful interpretative analysis, future research could and should seek to capitalize on emerging methodologies in the study of law and politics.

Notes

* Assistant Professor of International Relations, Carnegie Mellon Institute for Strategy and Technology, Carnegie Mellon University: jcanfil@cmu.edu.

1 Dan Maurer, "A U.N. Security Council Permanent Member's De Facto Immunity from Article 6 Expulsion: Russia's Fact or Fiction?" Lawfare, April 15, 2022, https://www.lawfareblog.com/un-security-council-permanent-members-de-facto-immunity-article-6-expulsion-russias-fact-or-fiction.

2 UN General Assembly Emergency Meeting, 15:38 mark, https://www.youtube.com/watch?v=EbiikviPOa8&ab_channel=C-SPAN

3 China's seat formally switched to the People's Republic of China in 1971 after a General Assembly vote. See: United Nations General Assembly Session 26 Resolution 2758. Restoration of the lawful rights of the People's Republic of China in the United Nations A/RES/2758(XXVI), 1, October 25, 1971.

4 E.g. A/RES/47/221, admission of the Czech Republic to membership in the United Nations: resolution/adopted by the General Assembly 1993, https://digitallibrary.un.org/record/283095?ln=en; A/46/834, application of the Republic of Kazakhstan for admission to membership in the United Nations: note/by the Secretary-General. UN. Secretary-General 1992: https://digitallibrary.un.org/record/135248?ln=en.

5 Ian Hurd, "Ian Hurd: Read the Words as They Appear – Russia Is Not a Member of the United Nations Security Council," Chicago Tribune, February 28, 2022, https://www.chicagotribune.com/opinion/commentary/ct-opinion-russia-un-security-council-ukraine-20220228–5ftidozwlbdx5k2ex6qera5skm-story.html; Andrew MacLeod, "Ukraine Invasion: Should Russia Lose Its Seat on the UN Security Council?" The Conversation, accessed April 27, 2022, http://theconversation.com/ukraine-invasion-should-russia-lose-its-seat-on-the-un-security-council-177870; Dan Maurer, "A U.N. Security Council Permanent Member's De Facto Immunity From Article 6 Expulsion: Russia's Fact or Fiction?" Lawfare, April 15, 2022, https://www.lawfareblog.com/un-security-council-permanent-members-de-facto-immunity-article-6-expulsion-russias-fact-or-fiction.

6 UN News, "UN General Assembly Votes to Suspend Russia from the Human Rights Council," April 7, 2022, https://news.un.org/en/story/2022/04/1115782.

7 Jill Goldenziel, "Ukraine Is Weaponizing Corporations Against Russia – Using Lawfare," Forbes, February 20, 2022, https://www.forbes.com/sites/jillgoldenziel/2022/02/20/ukraine-weaponizes-corporations-to-surround-russia-using-lawfare/.

8 Orde F. Kittrie, Lawfare: Law as a Weapon of War, 1 edition (Oxford; New York, NY: Oxford University Press, 2016).

9 E.g. Jonathan G. Odom, "Understanding China's Legal Gamesmanship in the Rules-Based Global Order," in S. D. McDonald and M. C. Burgoyne (Eds.), China's Global Influence: Perspectives and Recommendations (Daniel K. Inouye, Asia-Pacific Center for Security Studies, 2019); Yaroslav Trofimov, Drew Hinshaw and Kate O'Keeffe, "How China Is Taking Over International Organizations, One Vote at a Time," Wall Street Journal, September 29, 2020, sec. World, https://www.wsj.com/articles/how-china-is-taking-over-international-organizations-one-vote-at-a-time-11601397208.

10 'They Call It Lawfare': Trump Faces Sentencing.

11 Bryan York, "Democratic Lawfare vs. Donald Trump," Washington Examiner, December 20, 2023, https://www.washingtonexaminer.com/daily-memo/2644325/democratic-lawfare-vs-donald-trump/

12 Andrew Cherkasky and Katie Cherkasky, "Trump Is Finally – and Rightly – Winning the 'Lawfare' Battle," The Hill, July 5, 2024, https://thehill.com/opinion/judiciary/4755935-trump-is-finally-and-rightly-winning-the-lawfare-battle/

13 Justin Key Canfil, "Yesterday's Reach: How Legal Institutions Keep Pace with Technological Change," SSRN Scholarly Paper (Rochester, NY: Social Science Research Network, January 2, 2020), https://papers.ssrn.com/abstract=3684991.

14 Ronald R. Krebs and Patrick Thaddeus Jackson, "Twisting Tongues and Twisting Arms: The Power of Political Rhetoric," European Journal of International Relations 13, no. 1 (March 2007): 35–66, 147; Thomas Risse, "'Let's Argue!': Communicative Action in World Politics," International Organization 54, no. 1 (2000): 1–39.
15 Thomas M. Franck, "The Power of Legitimacy and the Legitimacy of Power: International Law in an Age of Power Disequilibrium," The American Journal of International Law 100, no. 1 (2006): 88–106, https://doi.org/10.2307/3518832.
16 Justin Key Canfil, "System Shocks: Technology and Ambiguity in International Law and Arms Control," Columbia University, 2021, https://doi.org/10.7916/d8-wf9g-fb06.
17 Ronald R. Krebs and Patrick Thaddeus Jackson, "Twisting Tongues and Twisting Arms: The Power of Political Rhetoric," European Journal of International Relations 13, no. 1 (2007): 35–66.
18 Joseph Raz, "Legal Principles and the Limits of Law," The Yale Law Journal 81, no. 5 (1972): 823–54, https://doi.org/10.2307/795152.
19 See Allison Carnegie and Austin Carson, "Reckless Rhetoric? Compliance Pessimism and International Order in the Age of Trump," The Journal of Politics 81, no. 2 (April 1, 2019): 739–46, https://doi.org/10.1086/702232.
20 Annual Report to Congress: Military and Security Developments Involving the People's Republic of China 2011. Report to Congress Pursuant to the National Defense Authorization Act for Fiscal Year 2000," US Department of Defense. Office of the Secretary of Defense.
21 Michael Glennon, The Fog of Law: Pragmatism, Security, and International Law. 1 edition (Washington, DC; Stanford, CA: Stanford Law Books, 2010).
22 Benjamin Allen Coates, Legalist Empire: International Law and American Foreign Relations in the Early Twentieth Century (Oxford University Press, 2016).
23 Gillian K. Hadfield and Barry R. Weingast, "What Is Law? A Coordination Model of the Characteristics of Legal Order," Journal of Legal Analysis 4, no. 2 (December 1, 2012): 471–514, https://doi.org/10.1093/jla/las008.
24 Andrew T. Guzman, How International Law Works: A Rational Choice Theory (Oxford University Press, 2008).
25 Thomas Erskine Holland, Elements of Jurisprudence (St. Paul: West Publishing Co., 1896).
26 Gillian K. Hadfield and Barry R. Weingast, "Microfoundations of the Rule of Law," Annual Review of Political Science 17, no. 1 (May 11, 2014): 21–42, https://doi.org/10.1146/annurev-polisci-100711-135226.
27 Louis Henkin, How Nations Behave. 2 edition (New York: Columbia University Press, 1979).
28 Abram Chayes and Antonia Handler Chayes, "On Compliance," International Organization 47, no. 2 (1993): 175–205.
29 Alastair Iain Johnston, "Treating International Institutions as Social Environments," International Studies Quarterly 45, no. 4 (December 1, 2001): 487–515, https://doi.org/10.1111/0020-8833.00212; Poznansky, Michael. In the shadow of international law: Secrecy and regime change in the postwar world. Oxford University Press, 2020.
30 George W. Downs, David M. Rocke and Peter N. Barsoom, "Is the Good News about Compliance Good News about Cooperation?" International Organization 50, no. 3 (1996): 379–406.
31 Beth A. Simmons, "Compliance with International Agreements," Annual Review of Political Science 1, no. 1 (1998): 75–93, https://doi.org/10.1146/annurev.polisci.1.1.75; Beth A. Simmons, "International Law and State Behavior: Commitment and Compliance in International Monetary Affairs," American Political Science Review 94, no. 4 (December 2000): 819–35, https://doi.org/10.2307/2586210; Beth A. Simmons and Daniel J. Hopkins, "The Constraining Power of International Treaties: Theory and Methods," American Political Science Review 99, no. 4 (November 2005): 623–31, https://doi.org/10.1017/S0003055405051920; Jana Von Stein, "Do Treaties Constrain

or Screen? Selection Bias and Treaty Compliance," The American Political Science Review 99, no. 4 (2005): 611–22; Matthew Fuhrmann and Yonatan Lupu, "Do Arms Control Treaties Work? Assessing the Effectiveness of the Nuclear Nonproliferation Treaty," International Studies Quarterly 60, no. 3 (September 1, 2016): 530–39, https://doi.org/10.1093/isq/sqw013.
32 Lisa L. Martin, "Against Compliance," in Jeffrey L. Dunoff and Mark A. Pollack (Eds.), Interdisciplinary Perspectives on International Law and International Relations: The State of the Art (Cambridge: Cambridge University Press, 2012), 591–610, https://doi.org/10.1017/CBO9781139107310.029.
33 Ronald B. Mitchell, "Regime Design Matters: Intentional Oil Pollution and Treaty Compliance," International Organization 48, no. 3 (1994): 425–58; Kenneth W. Abbott, Robert O. Keohane, Andrew Moravcsik, Anne-Marie Slaughter and Duncan Snidal, "The Concept of Legalization," International Organization 54, no. 3 (ed 2000): 401–19, https://doi.org/10.1162/002081800551271; Kenneth W. Abbott and Duncan Snidal, "Hard and Soft Law in International Governance," International Organization 54, no. 3 (2000): 421–56.
34 Zoltán I. Buzas and Erin R. Graham, "Emergent Flexibility in Institutional Development: How International Rules Really Change," International Studies Quarterly 64, no. 4 (December 7, 2020): 821–33, https://doi.org/10.1093/isq/sqaa049; Brian Bix, "H. L. A. Hart and the 'Open Texture' of Language," Law and Philosophy 10, no. 1 (1991): 51–72, https://doi.org/10.2307/3504835.
35 Ian Hurd, How to Do Things with International Law (Princeton, NJ: Princeton University Press, 2017).
36 Zoltan Buzas, "Evading International Law: How Agents Comply with the Letter of the Law but Violate Its Purpose," European Journal of International Relations 23, no. 4 (December 1, 2017): 857–83, https://doi.org/10.1177/1354066116679242; Zoltan Buzas, "Is the Good News About Law Compliance Good News About Norm Compliance? The Case of Racial Equality," International Organization 72, no. 2 (ed 2018): 351–85, https://doi.org/10.1017/S0020818318000024.
37 Kittrie, 2016, 4–5.
38 Karen J. Alter, The New Terrain of International Law: Courts, Politics, Rights (Princeton, NJ: Princeton University Press, 2014).
39 Tonya L. Putnam, "Mingling and Strategic Augmentation of International Legal Obligations," International Organization 74, no. 1 (ed 2020): 31–64, https://doi.org/10.1017/S0020818319000377.
40 See United States v. Alvarez- Machain, 504 U.S. 655 (1992); Kerv. People, 110 111. 627 (1884), affd 119 U.S. 436(1886).
41 Tonya L. Putnam, Courts without Borders: Law, Politics, and US Extraterritoriality, Reprint edition (Cambridge: Cambridge University Press, 2017).
42 Garrett Hinck and Tim Maurer, "Persistent Enforcement: Criminal Charges as a Response to Nation-State Malicious Cyber Activity," Journal of National Security Law and Policy 10, no. 3 (2020): 525–62.
43 Kittrie, 2016, 29–30.
44 Ben Westcott, "Hong Kong Was a Home for Beijing Critics. Now They Might Not Even Be Safe at the Airport," CNN, July 3, 2020, https://www.cnn.com/2020/07/03/asia/hong-kong-airport-arrests-national-security-intl-hnk/index.html.
45 Will Oremus, "In Putin's Russia, 'Fake News' Now Means Real News," Washington Post, March 11, 2022, https://www.washingtonpost.com/technology/2022/03/11/russia-fake-news-law-misinformation/.
46 Michael Scharf and Shannon Pagano, "Foreward: Lawfare!" Case Western Reserve Journal of International Law 43, no. 1 (January 1, 2010): 1.
47 Kittrie, 2016.

48 Dunlap, Charles J. Jr., "Lawfare Today: A Perspective," Yale Journal of International Affairs 3 (2008): 146.
49 Dunlap, Charles J. Jr., "Does Lawfare Need an Apologia," Case Western Reserve Journal of International Law 43 (2010): 121.
50 Dunlap, Charles J. Jr., "Law and Military Interventions: Preserving Humanitarian Values in 21st Conflicts," (2001).
51 Kittrie, 2016, 12.
52 Charles Dunlap, "Lawfare Today . . .and Tomorrow," in Raul A. "Pete" Pedrozo and Daria P. Wollschlaeger (Eds.), International Law and the Changing Character of War, Vol. 87 (US Naval War College International Law Studies, 2011), 315–25, https://scholarship.law.duke.edu/faculty_scholarship/2465.
53 "Lawfare is almost always less deadly . . . [and] almost always less financially costly" than traditional warfare, and it "can sometimes also be more effective," See: Kittrie, 2016, 3.
54 Kittrie, 2016, 1.
55 Kittrie, 2016, referencing evidence in Dong Wang, China's Unequal Treaties: Narrating National History (Lexington Books, 2005).
56 Works by other Chinese authors that further developed this concept include Analysis of 100 cases of Legal Warfare (2004) by Cong Wensheng et al., Legal Warfare in the Modern Era (2004) by Xun Hengdong, Lawfare in Modern War (2005) also by Xun Hengdong, and Under Informatized Conditions (2007) by Song Yuxia ed. See Kittrie, 2016.
57 Coates, 2016.
58 Lee A. Casey and David B. Rivkin, "The Rocky Shoals of International Law," in The National Interest on International Law & Order (Routledge, 2003).
59 David A. Koplow, "Indisputable Violations: What Happens When the United States Unambiguously Breaches a Treaty," Fletcher Forum of World Affairs 37, no. 1 (2013): 53–76.
60 Andrea Schneiker, "Norm Sabotage: Conceptual Reflection on a Phenomenon That Challenges Well-Established Norms," International Studies Perspectives 22, no. 1 (January 30, 2021): 106–23, https://doi.org/10.1093/isp/ekaa003; Woodward, Angela, "National Implementing Laws for Arms Control and Disarmament Treaties," Verification Yearbook, December 10, 2003, 151–68.
61 Fernando G. Nuñez-Mietz, "Lawyering Compliance with International Law: Legal Advisers in the 'War on Terror.'" European Journal of International Security 1, no. 2 (July 2016): 215–38, https://doi.org/10.1017/eis.2016.9; Zoltan Buzas, "Is the Good News About Law Compliance Good News About Norm Compliance? The Case of Racial Equality," International Organization 72, no. 2 (ed 2018): 351–85, https://doi.org/10.1017/S0020818318000024; Daniel Ghezelbash, "Hyper-Legalism and Obfuscation: How States Evade Their International Obligations Towards Refugees," The American Journal of Comparative Law 68, no. 3 (September 1, 2020): 479–516, https://doi.org/10.1093/ajcl/avaa019.
62 Giovanni Mantilla, "Deflective Cooperation: Social Pressure and Forum Management in Cold War Conventional Arms Control," International Organization (January 10, 2023): 1–35, https://doi.org/10.1017/S0020818322000364; Paul K. MacDonald, "Most Potent and Irresistible Moral Influence: Public Opinion, Rhetorical Coercion, and the Hague Conferences," Global Policy 11, no. S3 (2020): 104–14, https://doi.org/10.1111/1758-5899.12847; Stacie E. Goddard, "Revolution from the Inside: Institutions, Legitimation Strategies, and Rhetorical Pathways of Institutional Change," Global Policy 11, no. S3 (2020): 83–92, https://doi.org/10.1111/1758-5899.12853; Radoslav S. Dimitrov, "Empty Institutions in Global Environmental Politics," International Studies Review 22, no. 3 (September 1, 2020): 626–50, https://doi.org/10.1093/isr/viz029.

63 Courtney J. Fung, "Rhetorical Adaptation, Normative Resistance and International Order-Making: China's Advancement of the Responsibility to Protect," Cooperation and Conflict 55, no. 2 (June 1, 2020): 193–215, https://doi.org/10.1177/0010836719858118.
64 Tonya L. Putnam, Courts without Borders: Law, Politics, and US Extraterritoriality. Reprint edition (Cambridge: Cambridge University Press, 2017).
65 Mario Daniels and John Krige, Knowledge Regulation and National Security in Postwar America (University of Chicago Press, 2022).
66 Justin Key Canfil, "The Illogic of Plausible Deniability: Why Proxy Conflict in Cyberspace May No Longer Pay," Journal of Cybersecurity 8, no. 1 (January 1, 2022): tyac007, https://doi.org/10.1093/cybsec/tyac007; Marcelo Leal and Paul Musgrave, "Cheerleading in Cyberspace: How the American Public Judges Attribution Claims for Cyberattacks," Foreign Policy Analysis 18, no. 2 (April 1, 2022): orac003, https://doi.org/10.1093/fpa/orac003; Garrett Hinck and Tim Maurer, "Persistent Enforcement: Criminal Charges as a Response to Nation-State Malicious Cyber Activity," Journal of National Security Law and Policy 10, no. 3 (2020 2019): 525–62.
67 Isaac B. Kardon, China's Law of the Sea: The New Rules of Maritime Order (New Haven: Yale University Press, 2023).
68 See footnote 60.
69 Abbott et al., 2000; Mark Raymond, "Social Practices of Rule-Making for International Law in the Cyber Domain," Journal of Global Security Studies. Accessed December 3, 2020, https://doi.org/10.1093/jogss/ogz065.
70 Kenneth W. Abbott and Duncan Snidal, "Hard and Soft Law in International Governance," International Organization 54, no. 3 (2000): 421–56.
71 Xinyuan Dai, "Information Systems in Treaty Regimes," World Politics 54, no. 4 (2002): 405–36.
72 Allison Carnegie and Austin Carson, "The Disclosure Dilemma: Nuclear Intelligence and International Organizations," American Journal of Political Science 63, no. 2 (2019): 269–85, https://doi.org/10.1111/ajps.12426; "The Spotlight's Harsh Glare: Rethinking Publicity and International Order," International Organization 72, no. 3 (ed 2018): 627–57, https://doi.org/10.1017/S0020818318000176.
73 Laurence R. Helfer and Erik Voeten, "International Courts as Agents of Legal Change: Evidence from LGBT Rights in Europe," SSRN Scholarly Paper (Rochester, NY: Social Science Research Network, February 13, 2014), http://papers.ssrn.com/abstract=1850526; Paul K. Huth, Sarah E. Croco and Benjamin J. Appel, "Bringing Law to the Table: Legal Claims, Focal Points, and the Settlement of Territorial Disputes Since 1945," American Journal of Political Science 57, no. 1 (January 1, 2013): 90–103, https://doi.org/10.1111/j.1540-5907.2012.00607.x; Robert O. Keohane, Andrew Moravcsik and Anne-Marie Slaughter, "Legalized Dispute Resolution: Interstate and Transnational," International Organization 54, no. 3 (July 2000): 457–88, https://doi.org/10.1162/002081800551299; Eric A. Posner and Miguel de Figueiredo, "Is the International Court of Justice Biased?" Journal of Legal Studies (2005), http://papers.ssrn.com/abstract=642581.
74 Michael N. Barnett and Martha Finnemore, "The Politics, Power, and Pathologies of International Organizations," International Organization 53, no. 4 (1999): 699–732.
75 Marc L. Busch, "Overlapping Institutions, Forum Shopping, and Dispute Settlement in International Trade," International Organization 61, no. 4 (October 2007): 735–61, https://doi.org/10.1017/S0020818307070257; Emilia Justyna Powell and Sara McLaughlin Mitchell, "Forum Shopping for the Best Adjudicator: Dispute Settlement in the United Nations Convention on the Law of the Sea," Journal of Territorial and Maritime Studies 9, no. 1 (Winter/Spring 2022): 7–33, https://doi.org/https://doi.org/10.2307.
76 Tonya L. Putnam, "Mingling and Strategic Augmentation of International Legal Obligations," 2020.

77 For earlier use of the term "fog of law" in a separate context, see: Michael Glennon, The Fog of Law: Pragmatism, Security, and International Law, 1 edition (Washington, DC; Stanford, CA: Stanford Law Books, 2010).
78 Courtney J. Fung, "Rhetorical Adaptation, Normative Resistance and International Order-Making: China's Advancement of the Responsibility to Protect," Cooperation and Conflict 55, no. 2 (June 1, 2020): 193–215, https://doi.org/10.1177/0010836719858118; Stacie E. Goddard, "Revolution from the Inside: Institutions, Legitimation Strategies, and Rhetorical Pathways of Institutional Change," Global Policy 11, no. S3 (2020): 83–92, https://doi.org/10.1111/1758-5899.12853; Paul K. MacDonald, "'Most Potent and Irresistible Moral Influence': Public Opinion, Rhetorical Coercion, and the Hague Conferences," Global Policy 11, no. S3 (2020): 104–14, https://doi.org/10.1111/1758-5899.12847; Alexander Thompson, "Emerging Powers and Differentiation in Global Climate Institutions," Global Policy 11, no. S3 (2020): 61–72, https://doi.org/10.1111/1758-5899.12883.
79 Lori Damrosch and Sean Murphy, International Law, 6 edition (St. Paul, MN: West Academic Publishing, 2014).
80 Garrett Hinck and Tim Maurer, "Persistent Enforcement: Criminal Charges as a Response to Nation-State Malicious Cyber Activity," Journal of National Security Law and Policy 10, no. 3 (2020): 525–62.
81 Gillian K. Hadfield, "Don't Forget the Lawyers: The Role of Lawyers in Promoting the Rule of Law in Emerging Market Democracies," DePaul Law Review 56, no. 2 (Winter 2007).
82 Tyler Jost, Joshua D. Kertzer, Eric Min and Robert Schub, "Advisers and Aggregation in Foreign Policy Decision-Making," n.d., 92.
83 The role of domestic lawyers has been considered extensively in the political science and sociology literature, as have transnational advocates, but foreign policy lawyers per se have apparently received less attention despite their important role. Legal scholarship may offer exceptions. C.f. Fernando G. Nuñez-Mietz, "Lawyering Compliance with International Law: Legal Advisers in the 'War on Terror.'" European Journal of International Security 1, no. 2 (July 2016): 215–38, https://doi.org/10.1017/eis.2016.9.
84 C.f. Lauri Mälksoo, Russian Approaches to International Law, Reprint edition (Oxford: Oxford University Press, 2017); William E. Pomeranz, Jonathan Smele and Michael Melancon, Law and the Russian State: Russia's Legal Evolution from Peter the Great to Vladimir Putin (London, Oxford, New York, New Delhi and Sydney: Bloomsbury Academic, 2018); Michael P. Scharf and Colin McLaughlin, "Foreign Legal Advisors' Roundtable," Proceedings of the Annual Meeting (American Society of International Law) 99 (2005): 161–69.
85 For illustrative examples, see: Francine Hirsch, Soviet Judgment at Nuremberg: A New History of the International Military Tribunal after World War II (New York: Oxford University Press, 2020).
86 Lawrence J. Liu, "The Rules of the (Belt and) Road: How Lawyers Participate in China's Outbound Investment and Infrastructure Initiatives," Yale Journal of International Law 46 (July 31, 2021); Jun Mai, "Xi Jinping Says China Has a Legal Problem: Finding the Lawyers to Defend Its Interests Abroad," South China Morning Post, March 1, 2021, https://www.scmp.com/news/china/politics/article/3123130/xi-jinping-says-china-has-legal-problem-finding-lawyers-defend.
87 Maya Wang, "Xi Is Bending Chinese Law to His Will," November 25, 2021, https://www.foreignaffairs.com/articles/china/2021-11-24/xi-bending-chinese-law-his-will.
88 See e.g. Richard A. Posner, Overcoming Law (Harvard University Press, 1995).
89 Roger Cramton, "The Ordinary Religion of the Law School Classroom," Cornell Law Faculty Publications, January 1, 1978, https://scholarship.law.cornell.edu/facpub/972.
90 Brian Z. Tamanaha, "How an Instrumental View of Law Corrodes the Rule of Law," DePaul Law Review 56 (2006): 469.

91 "The United States should consider whether a bilateral or multilateral agreement limiting in some ways the waging of lawfare between the United States and PRC would be feasible and useful," Kittrie, 2016, 194–95.
92 Robert W. Gordon, "The Return of the Lawyer-Statesman," Stanford Law Review 69, no. 6 (2017): 1731–64.
93 Under the "advocacy ideal," according to Kittrie, "attorneys should not tell outright lies to judges or fabricate evidence . . . they may, and if it will serve their clients' interest must, exploit any gap, ambiguity, technicality or loophole, any not-obviously-and-totally-implausible interpretation of the law or facts," Kittrie, 2016.
94 Gordon, 2017.
95 Canfil, "Yesterday's Reach: How Legal Institutions Keep Pace with Technological Change," 2020.
96 Thomas M. Franck, "The Power of Legitimacy and the Legitimacy of Power: International Law in an Age of Power Disequilibrium," American Journal of International Law 100, no. 1 (2006): 88–106.
97 Franck, 2006.
98 James D. Morrow, Order within Anarchy: The Laws of War as an International Institution (Cambridge University Press, 2014).
99 Judith A. McMorrow, Sida Liu and Benjamin van Rooij, "Lawyer Discipline in a Authoritarian Regime: Empirical Insights from Zhejiang Province, China," Georgetown Journal of Legal Ethics 30 (2017): 267; Laura A. Dickinson, "Military Lawyers on the Battlefield: An Empirical Account of International Law Compliance," American Journal of International Law 104, no. 1 (January 2010): 1–28, https://doi.org/10.5305/amerjintelaw.104.1.0001.

Bibliography

Abbott, K. W., & Snidal, D. (2000). Hard and soft law in international governance. *International Organization, 54*(3), 421–456.

Abbott, K. W., Keohane, R. O., Moravcsik, A., Slaughter, A.-M., & Snidal, D. (2000). The concept of legalization. *International Organization, 54*(3), 401–419.

Alter, K. J. (2014). *The new terrain of international law: Courts, politics, rights*. Princeton University Press.

Barnett, M. N., & Finnemore, M. (1999). The politics, power, and pathologies of international organizations. *International Organization, 53*(4), 699–732.

Bix, B. (1991). H. L. A. Hart and the 'open texture' of language. *Law and Philosophy, 10*(1), 51–72.

Busch, M. L. (2007). Overlapping institutions, forum shopping, and dispute settlement in international trade. *International Organization, 61*(4), 735–761.

Buzas, Z. I. (2017). Evading international law: How agents comply with the letter of the law but violate its purpose. *European Journal of International Relations, 23*(4), 857–883.

Buzas, Z. I. (2018). Is the good news about law compliance good news about norm compliance? The case of racial equality. *International Organization, 72*(2), 351–385.

Buzas, Z. I., & Graham, E. R. (2020). Emergent flexibility in institutional development: How international rules really change. *International Studies Quarterly, 64*(4), 821–833.

Canfil, J. K. (2020). *Yesterday's reach: How legal institutions keep pace with technological change*. Social Science Research Network.

Canfil, J. K. (2021). *System shocks: Technology and ambiguity in international law and arms control*. Columbia University.

Canfil, J. K. (2022). The illogic of plausible deniability: Why proxy conflict in cyberspace may no longer pay. *Journal of Cybersecurity, 8*(1).

Carnegie, A., & Carson, A. (2018). The spotlight's Harsh Glare: Rethinking publicity and international order. *International Organization, 72*(3), 627–657.

Carnegie, A., & Carson, A. (2019a). Reckless rhetoric? Compliance pessimism and international order in the age of Trump. *The Journal of Politics*, *81*(2), 739–746.
Carnegie, A., & Carson, A. (2019b). The disclosure dilemma: Nuclear intelligence and international organizations. *American Journal of Political Science*, *63*(2), 269–285.
Casey, L. A., & Rivkin, D. B. (2003). The rocky shoals of international law. In *The national interest on international law & order*. Routledge.
Chayes, A., & Chayes, A. H. (1993). On compliance. *International Organization*, *47*(2), 175–205.
Cherkasky, A., & Cherkasky, K. (2024, July 5). Trump is finally – and rightly – winning the 'lawfare' battle. *The Hill*. https://thehill.com/opinion/judiciary/4755935-trump-is-finally-and-rightly-winning-the-lawfare-battle/
Coates, B. A. (2016). *Legalist empire: International law and American foreign relations in the early twentieth century*. Oxford University Press.
Cramton, R. C. (1978). *The ordinary religion of the law school classroom*. Cornell Law Faculty Publications.
Cramton, R. C. (1994). Beyond the ordinary religion, law, and negotiation. *Harvard Negotiation Law Review*, *39*(2), 39–61.
Dai, X. (2002). Information systems in treaty regimes. *World Politics*, *54*(4), 405–436.
Damrosch, L., & Murphy, S. (2014). *International law* (6th ed.). West Academic Publishing.
Daniels, M., & Krige, J. (2022). *Knowledge regulation and national security in postwar America*. University of Chicago Press.
Dickinson, L. A. (2010). Military lawyers on the battlefield: An empirical account of international law compliance. *American Journal of International Law*, *104*(1), 1–28.
Dimitrov, R. S. (2020). Empty institutions in global environmental politics. *International Studies Review*, *22*(3), 626–650.
Downs, G. W., Rocke, D. M., & Barsoom, P. N. (1996). Is the good news about compliance good news about cooperation? *International Organization*, *50*(3), 379–406.
Dunlap, C. J. (2001). Law and military interventions: Preserving humanitarian values in 21st conflicts. In *Paper presented to humanitarian challenges in Military intervention conference, Carr center for Human Rights policy*. Kennedy School of Government, Harvard University, November 29, 2001.
Dunlap, C. J. (2008). Lawfare today: A perspective. *Yale Journal of International Affairs*, *3*, 146.
Dunlap, C. J. (2010a). Does lawfare need an apologia. *Case Western Reserve Journal of International Law*, *43*, 121.
Dunlap, C. J. (2010b). Lawfare today . . . and tomorrow. In *International law and the changing character of war* (Vol. 87). US Naval War College International Law Studies.
Franck, T. M. (1990). *The power of legitimacy among nations*. Oxford University Press.
Franck, T. M. (2006). The power of legitimacy and the legitimacy of power: International law in an age of power disequilibrium. *The American Journal of International Law*, *100*(1), 88–106.
Fuhrmann, M., & Lupu, Y. (2016). Do arms control treaties work? Assessing the effectiveness of the nuclear nonproliferation treaty. *International Studies Quarterly*, *60*(3), 530–539.
Fung, C. J. (2020). Rhetorical adaptation, normative resistance and international order-making: China's advancement of the responsibility to protect. *Cooperation and Conflict*, *55*(2), 193–215.
Ghezelbash, D. (2020). Hyper-legalism and obfuscation: How states evade their international obligations towards refugees. *The American Journal of Comparative Law*, *68*(3), 479–516.
Glennon, M. (2010). *The fog of law: Pragmatism, security, and international law* (1st ed.). Stanford Law Books.
Goddard, S. E. (2020). Revolution from the inside: Institutions, legitimation strategies, and rhetorical pathways of institutional change. *Global Policy*, *11*(S3), 83–92.

Goldenziel, J. (2022, February 20). Ukraine is weaponizing corporations against Russia – using lawfare. *Forbes*. https://www.forbes.com/sites/jillgoldenziel/2022/02/20/ukraine-weaponizes-corporations-to-surround-russia-using-lawfare/

Gordon, R. W. (2017). The return of the lawyer-statesman? *Law and Contemporary Problems*, *80*, 7–20.

Guzman, A. T. (2008). *How international law works: A rational choice theory*. Oxford University Press.

Hadfield, G. K. (2007). Don't forget the lawyers: The role of lawyers in promoting the rule of law in emerging market democracies. *DePaul Law Review*, *56*(2).

Hadfield, G. K., & Weingast, B. R. (2012). What is law? A coordination model of the characteristics of legal order. *Journal of Legal Analysis*, *4*(2), 471–514.

Hadfield, G. K., & Weingast, B. R. (2014). Microfoundations of the rule of law. *Annual Review of Political Science*, *17*(1), 21–42.

Helfer, L. R., & Voeten, E. (2014). *International courts as agents of legal change: Evidence from LGBT rights in Europe*. Social Science Research Network.

Henkin, L. (1979). *How nations behave: Law and foreign policy*. Columbia University Press.

Hinck, G., & Maurer, T. (2019). Persistent enforcement: Criminal charges as a response to nation-state malicious cyber activity. *Journal of National Security Law and Policy*, *10*(3), 525–562.

Hirsch, F. (2020). *Soviet judgment at Nuremberg: A new history of the international military tribunal after World War II*. Oxford University Press.

Holland, T. E. (1896). *Elements of jurisprudence*. West Publishing Co.

Hurd, I. (2017). *How to do things with international law*. Princeton University Press.

Hurd, I. (2022, February 28). Ian Hurd: Read the words as they appear – Russia is not a member of the United Nations Security Council. *Chicago Tribune*. https://www.chicagotribune.com/opinion/commentary/ct-opinion-russia-un-security-council-ukraine-20220228–5ftidozwlbdx5k2ex6qera5skm-story.html

Huth, P. K., Croco, S. E., & Appel, B. J. (2013). Bringing law to the table: Legal claims, focal points, and the settlement of territorial disputes since 1945. *American Journal of Political Science*, *57*(1), 90–103.

Johnston, A. I. (2001). Treating international institutions as social environments. *International Studies Quarterly*, *45*(4), 487–515.

Jost, T., Kertzer, J. D., Min, E., & Schub, R. (2024). Advisers and aggregation in foreign policy decision-making. *International Organization*, *78*(1), 1–37.

Kardon, I. B. (2023). *China's law of the sea: The new rules of maritime order*. Yale University Press.

Keohane, R. O., Moravcsik, A., & Slaughter, A.-M. (2000). Legalized dispute resolution: Interstate and transnational. *International Organization*, *54*(3), 457–488.

Kittrie, O. F. (2016). *Lawfare: Law as a weapon of war* (1st ed.). Oxford University Press.

Koplow, D. A. (2013). Indisputable violations: What happens when the United States unambiguously breaches a treaty. *Fletcher Forum of World Affairs*, *37*(1), 53–76.

Krebs, R. R., & Jackson, P. T. (2007). Twisting tongues and twisting arms: The power of political rhetoric. *European Journal of International Relations*, *13*(1), 35–66.

Leal, M., & Musgrave, P. (2022). Cheerleading in cyberspace: How the American public judges attribution claims for cyberattacks. *Foreign Policy Analysis*, *18*(2).

Liu, L. J. (2021). The rules of the (belt and) road: How lawyers participate in China's outbound investment and infrastructure initiatives. *Yale Journal of International Law*, *46*.

MacDonald, P. K. (2020). Most potent and irresistible moral influence: Public opinion, rhetorical coercion, and the Hague conferences. *Global Policy*, *11*(S3), 104–114.

MacLeod, A. (2022). Ukraine invasion: Should Russia lose its seat on the UN security council? *The Conversation*. https://theconversation.com/ukraine-invasion-should-russia-lose-its-seat-on-the-un-security-council-177870

Mai, J. (2021, March 1). Xi Jinping says China has a legal problem: Finding the lawyers to defend its interests abroad. *South China Morning Post*. https://www.scmp.com/news/china/politics/article/3123130/xi-jinping-says-china-has-legal-problem-finding-lawyers-defend

Mälksoo, L. (2017). *Russian approaches to international law* (Reprint ed.). Oxford University Press.

Mantilla, G. (2023). Deflective cooperation: Social pressure and forum management in cold war conventional arms control. *International Organization*, 1–35.

Martin, L. L. (2012). Against compliance. In *Interdisciplinary perspectives on international law and international relations: The state of the art*. Cambridge University Press.

Maurer, D. (2022). A U.N. security council permanent member's De Facto immunity from Article 6 expulsion: Russia's fact or fiction? *Lawfare*. https://www.lawfaremedia.org/article/un-security-council-permanent-members-de-facto-immunity-article-6-expulsion-russias-fact-or-fiction

McMorrow, J. A., Liu, S., & van Rooij, B. (2017). Lawyer discipline in a authoritarian regime: Empirical insights from Zhejiang Province, China. *Georgetown Journal of Legal Ethics*, *30*, 267.

Mitchell, R. B. (1994). Regime design matters: Intentional oil pollution and treaty compliance. *International Organization*, *48*(3), 425–458.

Morrow, J. D. (2014). *Order within anarchy: The laws of war as an international institution*. Cambridge University Press.

Nuñez-Mietz, F. G. (2016). Lawyering compliance with international law: Legal advisers in the 'war on terror.' *European Journal of International Security*, *1*(2), 215–238.

O'Keeffe, K., Trofimov, Y., & Hinshaw, D. (2020, September 29). How China is taking over international organizations, one vote at a time. *Wall Street Journal*. https://www.wsj.com/articles/how-china-is-taking-over-international-organizations-one-vote-at-a-time-11601397208

Odom, J. G. (2019). Understanding China's legal gamesmanship in the rules-based global order. In *China's global influence: Perspectives and recommendations*. Daniel K. Inouye Asia-Pacific Center for Security Studies.

Oremus, W. (2022, March 11). In Putin's Russia, 'fake news' now means real news. *Washington Post*. https://www.washingtonpost.com/technology/2022/03/11/russia-fake-news-law-misinformation/

Pomeranz, W. E., Smele, J., & Melancon, M. (2018). *Law and the Russian state: Russia's legal evolution from Peter the great to Vladimir Putin*. Bloomsbury Academic.

Posner, E. A., & de Figueiredo, M. (2005). Is the international court of justice biased? *Journal of Legal Studies*, *34*(2), 599–630.

Posner, R. A. (1995). *Overcoming law*. Harvard University Press.

Powell, E. J., & Mitchell, S. M. (2022). Forum shopping for the best adjudicator: Dispute Settlement in the United Nations convention on the law of the sea. *Journal of Territorial and Maritime Studies*, *9*(1), 7–33.

Poznansky, M. (2020). *In the shadow of international law: Secrecy and regime change in the postwar world*. Oxford University Press.

Putnam, T. L. (2017). *Courts without borders: Law, politics, and US extraterritoriality* (Reprint ed.). Cambridge University Press.

Putnam, T. L. (2020). Mingling and strategic augmentation of international legal obligations. *International Organization*, *74*(1), 31–64.

Raymond, M. (2020). Social practices of rule-making for international law in the cyber domain. *Journal of Global Security Studies*, *6*(2).

Raz, J. (1972). Legal principles and the limits of law. *The Yale Law Journal*, *81*(5), 823–854.

Risse, T. (2000). 'Let's argue!' Communicative action in world politics. *International Organization*, *54*(1), 1–39.

Scharf, M. P., & McLaughlin, C. (2005). Foreign legal advisors' roundtable. *Proceedings of the Annual Meeting (American Society of International Law)*, *99*, 161–169.

Scharf, M. P., & Pagano, S. (2010). Foreword: Lawfare! *Case Western Reserve Journal of International Law*, *43*(1).

Schneiker, A. (2021). Norm sabotage: Conceptual reflection on a phenomenon that challenges well-established norms. *International Studies Perspectives*, *22*(1), 106–123.
Simmons, B. A. (1998). Compliance with international agreements. *Annual Review of Political Science*, *1*(1), 75–93.
Simmons, B. A. (2000). International law and state behavior: Commitment and compliance in international monetary affairs. *American Political Science Review*, *94*(4), 819–835.
Simmons, B. A., & Hopkins, D. J. (2005). The constraining power of international treaties: Theory and methods. *American Political Science Review*, *99*(4), 623–631.
Tamanaha, B. Z. (2006a). How an instrumental view of law corrodes the rule of law. *DePaul Law Review*, *56*, 469.
Tamanaha, B. Z. (2006b). *Law as a means to an end: Threat to the rule of law*. Cambridge University Press.
Thompson, A. (2020). Emerging powers and differentiation in global climate institutions. *Global Policy*, *11*(S3), 61–72.
UN General Assembly Emergency Meeting. (2022). https://www.youtube.com/watch?v=EbiikviPOa8&ab_channel=C-SPAN
UN News. (2022, April 7). UN general assembly votes to suspend Russia from the human rights council. *UN News*. https://news.un.org/en/story/2022/04/1115782
United Nations General Assembly Session 26 Resolution 2758. (1971). *Restoration of the lawful rights of the People's Republic of China in the United Nations* A/RES/2758 (XXVI). https://digitallibrary.un.org/record/192054?v=pdf
United States Department of Defense. (2011). Annual report to congress: Military and security developments involving the People's Republic of China 2011. In *Report to congress pursuant to the National Defense Authorization Act for fiscal year 2000*. Office of the Secretary of Defense.
Von Stein, J. (2005). Do treaties constrain or screen? Selection bias and treaty compliance. *The American Political Science Review*, *99*(4), 611–622.
Wang, D. (2005). *China's unequal treaties: Narrating national history*. Lexington Books.
Wang, M. (2021, November 25). Xi is bending Chinese law to his will. *Human Rights Watch*. https://www.hrw.org/news/2021/11/30/xi-bending-chinese-law-his-will
Westcott, B. (2020, July 3). Hong Kong was a home for Beijing critics. Now they might not even be safe at the airport. *CNN*. https://edition.cnn.com/2020/07/03/asia/hong-kong-airport-arrests-national-security-intl-hnk/index.html
Woodward, A. (2003). National implementing laws for arms control and disarmament treaties. In T. Finlay (Ed.), *Verification yearbook* (pp. 151–168). The Verification Research, Training and Information Centre.
York, B. (2023, December 20). Democratic lawfare vs. Donald Trump. *Washington Examiner*. https://www.washingtonexaminer.com/daily-memo/2644325/democratic-lawfare-vs-donald-trump/

8 Science Securitized, Technology Transformed

Innovation and Exploitation in Non-Military Warfare

*Elsa B. Kania**

Science and Technology in Non-Military Warfare

Throughout history, advances in science and technology have revolutionized warfare; so too, in competition, innovation has been foundational to national competitiveness. As the character of conflict continues to change, including with the increased prominence of non-military warfare, science and technology have become not only an arena for intensifying competition but also a means of coercive leverage when dependencies on technology or the underlying resources and components can be exploited. The pursuit of progress in science and technology was once regarded primarily as an opportunity for international cooperation, centering on diplomacy and promotion of soft power on the world stage, yet there are also aspects of science diplomacy that can be closely linked to the exercise of hard power and coercion.[1] Innovations have driven growth in the global economy for decades, accelerated by free flows of ideas, talent, and often technology across international innovation ecosystems. Previously, trade was often regarded as a factor promoting peace and a check against the outbreak of war. There have since been a range of refutations of that belief in favor of more complex arguments; and to the contrary, interdependence has also been identified as a factor that can create frictions or even increase the risks of conflict, especially when leaders have negative expectations about the future environment.[2] Increasingly, economic interdependence has come to be regarded both by states and scholars as creating vulnerabilities that can be exploited, even weaponized.[3] In parallel, emerging technological developments have accelerated in ways that have far-reaching implications for national power.

While economic networks can be weaponized, so too can the global networks underpinning scientific cooperation and collaborative technological development be exploited to seek advantage. Increasingly, great powers explicitly recognize national security as encompassing economic security, which tends to motivate a drive for technological self-reliance and autonomy through investing in independent capabilities. Today, innovation is regarded increasingly as a national imperative with strategic consequences. These evolving dynamics are currently motivating a sweeping realignment of the international innovation ecosystem, based on exigencies of geopolitics. Whenever a state is dependent, especially asymmetrically so,

DOI: 10.4324/9781003616993-8

on other countries for critical technologies, that reliance can be risky. Ultimately, these trends are poised to reshape the global scientific and technological landscape. This chapter will examine several of these dynamics. In the process, I am informed by several perspectives from literatures on international relations, science and technology studies, and industry/business perspectives.

This chapter will explore three primary mechanisms for states that can come into play in an era of non-military warfare.[4] In a world of *competitive innovation*, states pursue proactive science, technology, and industrial policies with the objective to increase national capabilities in strategic domains of research and critical industries. States have become more deliberate in pursuing *chokepoint exploitation* by identifying particular technologies or the underlying resources and components where one state has an edge or unique position. Where such dependencies can be identified, states can start to impose costs on an adversary, as a means of coercive leverage, punitive measure, or even deterrent. In the process, states also identify and seek to mitigate risks that could result from their dependencies. And as ecosystems of academic research and commercial collaboration that were once highly globalized integrated begin to fracture and become fragmented, a trend of *innovation realignment* is occurring. States are pursuing new avenues of cooperation through policies intended to ensure technological partnerships are more closely aligned with their strategic interests and national values or ideologies. The sections that follow examine each of these dynamics, in turn, concentrating on interactions among the United States, the People's Republic of China (PRC), and the Russian Federation. These great powers have been engaged in long-term competition that can manifest in non-military warfare extending into the domain of science and technology, which will impact the future of innovation – and ultimately, global balances of power.

Competitive Innovation

Innovation is foundational to national power. Throughout history, the rise and fall of great powers has been influenced by adoption and diffusion of scientific and technological developments.[5] Mastering and realizing the full potential of a new innovation can change the global balance of power, whereas failure to keep pace can accelerate decline and raise the risks of foreign predation. For nations to maintain and enhance their competitiveness depends on not only innovation that usually arises within academia or industry but also the capacity of governments to pursue policies that are effective in leveraging, accelerating, and integrating technological developments, including to facilitate diffusion of new innovations.[6] "As the basis of competition has shifted more and more to the creation and assimilation of knowledge, the role of the nation has grown," the scholar of business and innovation Michael Porter noted.[7]

However, the question of the role of the nation-state in the process of innovation has been contested and generated different national responses. Historically, the United States dedicated significant amounts of public funding to basic research. This core ethos can be traced back to arguments articulated originally by Dr.

Vannevar Bush, director of the Office of Scientific Research and Development, who argued in a report addressed to Present Roosevelt in the wake of World War II:

> There are areas of science in which the public interest is acute but which are likely to be cultivated inadequately if left without more support than will come from private sources. These areas . . . should be advanced by active Government support.[8]

Such support was foundational to enabling American innovation advantage in the decades thereafter, including the origins and early successes of Silicon Valley.[9]

During the Cold War, U.S. military and government support for research contributed considerably to establishing the foundations for a robust American innovation ecosystem. Thereafter, the notion of state support to promote innovation came to be regarded unfavorably; industrial policies were caricatured and conflated with rigid approaches to state planning in managed economies that were regarded as relegated to irrelevance with the collapse of the Soviet Union.[10] However, such previous conventional wisdoms in a U.S. context often dismissed various examples of success where states have taken targeted, deliberate approaches to upgrading their capacity for scientific and technological development in alignment with national priorities. The literature on late industrializing economies has highlighted examples of developmental states that effectively catalyzed advances through technocratic approaches, including Taiwan, Singapore, and South Korea.[11]

Since then, China has become the prime example of a great power that has risen through a model of "innovation-driven" development.[12] The state has acted as a core driver and major enabler of advances in innovation, serving as a "champion" for science and technology, while leveraging business sectors as a force multiplier. This approach has centered upon effective implementation of technology policies that balanced between acquiring technologies, whether through licit or illicit means, and indigenous development. Meanwhile, seeing relative declines in its scientific prowess and technological capabilities, Russia had nonetheless launched new initiatives seeking to reverse those trends and establish its own alternatives and competitive advantages but has seen damage to those objectives since its invasion of Ukraine. And the United States, where industrial policy was once anathema, motivated by concerns about perceived declines in its own competitiveness, has introduced new initiatives and investments that are intended to maintain leadership and rebuild industrial capabilities that eroded for years prior.

The People's Republic of China

China today is recognized to have emerged as a global powerhouse in science and technology. By any metrics, China's rise as a would-be innovation superpower has been rapid and dramatic, yet often uneven, insofar as striking successes are often juxtaposed with persistent problems. Increasingly, a paradigm of "innovation-driven" development has been positioned at the center of China's national strategy, a feature of national policies reaffirmed across multiple five-year plans.[13] Xi Jinping has reiterated the importance of national strength in science and technology for China's rise, saying: "Science and technology are the nation's sharp weapons, [which] the country depends upon to strengthen, enterprises depend upon to succeed, and people's lives depend upon to be good."[14] From this vantage point, science and technology are essential to a nation's success, even viability, in development, as well as for national security.

China's agenda for science and technology has evolved and adapted over time along with national priorities. The PRC state-driven approach has leveraged large-scale mobilization of resources, demonstrating strengths and limitations. Over the decades, China's drive to "catch up and overtake" (赶超) and pursuit of "leapfrog development" (跨越发展) in science and technology has proven effective, though sometimes inefficient.[15] Underpinning China's rise and relative success has been a strategy combining state guidance, direction, and incentives, often in the form of now-infamous industrial policies, such as "Made in China 2025." The high-level guidance from China's Party-state, executed through local policies and initiatives, can mobilize business resources aligned with national objectives. Even as priorities are set from the top, underlying those advances have been the energy and dynamism of commercial enterprises operating within a fiercely competitive ecosystem.

The PRC has established economies of scale and positioned itself at key nodes within global supply chains and in critical resources. In the process, China's efforts to close the gap have often leveraged both licit and illicit transfers of scientific knowledge and intellectual property. To that end, China has dedicated considerable resources to open-source scientific and technical intelligence, as well as cyber espionage, industrial espionage, and targeted acquisitions of sensitive technologies.[16] PRC reliance upon expropriation and incorporation of foreign technologies, often characterized euphemistically as "re-innovation" (再创新), has been variously an impediment and accelerant for progress.[17]

For China, considerable investments in science and education have served as critical enablers to accelerate advances. Over time, China's spending on science across multiple disciplines and allocations of funding for applied research and development have increased considerably, gradually approaching and seemingly on track to surpass levels of U.S. expenditures.[18] In 2024, China's R&D funding surpassed 3.6 trillion RMB ($496 billion), continuing a steady trend of increases yet by a lower rate than prior year due to budget constraints.[19] Whereas there was initially an emphasis in China's budget on funding primarily applied research, the support to basic research has also increased over time; the expenditures on science and technology reached 328 billion RMB ($47.55 billion) in 2023 and 371 billion ($52 billion) in 2024.[20]

By most metrics, these investments are contributing to significant advancements across multiple disciplines. In such priority disciplines as artificial intelligence, the surge in the quantity of research produced by Chinese scientists has often been a leading indicator; the increases in the quantity of outputs, such as patents and publications, have contributed to eventual progress in quality as well. In AI as in other disciplines, the representation of Chinese scientists/institutions in top-cited papers has also risen in turn.[21] Indeed, by one estimate, China had reached first in most-cited papers in S&T overall as of mid-2021.[22] However, issues of scientific integrity, including plagiarism and falsification of results, have increased given the pressures and incentives for academic production, which the Chinese government has sought to remedy through improved oversight and enforcement.[23]

For China's leaders today, the relationship between technology and national security is essentially inextricable. The "comprehensive national security concept" introduced by Xi Jinping explicitly incorporates "S&T security" among the facets of concern.[24] In fact, China today can be regarded as a "techno-security state," as categorized by scholar Tai Ming Cheung, a transformation of the regime that has accelerated under Xi Jinping's leadership.[25] These issues are heightened as innovation has accelerated with breakthroughs and convergences. PRC leaders fear the consequences of failing to overcome current impediments to independent innovation. As China has faced a slew of policies that Beijing perceives to be intensifying technological "blockade" or "containment" orchestrated by the United States, these imperatives have become even more acute, accelerating its agenda for self-reliance and self-sufficiency in innovation.[26]

The extensive connectivity that previously extended between the United States and China has started to fracture. This trend has accelerated since the U.S.-China trade war initially started during the first Trump administration and continued since the U.S. government has introduced new export controls, especially restricting access to semiconductors and semiconductor manufacturing equipment, across the Biden and second Trump administrations. As a result, China's leaders have prioritized developing "independent, controllable, safe, and reliable" supply chains in domains connected to national security.[27] "Amid fierce international competition, unilateralism and protectionism, we must work out our own way to upgrade, become more innovative and make more breakthroughs from scratch," Xi Jinping urged at a 2020 symposium of scientists.[28] He continued, "For technologies where there can be imminent breakthroughs and solutions, you must quickly move forward; for strategic technologies that take time, you must deploy resources in advance."[29] With the de facto blockade that Beijing believes the United States to be undertaking through a range of methods that restrict access to American technologies, the tactics, techniques, and procedures associated are likely to evolve in turn, including attempts to find workarounds, such as smuggling, to access chips illicitly.[30]

Beyond catching up and closing the gap relative to the United States, the aspiration espoused in China's plans and policies is apparently to be not merely a fast follower but rather a true global leader in new frontiers of science and technology. Among the major S&T projects launched, in 2016 under the 13th Five-Year Plan and continued through 2030, are brain science, quantum computing and communications, and cyber security, among other priority projects.[31] The 14th Five-Year Plan (2021–2025), launched in March 2021, has included as core themes the pursuit of "self-reliance," self-strengthening (自立自强), and technological independence as a "strategic support" of national development.[32] Xi Jinping reaffirmed these themes in his remarks at the 20th Party Congress in October 2022, which again highlighted the aspiration for China to "join the ranks of the world's most innovative countries."[33] A range of recent Chinese policies and priorities have highlighted the importance of more fully original and cutting-edge innovation beyond the more incremental innovation of its recent past.[34]

PRC leaders have prioritized improved integration of the national innovation ecosystem. During the "Two Sessions" in March 2023, which marked the official beginning of his latest term after a decade in power, Xi Jinping again emphasized the importance of developing an integrated national strategic system and capabilities.[35] Notably, the Chinese Communist Party has also undertaken a significant reorganization of its S&T leadership as of April 2023 by creating the Central Science and Technology Commission (中央科技委员会), which is expected to provide more forceful, high-level guidance for this S&T agenda.[36] The conclusion is clear: the Party-state will continue to direct and guide innovation through mobilizing the requisite resources at scale.[37]

The Russian Federation

Russia too has taken a relatively state-driven approach to advancing the development of science and technology in its recent history. The collapse of the Soviet Union had adversely impacted science in modern Russia, which atrophied in the decades following, in particular due to the massive exodus of talent.[38] In the years since, Russia started trying to revitalize its overall science and technology ecosystem, concentrating on rebuilding capacity that had been lost and investing in priority disciplines, such as information technology. In particular, the Russian government has set up new programs and introduced incentives for academics to increase scientific publications.[39] Of note, in 2014, the Russian government launched the National Technology Initiative to build up Russia's tech industry through supporting start-ups and creating new centers to provide resources, expertise, and infrastructure to facilitate advances in breakthrough technologies.[40] The Russian Science Foundation was also established in 2014 with a mandate to award grants for research competitively, yet its budget has been less than the U.S. National Science Foundation or other similar agencies.[41] As Russia's attention to emerging technologies has increased, the Moscow Institute of Physics and Technologies established the Center for Artificial Intelligence in 2017.[42] As of 2018, Putin introduced a new national research strategy that extended through 2024, which was intended to increase funding, support early career scientists, and establish new laboratories, including in robotics.[43] In December 2019, Russia also introduced a new national quantum research program that provided state support, and the Russian Quantum Center also engaged in partnerships with international scientists.[44] Although the quantity of research from Russia started to increase as a result of these policies and incentives, fraud has persisted, despite recurrent retractions and multiple investigations.[45]

Prior to its invasion of Ukraine, Russia had also relied upon scientific partnerships and technological collaboration to rebuild its capacity for innovation. In 2011, the Skolkovo Institute of Science and Technology was founded as a partnership with the Massachusetts Institute of Technology.[46] This institute contributed to the creation of the Skolkovo Innovation Center, which was intended to become a "Russian Silicon Valley."[47] However, after a number of years and millions later, however, Skolkovo is generally regarded to have failed to live up to its potential promise.[48] Even when the ingredients believed to be integral to innovation were combined and bolstered with ample funding – and augmented by access to world-class collaborators at the Massachusetts Institute of Technology (MIT) – the outcomes didn't achieve the success intended. Russia's efforts to recruit and attract talent also tended to be impeded by political and geopolitical circumstances that have exacerbated issues of large-scale brain drain. Since the invasion of Ukraine, the repercussions for Russia in terms of lessened funding, drains on human capital, and reduced options for scientific engagement have been severe and could prove enduring.

The United States

The United States has a long history of dedicating significant investments in science and technology. This has been a core concern for U.S. policy ever since Vannevar Bush had initially advocated, "science is a proper concern of government," calling for more active engagement by the U.S. government to build upon wartime efforts to continue advancing scientific progress in peacetime.[49] The National Science Foundation has a long tradition of robust support for basic research.[50] While the U.S. potential to attract world-class talent and of leading universities, could enable enduring advantages, the risks of shortfalls in funding, as well as immigration policies that can impede talented students and researchers from remaining, can cause damage. Moreover, funding for basic research had started to flatten over time, a trend that raises concerns about that American investments could be insufficient to sustain progress in new technologies. In 2013 and again in 2015, the share of U.S. federal spending on basic research fell below 50%, a low relative to historical levels.[51] This marked the first time since World War II that the U.S. government was no longer supporting the majority of basic research, and the outlook for research spending has become especially uncertain in the Trump administration. Although corporate spending on research has increased in parallel, the lessening proportion of federal funding for R&D has also raised concerns about the potential distortion that could ensue, considering the incentives of companies may be more aligned toward near-term profits.

In its modern history, the U.S. approach to innovation had centered on the dynamism of American academia and its private sector. The United States often eschewed industrial policies, instead pointing to the success of American science and leading companies as a testament of the strength of the U.S. model. Increasingly, the center of gravity for innovation had shifted to commercial enterprises, which creates new challenges for the government to capture, leverage, and promote partnerships. For American policymakers, there have been intense concerns that China was catching up through mobilizing at a scale beyond U.S. efforts. This outlook has provoked anxieties that China's rise could jeopardize American technological dominance. These trends have motivated a turn to industrial policies once been dismissed as outdated and counterproductive. Previously, U.S. concepts of national security didn't explicitly include economic security. However, economic and technological security has since become a more central feature of U.S. policy, overtly recognized in national security strategies as such starting under the Trump administration and continuing through the Biden administration.[52] In a speech in fall 2022, National Security Adviser Jake Sullivan declared, "Given the foundational nature of certain technologies, such as advanced logic and memory chips, we must maintain as large of a lead as possible."[53]

U.S. policies to reinvigorate science and technology have been undertaken in a relatively bipartisan fashion across the Trump and Biden administrations. In 2018, the National Quantum Initiative created an expansive program dedicated to advancing quantum information science, including setting up major research centers.[54] The National Artificial Intelligence Initiative Act of 2020 focused on expanding support for AI research and supporting efforts in data and standardization.[55] The Biden administration continued and expanded these initiatives, including supporting increased research and development in quantum technology. Notably, in August 2022, President Biden signed the CHIPS and Science Act, which dedicated $52 billion for research, training, and manufacturing in microelectronics. The CHIPS Act also introduced new measures for research security, including prohibitions against scientists' joining foreign talent recruitment programs and introducing new initiatives to use data analytics to identify gaps in reporting of foreign support.[56] The CHIPS Act has since been not only praised but also generated controversy, including because of the expense and potential for inefficiencies.[57] With the start of the second Trump administration, including a series of executive orders that included measures to undo major initiatives from the Biden administration, the trajectory of S&T policies is highly uncertain, yet there continues to be bipartisan consensus on the fundamental importance of contesting advantage in emerging technologies, especially AI.

At present, the long-term prospects for these efforts to be successful remain to be seen, as does the patience, consistency, and capacities through which new U.S. policies will be implemented. With such efforts as the CHIPS Act, the United States turn to selecting priorities in its policy approach could be construed as reflecting a trend that is long overdue – or alternatively, a direction that may not be as successful as initially intended, especially if implementation is not consistent or sustained. There are still risks of a gap between expectations and the feasible near-term returns on investment, in an environment where issues of funding can be highly politicized. Moreover, reporting as of early in the Trump administration about a freeze in grant review for the National Science Foundation to realign funding, as well as potential layoffs for its staff, is concerning, as is the potential politicization of the allocation of scientific funding.[58]

Chokepoint Exploitation

Pursuant to globalization has come increased scientific and technological connectivity through global supply chains and international partnerships in research and development, yet these entanglements can also become a source of friction. Indeed, these dynamics have been increasingly securitized; asymmetries in dependency can instead become chokepoints to be leveraged for strategic purposes, in ways that can enhance American power.[59] To undertake actions of non-military warfare through science and technology, states have been starting to employ more export controls and restrictions on cross-border investments and research partnerships. These trends raise a complex interplay among actors, including not only national governments but also academic institutions, enterprises, and scientists themselves. Once primarily instruments for soft power, diplomacy, and cooperative engagement, S&T issues have since come to be regarded more through the lens of hard power.[60] The recent trajectory of measures and countermeasures has demonstrated a progression of techniques, including mutual learning and mimicry among competitors. This chapter considers recent national approaches as case studies that may establish precedents that could result in adaptation of these tactics going forward.

Semiconductor Strangleholds

From Beijing's vantage point, "chokepoints" or "strangleholds" (卡脖子) occur when China faces an unfavorable situation where core technologies "are controlled by others," as Xi Jinping has warned in describing that dilemma.[61] Among the particular chokepoints of urgent concern to China today are semiconductors, aero-engines, and specialized materials.[62] In particular, semiconductors have come to be seen as a strategic weapon in U.S.-China competition as U.S. policies have recognized and sought to exploit that dependency. Despite China's successes on several fronts, shortcomings in a series of key technologies have provoked particular anxiety, and semiconductors have been consistently at the top of that list.[63]

Among the impetuses for the U.S. government's initial introduction of these measures was the success of Huawei, which was enabled in no small part by distortionary measures. Huawei, regarded previously as a juggernaut unstoppable in its advance as a global leader in 5G, faced near-term damages from U.S. policies limiting its access to semiconductors, yet has sought to develop indigenous alternatives in the mid to long term. The U.S. use of blacklists has adversely impacted a growing number of Chinese companies that have been specifically incorporated.[64] The Department of Commerce's Bureau of Industry and Security (BIS) has become a critical actor based on its role in devising and imposing these export controls across a range of rules and filings that seek to take a targeted approach.[65]

Chinese leaders recognize semiconductors are foundational to modern information technology; in particular, specialized semiconductors provide critical compute for AI systems. China's long-standing investments have brought only limited progress, and its continued dependence on foreign sources has been highly costly. However, even many leading Chinese companies have remained highly dependent on semiconductors produced by American companies, a dependence that the U.S. government has sought to exploit. With U.S. export controls on chips sustained and becoming increasingly restrictive since 2022, Chinese stakeholders have moved to stockpile chips. For instance, in 2024, China imported an estimated $385 billion in chips, an increase by 10.4%,[66] and its chip imports have consistently exceeded that spent on crude oil imports for years.[67] Beijing has since dedicated hundreds of billions further to overcoming dependence on such imports.

The U.S. government has leveraged a blacklisting mechanism known as the Entity List as a means of restriction. The U.S. Commerce Department added Huawei and its affiliats to its "Entity List" of persons, nations, etc., for which access to certain items would be limited. These actions were initially characterized as "death blows" to these companies given their dependency on U.S. chips.[68] As a result, Huawei saw a drop in revenue by 30% in 2021, damage that was characterized and, in some cases, criticized as a strangulation of the company at the discretion of the U.S. government.[69] The entity list has since been expanded repeatedly and dramatically, including in March and October 2023, and including 140 additional companies as of December 2024.[70] From Beijing's perspective, these policies have directly threatened China's economic and technological security. U.S. export controls that restricting China's access to advanced technologies have been regarded as tantamount to technological containment. In response to U.S. policies viewed as attempted technical "strangling," Chinese leaders have even more urgently promoted self-reliance and domestic innovation.[71]

However, in retrospect, there were gaps in the initial export controls, loopholes that prompted adaptation and countermeasures in response. Chinese companies sought to take advantage of countermeasures, such through renting access to chips through cloud service providers or through using intermediaries.[72] PRC AI companies involved in surveillance applications have often leveraged front companies or intermediaries to access U.S. AI chips and components to which U.S. policies had sought to restrict their access. iFlytek, a company that has also collaborated with the Chinese military, has also rented NVIDIA chips to circumvent the restrictions on purchasing them outright.[73] In October 2024, the Taiwan Semiconductor Manufacturing Company (TSMC) curtailed shipments of chips to a client and notified U.S. officials after discovering that those semiconductors were being routed to Huawei.[74] After the reports starting in December 2024 about DeepSeek's advances, which initially appeared to present a significant challenge to American leadership in AI, the question of how best to reduce those loopholes, without damaging American companies, has again become a subject of active debate as of early 2025.[75]

Increasingly, the TSMC, situated in a pivotal position in the semiconductor supply chain, has come to be regarded among the most important companies in the world and potentially a flashpoint in its own right. TSMC is situated at a uniquely strategic position within the chip industry, as the world's largest foundry and seemingly unassailable in its capabilities.[76] The history of TSMC and the means by which the company came to reach this strategic position is complex, reflecting how Taiwan's state policies were influential in creating conditions where a unique capacity and ecosystem around it could be developed.[77] While its importance was once more obscured or at least less prominent in popular discourse, concerns about TSMC have since reached such a fever pitch that concerned American commentators even raised the possibility that China could be further incentivized to target or invade Taiwan for the purpose of seizing control of chips. For instance, a U.S. think-tank arranged a war-game centered upon the semiconductor sector as a flashpoint.[78] Several defense academics have proposed the threat of sabotage to TSMC as a means of deterring a Chinese invasion.[79] Could such a threat have any deterrent impacts, and how might that risk increase the likelihood that the United States and allied governments would intervene? These questions highlight the enormous consequence that a private company can have when the technology it has mastered becomes a potential flashpoint in its own right. So too, TSMC's decision to halt chip sales to Russia has been significant, weaponizing that dependency to the Russian military's detriment.[80]

U.S. measures forcing such a choice upon American companies, as well as those foreign companies leveraging American-origin technologies, will have second-order effects. Leading American companies, including NVIDIA, that had looked to opportunities in China's markets, including to generate profits that enable their own investments, have been adversely impacted and looked for workarounds, such as increasing their sales of less advanced chips to China.[81] The latest measures, under the Biden administration, leveraged "foreign produced direct product rules," which exploits the current reality that despite the United States' lack of its own indigenous capability for manufacturing semiconductors domestically, many, even most, chip factories leverage U.S.-designed software or equipment.[82] Regarding the United States, observers in Beijing believes, "The fundamental purpose is still to contain China with 'chips,' with the intent to build its own 'chip hegemony,' and then create a global semiconductor supply chain led by the U.S.," by the characterization of one commentary in PRC state media.[83]

However, the capacity to exploit superiority in semiconductors may start to change within the foreseeable future. The demonstrated destructive potential of current consolidation may create further impetus for growing investments in alternatives. There are efforts underway worldwide to diversify semiconductors manufacturing to avoid reliance upon limited sources and locations. TSMC has attempted to build a U.S. factory in Arizona with only marginal success to date.[84] Current attempts to replicate the unique conditions that have allowed TSMC to be so successful in Taiwan are likely to confront challenges. Going forward, the export controls may cause short-term pain but catalyze long-term progress, and Beijing is also looking to learn lessons from these efforts for its own future efforts to develop indigenous semiconductors at scale, in ways that could enable China to capture more market share, such as it currently commands in chips for more everyday applications, in the future.[85] The massive amounts of resources dedicated to that effort have contributed to China's progress toward more advanced semiconductors. However, the actual return on investment appears to have been uneven or moderate, and the overall funding and approach have since been reevaluated, including with investigation of one of the main funds involved over potential corruption.[86] China's "Big Fund," a state-sponsored investment fund, for semiconductor development has progressed across three rounds so far, the latest of which was 344 billion RMB ($47.5 billion); in total, the "big fund" has so far totaled an estimated $95 billion in equity investments.[87] However, China's fund for semiconductor investments has been the focus of a Chinese government anticorruption investigation.[88]

China has in turn started to develop its own regime of export controls and restrictions on the sales of critical minerals. In December 2024, Beijing banned the export of certain critical materials, including gallium, antimony, and germanium, that are used in the production of semiconductors.[89] In January 2025, Beijing announced plans to curb the export of battery cathode technology and other enabling technologies.[90] And in February, after the Trump administration started to threaten tariffs against China, the PRC announced further export controls of its own that restricted critical minerals relevant for clean energy and defense, including tungsten and indium.[91] Beijing's capacity to exploit mineral resources as a source of leverage or for punitive measures has not yet been exercised to the fullest.

Technological Denial

Since the Russian invasion of Ukraine, the United States and allied efforts to restrict Russia's access to science and technology have been a critical element of the coordinated campaign undertaken in response. According to Secretary of Commerce Raimondo, U.S. tech exports to Russia were reduced by 70% after February 2022 as of mid-2022, or by as much as 85% by another estimate.[92] According to a fact sheet released by the White House, "Our unprecedented sanctions are already exacting an immense toll on Russia's economy and our export controls have strangled Russia's access to critical technology and the supply chains it needs to sustain its military ambitions."[93] Through the use of export controls, "Our approach was to deny Russia technology – technology that would cripple their ability to continue a military operation. And that is exactly what we are doing," Secretary Raimondo emphasized.[94] Of note, chips that used to be manufactured by TSMC but were designed in Russia and branded as Elbrus were used previously by Russian military and security services, such that the application of this rule was expected to prove "devastating," according to one expert analysis.[95] The primary providers of chips in Russia, Baikal and Elbrus, had relied on TSMC for the fabrication.[96]

Since these restrictions have been introduced, Russia has been forced to resort to creative scavenging to access the semiconductors required for modern weaponry. Russian military equipment has even been found to have chips scavenged from kitchen appliances.[97] So too, at later stages in the conflict, the electronic components of a number of Russian drones that were shot down over Ukraine proved to be derived from Western technology, according to a technical analysis produced by the Conflict Armament Research group, which dissected the drones.[98] However, recent analyses have concluded that the impact of these export controls has been limited, and access to American technology has continued to be an enabler for Russian military capabilities.[99] Despite the myriad sanctions and export restrictions imposed on Russia, its capacity for weapons production has increased nonetheless, according to an analysis from UK think-tank RUSI.[100]

In recent history, U.S. sanctions and export controls on Russia have been impactful, yet these measures have failed to change the calculus of Russian decision-making. While essentially unprecedented in scope, these efforts to restrict the transfer of sensitive technologies have pertinent historical antecedents.[101] Today, American policymakers are looking to exercise further leverage through extending the reach of these restrictions through multinational coordination. Along the way, the U.S. government has also worked to build up capacity to implement export controls, and the latest budget provides additional funding to the BIS at the Department of Commerce. This significant increase in resources is intended to increase its capacity to "apply and enforce export controls to advance our national security objectives."[102] As in the case of Huawei, the U.S. government applied the foreign direct product rule to Russia, thereby requiring any company using U.S. software or manufacturing equipment abide by the same policies.[103] Among the items covered by these measures were semiconductors and telecommunications equipment, as well as lasers and avionics.[104] The relative success of these measures reflected the multilateral character, involving over three dozen countries coordinating and implementing in parallel.[105]

While Russia has been cognizant of these risks, the capacity to course-correct or introduce changes to mitigate the vulnerability to these measures has been limited to date. Russia must "radically reduce the risks associated with the use of foreign programmes, computer technology and telecommunications equipment," Putin had declared at a Security Council meeting in May 2022.[106] He continued, "In order to strengthen our technological sovereignty, the government needs to create a modern Russian electronic component base in the shortest time possible."[107] Turning to China for chips is an obvious alternative, yet made-in-China chips cannot completely substitute for TSMC's capacity or quality. And despite pressure on Beijing to distance itself from Moscow, "China attaches great importance to cooperation in this field and will work together with the Russian side to promote military-technical cooperation between the two countries to a higher level and a broader field," declared China's ambassador to Russia, Zhang Hanhui, in May 2022.[108]

Even as Beijing has publicly opposed and condemned Western sanctions, pursuing official policies deepening scientific engagement with Russia, several Chinese companies agreed to adhere to technological restrictions on Russia due to fears of secondary sanctions.[109] Within Russia's market, Chinese technology companies have been relatively dominant. For instance, Lenovo and Xiaomi both limited or curtailed sales/shipments in Russia according to one report.[110] However, relative to Western technology companies, most Chinese companies didn't publicly discuss or release statements regarding their decisions to withdraw from the Russian market. Notably, DJI announced its intention to suspend its business activities in Russia ahead of a compliance review.[111] A U.S. official confirmed when asked, "We have not seen the provision of military equipment," as of mid-2022, but later reporting raised the question of whether quieter cooperation could continue.[112] However, the extent of that cooperation is difficult to ascertain from the available information, and the long-term trajectory of Sino-Russian innovation collaboration will also reflect the degree of secondary leverages that U.S. policy can enact.

Innovation Realignments

Maneuvering to contest advantage, great powers have pursued new partnerships and deepened existing channels of cooperation in science and technology. Once, science and technology were regarded as a highly globalized domain of cooperation that was relatively independent from geopolitics. However, that utopian notion of innovation as oriented toward global public goods has not been completely sustained against the backdrop of the recent intensification of great power rivalries. Previously, U.S.-China competition was juxtaposed with the reality of extensive cooperation between American and Chinese scientists, universities, and companies. Across most critical sectors, to include biotechnology and artificial intelligence, and by most metrics, whether that be co-authorship, co-founding of start-ups, and cross-border venture capital, ties between American and Chinese companies were strong and initially remained robust even as the political and military dimensions of the relationship became fraught.[113] The question of which country drew greater benefits from these dynamics was difficult at best to evaluate, yet American officials became concerned that the benefits were asymmetric in Beijing's favor.[114] The Chinese government often leveraged research and scientific partnerships as a means of accessing sensitive technologies. In that regard, even when cooperation was regarded as relatively independent of security issues, those problems were underlying though often inconsistently appreciated.

The United States has looked to deepen engagement on technology with allies, partners, and like-minded countries on several fronts; meanwhile, facing pressures from the United States and other Western restrictions, China and Russia have continued to deepen technological collaboration on scientific research and technological development. For the United States and USSR, even at the height of the Cold War, scientific cooperation, at least on a limited number of issues, was able to continue, nonetheless. In that era, consistent engagement among scientists was perceived as potentially advantageous and conducive to stabilizing the relationship. Today, that dynamic has changed drastically, with U.S.-Russia scientific cooperation severely disrupted and that between the United States and China significantly diminished and constrained. Resultantly, the costs and consequences to open science and the creation of knowledge, as well as the loss of options for mutual understanding and creative diplomacy that scientific collaboration once enabled, may be grave, especially at a moment when the world confronts severe, systemic challenges, especially in pandemic prevention and climate change.

Scientific Cold War?

The isolation of Russia resulted in bifurcation of the global scientific ecosystem, to an extent that has been unprecedented in recent history. The vast majority of scientific collaborations between Russian scientists and Western counterparts were curtailed or outright canceled across multiple disciplines and contexts. As Russian scientist and dissident Andronick Arutyunov observed, "The divide between Russian sciences and global sciences is happening. . . . Collaboration projects with foreign scientists are finished and I fear they will not come here any more."[115] A number of countries entirely curtailed all scientific projects with Russia.[116] Of note, the G7 Science Ministers' Communiqué declared, in June 2022,

> We strongly condemn Russia's assault on the fundamental principles of freedom, democracy and self-determination. These fundamental principles

> provide the basis for academic freedom and for scientific exchange and cooperation. We are restricting, as appropriate, government funded research projects and programs involving Russian government participation.[117]

Even research initiatives that had originated under fraught conditions in the Cold War, as in basic research and space science, have adversely impacted. MIT has halted entirely its engagement with the Skolkovo Institute.[118] The European Space Agency canceled a planned mission to explore Mars along with Russia's ExoMars-22.[119] These measures, among many others, have imposed symbolic and reputational damages, more so than substantive consequences, and yet the freezing of scientific engagement has marked a new low below the chills of the Cold War. Russian researchers lost access to a range of resources, from databases to cloud infrastructure, and a number of partnerships were curtailed, in ways that risked undermining scientific cooperation on important challenges, including climate change.[120] Although the Russian invasion was initially expected to result in near-complete isolation of Russian science, later analyses based on patterns of publications showed that certain cooperative relationships remained relatively stable.[121]

Sino-Russian Innovation Collaboration

There is a long history of scientific and technological cooperation between China and Russia, and recent deepening of their geopolitical alignment may reinforce these trends.[122] In the PRC's early history, the flow of technology transfers occurred primarily from the Soviet Union to China, whether in arms sales or training. From the impetus of aligned interests and common concerns has come a deepening alignment of efforts for mutual benefits that may mitigate the damage imposed by Western technology restrictions.[123] Ultimately, "It's mutual interest – and not money – that really counts" for China-Russia cooperation in artificial intelligence, according to Russian President Putin.[124] While mutual interests matter, money conveys importance of its own, especially as Russia confronts acute constraints on resources. Facing scientific isolation and the loss of foreign funding, Russia has responded by seeking out new options for scientific collaboration beyond the United States and Europe, including with India and China.[125]

For China and Russia, respective comparative advantages have become increasingly complementary. While China's market and resources have outpaced those that Russia can generate today, Russia still possesses certain technical expertise that China lacks and is trying to generate.[126] "We can use our best qualities, expanding our technological potential and competitiveness," declared China's Foreign Ministry spokesman Geng Shuang in 2019.[127] Putin has emphasized Russian advantages in "mathematics and everything related to it."[128] The Chinese and Russian Academies of Sciences have expanded their cooperation, including in biotechnology and neuroscience, which has become an important element of this partnership.[129] For 2020 and 2021, which were deemed 'years of innovation cooperation,' joint activities included bilateral academic seminars and a range of S&T exchanges.[130] Despite disruption due to the pandemic and the quieting of official initiatives with dual-use potential since the Russian invasion of Ukraine likely because of Beijing's concerns about the potential for U.S. pressure against China, Sino-Russian innovation cooperation should be expected to be sustained and probably expanded for the mid/long term. Among the priority domains or cooperation has been Arctic research. So too, as of 2025, after DeepSeek's rise to prominence internationally, Russia's Sberbank announced plans to expand joint AI research.[131]

Democratic Technology Collaboration

America has started to focus on engaging more proactively with allies and partners to create new partnerships to promote scientific progress among stakeholders with shared values. In American debates on the topic, there have been calls for technology alliances among democracies.[132] This vision is starting to become a reality through several recent "mini-lateral" initiatives, including the Quad, which consists of the United States, India, Japan, and Australia, and AUKUS, among the United States, the United Kingdom, and Australia. To date, the Quad has sought to promote the concept of "Open Science." For instance, initial initiatives concentrated on options for sharing space-based civil Earth observation data as a common resource.[133] The Quad also established the Critical and Emerging Technologies Working Group as of March 2021, through which partners have "mapped collective capacity and vulnerabilities in global semiconductor supply chains" and agreed to common principles on supply chain resilience.[134] The promotion of innovation and standardization in telecommunications was a particular emphasis in initial efforts. Meanwhile, while AUKUS has been most prominent for its focus on the collaborative development of nuclear submarines for introduction into Australia's Navy, it has also launched several nascent efforts focused on other emerging technologies, including artificial intelligence, undersea capabilities, and quantum science.[135] The latest stages of AUKUS have concentrated on expanding research engagement among the partners, including joint projects and exchanges of talent.[136] While these efforts remain nascent, the early initiatives are evidence of a paradigm within which scientific engagement is more closely realigned with considerations of values and geopolitics than previously.

Conclusion

Across history, innovation has been a new frontier for human progress; this is also a fraught front for dynamics of non-military warfare. The securitization of science and technology in confrontation among major powers today is hardly surprising at a moment when the capacity to leverage technology to enhance national power has become increasingly consequential, relative to the centrality of natural resources in the past. The world's leading companies and those most prominent on the world stage are primarily technology companies that are often transnational in operations, yet subject to national jurisdictions. To some extent, this exploitation of science and technology to contest strategic advantage could be regarded as interrelated with economic warfare. However, these measures have been more targeted to focus on the control of specific sensitive or strategic technologies with targeted effects. Moreover, the potential to augment or undercut military capabilities, as has become apparent in the campaign to hinder Russia's ongoing military operations through denying it access to the requisite technologies, is a more unique feature of modern non-military warfare.

Ultimately, in an era of intensifying competition among great powers, the capacity to achieve advantage through innovation, exploit adversaries' technological dependences, and realign scientific engagement and collaborative technological development with strategic imperatives can be considered a new modality for non-military warfare. The prior decades of globalization of innovation ecosystems had created a complex landscape with ample space for maneuver. However, as states become more proficient in mapping and anticipating their vulnerabilities to identify options to mitigate and build domestic capacity, certain of such measures may have diminishing returns. Despite the recurrent tendencies toward idealism about a world of open science, these tendencies appear likely to continue recent trends of fracturing supply chains, disrupting previously far-reaching scientific partnerships, and undermining collaborative technological development. In some cases, confrontation may spur competition and innovations, yet the associated frictions can undermine scientific progress as well. The relationships among the relevant stakeholders across government and enterprises will continue to evolve. Based on current trends, the future global innovation ecosystem could be more centered on a combination of allied and indigenous capabilities, and competition may overcome cooperation for the foreseeable future.

Notes

* The opinions expressed in this article are the author's alone.

1 Luk Van Langenhove, Luc Soete, and Eric Piaget, "Science Diplomacy in the Age of War," Policylabs, May 5, 2022, https://policylabs.frontiersin.org/content/science-diplomacy-in-the-age-of-war

2 There is an extensive literature on the topic. See, for instance: Katherine Barbieri, The Liberal Illusion: Does Trade Promote Peace? (Ann Arbor: University of Michigan Press, 2002). Dale C. Copeland, Economic Interdependence and War. (Princeton: Princeton University Press, 2015).

3 For instance: Henry Farrell and Abraham L. Newman, "Weaponized Interdependence: How Global Economic Networks Shape State Coercion," International Security 44, no. 1 (2019): 42–79. See also: Daniel W. Drezner, Henry Farrell and Abraham L. Newman (Eds.), The Uses and Abuses of Weaponized Interdependence (Brookings Institution Press, 2021).

4 I choose to use these phrases to attempt to capture and characterize specific phenomena that are relevant to understanding great power rivalries and non-military warfare today. These concepts are not wholly unique. For instance, the term "competitive innovation" has also been used in the business literature. This phenomenon that I attempt to capture in discussing the exploitation of entanglements in science and technology builds upon ideas of targeted decoupling or weaponized interdependence, though these dynamics are not limited to networks or supply chain vulnerabilities. This phrasing is intended to capture other types of engagement, such as exchanges of talent and diffusion of knowledge through research, that can be exploited. The idea of "innovation realignment" is relatively analogous to the concept of technology alliances that has been

promoted by several think-tanks and policymakers. However, I chose the phrase "realignment" rather than "alliance" to capture that these trends are shifting global patterns of scientific engagement and technological collaboration; moreover, the activities and stakeholders involved in scientific and technological development expand beyond the bounds of those typically considered in the context of an alliance.

5 For a discussion on how this dynamic has played out in historical perspective, see, for instance: Jeffrey Ding, "The Rise and Fall of Technological Leadership: General-purpose Technology Diffusion and Economic Power Transitions," International Studies Quarterly 68, no. 2 (2024): sqae013. See also: Jeffrey Ding, "The Diffusion Deficit in Scientific and Technological Power: Re-assessing China's Rise," Review of International Political Economy 31, no. 1 (2024): 173–98.

6 For one perspective on the topic, see for instance: Thomas S. Robertson, "The Process of Innovation and the Diffusion of Innovation," Journal of Marketing 31, no. 1 (1967): 14–19.

7 Michael E. Porter, "The Competitive Advantage of Nations," Harvard Business Review, March/April 1990, https://hbr.org/1990/03/the-competitive-advantage-of-nations

8 Vannevar Bush, Science, the Endless Frontier: A Report to the President by Vannevar Bush, Director of the Office of Scientific Research and Development, July 1945 (Washington: United States Government Printing Office, 1945), https://www.nsf.gov/od/lpa/nsf50/vbush1945.htm

9 See, for instance: Thomas Heinrich, "Cold War Armory: Military Contracting in Silicon Valley," Enterprise & Society 3, no. 2 (2002): 247–84. Margaret O'Mara, The Code: Silicon Valley and the Remaking of America (Penguin, 2020).

10 For general reference, see, for instance: Ivan T. Berend, "Industrial Policy and Its Failure in the Soviet Bloc," in Industrial Policy in Europe after 1945 (London: Palgrave Macmillan, 2014), 279–99.

11 There is an extensive literature on these topics. See, for instance: Sanjaya Lall, Learning from the Asian Tigers: Studies in Technology and Industrial Policy (Springer, 1996). Peilei Fan and Chihiro Watanabe, "Promoting Industrial Development Through Technology Policy: Lessons from Japan and China," Technology in Society 28, no. 3 (2006): 303–20.

12 "Xi Jinping: Comprehensively Implement an Innovation-Driven Development Strategy; Promote the Achievement of New Leaps in National Defense and Military Construction" [习近平：全面实施创新驱动发展战略　推动国防和军队建设实现新跨越], Xinhua, March 13, 2016, http://news.xinhuanet.com/politics/2016lh/2016-03/13/c_1118316426.htm.

13 "The CCP Central Committee and State Council Release the "Outline of the National Innovation-Driven Development Strategy" [中共中央　国务院印发《国家创新驱动发展战略纲要》], Xinhua, May 19, 2016, http://news.xinhuanet.com/politics/2016-05/19/c_1118898033.htm.

14 "Xi Jinping: Striving to Build a Global S&T Power" [习近平：为建设世界科技强国而奋斗], Xinhua, May 30, 2016, http://news.xinhuanet.com/politics/2016-05/31/c_1118965169.htm.

15 For further context, see: Barry Naughton, Siwen Xiao and Yaosheng Xu, "The Trajectory of China's Industrial Policies," IGCC Working Paper no. 6 (June 2, 2023): 1–34, https://ucigcc.org/publication/working-paper/the-trajectory-of-chinas-industrial-policies/.

16 See, for reference: William C. Hannas and Didi Kirsten Tatlow (Eds.), China's Quest for Foreign Technology: Beyond Espionage (Routledge, 2020).

17 Tai Ming Cheung, "Innovation in China's Defense Technology Base: Foreign Technology and Military Capabilities," Journal of Strategic Studies 39, no. 5–6 (2016): 728–61.

18 See: Holly Chik, "China Makes Big Leap Forward in Basic Research Spending," South China Morning Post, August 31, 2022, https://www.scmp.com/news/china/science/article/3190874/china-makes-big-leap-forward-basic-research-spending.

19 Victoria Bela, "China's R&D Spending Passes US$496 Billion in Steady Rise to Hi-tech Dominance," South China Morning Post, January 24, 2025, https://www.scmp.com/news/china/science/article/3296098/chinas-rd-growth-passes-us496-billion-steady-rise-hi-tech-dominance

20 Echo Xie, "China's Two Sessions 2023: 2% Budget Boost for Science and Technology," South China Morning Post, March 5, 2023, https://www.scmp.com/news/china/science/article/3212432/chinas-two-sessions-2023-2-cent-budget-boost-science-and-technology. "China Promises More Money for Science in 2024," Nature, March 8, 2024, https://www.nature.com/articles/d41586-024-00695-4#:~:text=The%20government%20will%20spend%20371,by%20China's%20Ministry%20of%20Finance.

21 See multiple accounts from media reporting and bibliometric analysis: Tom Simonite, "China Is Catching Up to the US in AI Research – Fast," Wired, March 13, 2019, https://www.wired.com/story/china-catching-up-us-in-ai-research/. Akira Oikawa and Yuta Shimono, "China Overtakes US in AI Research," Nikkei Asia, August 10, 2021, https://asia.nikkei.com/Spotlight/Datawatch/China-overtakes-US-in-AI-research.

22 Jeffrey Brainard and Dennis Normilel, "China Rises to First Place for Most-Cited Papers," Science, September 16, 2021, https://www.science.org/content/article/china-rises-first-place-most-cited-papers.

23 Smriti Mallapaty, "China's Research-Misconduct Rules Target 'Paper Mills' that Churn out Fake Studies," Nature, August 21, 2020, https://www.nature.com/articles/d41586-020-02445-8.

24 See, for instance: "The '16 Types of Security' of the Overall National Security Concept" [总体国家安全观的"16种安全], S&T Daily, April 14, 2021.

25 See again for comparison: Tai Ming Cheung, Innovate to Dominate: The Rise of the Chinese Techno-Security State (Cornell University Press, 2022).

26 Chi Zhipei [池志培], "The Implementation and Restrictions of U.S S&T Containment Strategy against China" [美国对华科技遏制战略的实施与制约], *Pacific Journal* 《太平洋学报》, 2020.

27 "Xi Jinping: Several Major Issues in the National Medium-Term and Long-Term Economic and Social Development Strategy" [习近平：国家中长期经济社会发展战略若干重大问题], Seeking Truth [求是], October 31, 2020, http://www.ccdi.gov.cn/ldhd/gcsy/202011/t20201103_229231.html.

28 Liu Zhen, "Xi Jinping calls on science to solve the big problems choking China," South China Morning Post, 13 September 2020, https://www.scmp.com/news/china/science/article/3101376/xi-jinping-calls-science-solve-big-problems-choking-china.

29 Liu Zhen, "Xi Jinping calls on science to solve the big problems choking China," South China Morning Post, 13 September 2020, https://www.scmp.com/news/china/science/article/3101376/xi-jinping-calls-science-solve-big-problems-choking-china.

30 Qianer Liu, "Nvidia AI Chip Smuggling to China Becomes an Industry," The Information, August 12, 2024, https://www.theinformation.com/articles/nvidia-ai-chip-smuggling-to-china-becomes-an-industry.

31 State Council Notice on the Issuance of the 13th Five-Year Plan for National S&T Innovation [国务院关于印发"十三五"国家科技创新规划的通知], July 28, 2016, http://www.gov.cn/zhengce/content/2016-08/08/content_5098072.htm.

32 The 14th Five-Year Plan (2021–2025) has included as core themes the pursuit of "self-reliance," self-strengthening (自立自强), and technological independence as a "strategic support" of national development.

33 Low De Wei, "Full Text of Xi Jinping's Speech at China's Party Congress," Bloomberg, October 18, 2022, https://www.bloomberg.com/news/articles/2022-10-18/full-text-of-xi-jinping-s-speech-at-china-20th-party-congress-2022

34 See, for reference: "Xi Story: Cutting-Edge Tech Exemplifies China's Innovation Drive," Xinhua, August 27, 2022, https://english.cctv.com/2022/08/27/ARTICNiptxlpAvxKnl0tvpvo220827.shtml.

35 Smriti Mallapaty, "China is Mobilizing Science to Spur Development – and Self-Reliance," Nature, March 21, 2023, https://www.nature.com/articles/d41586-023-00744-4.
36 Eduardo Baptista, "China to Restructure Sci-Tech Ministry to Reach Self-Reliance Faster: State Media," Reuters, March 7, 2023, https://www.reuters.com/world/china/china-restructure-sci-tech-ministry-reach-self-reliance-faster-state-media-2023-03-07/.
37 Huang Qiang [黄强], "Strengthen the Coordination of Innovation Resources and the Organization of Forces" [加强创新资源统筹和力量组织], People's Daily [人民日报], August 23, 2024.
38 See, for instance: Ina Ganguli, "Scientific Brain Drain and Human Capital Formation After the End of the Soviet Union," International Migration 52, no. 5 (2014): 95–110.
39 Denis Kosyakov and Andrey Guskov. "Impact of National Science Policy on Academic Migration and Research Productivity in Russia." *Procedia Computer Science* 146 (2019): 60–71.
40 Vadim V. Medvedev, Alexander V. Ragutkin and Yulia A. Fedchenkova, "National Technology Initiative and Modern Economic Shocks," Вестник университета (2023): 93, https://doi.org/10.26425/1816-4277-2023-1-92-99
41 Vladimir Pokrovsky, "Putin Decree Shakes Russian Science Funding," Science, January 22, 2014, https://www.science.org/content/article/putin-decree-shakes-russian-science-funding.
42 Samuel Bendett, "Russia's National AI Center Is Taking Shape," Defense One, September27,2019,https://www.defenseone.com/technology/2019/09/russias-national-ai-center-taking-shape/160219/
43 Quirin Schiermeier, "Russia Aims to Revive Science After Era of Stagnation," Nature, March 18, 2020, https://www.nature.com/articles/d41586-020-00753-7
44 Quirin Schiermeier, "Russia Joins Race to Make Quantum Dreams a Reality," Nature, December 17, 2019, https://www.nature.com/articles/d41586-019-03855-z
45 Dalmeet Singh Chawla, "Russian Journals Retract More Than 800 Papers After 'Bombshell' Investigation," Science, January 8, 2020, https://www.science.org/content/article/russian-journals-retract-more-800-papers-after-bombshell-investigation.
46 "Collaboration: Skoltech and the Massachusetts Institute of Technology," https://www.skoltech.ru/en/about/mit/.
47 "US Fund To Invest $250 Million In Skolkovo, The 'Russian Silicon Valley,'" Tech Crunch, May 31, 2010, https://techcrunch.com/2010/05/31/us-fund-to-invest-250-million-in-skolkovo-the-russian-silicon-valley/
48 Justin Reynolds, "Skolkovo: The Moscow Suburb's Struggle to Survive," Harvard Kennedy School Student Policy Review, April 14, 2016, https://web.archive.org/web/0/https://studentreview.hks.harvard.edu/2016/04/14/skolkovo-the-moscow-suburbs-struggle-to-survive/
49 Vannevar Bush, Science, the Endless Frontier: A Report to the President by Vannevar Bush, Director of the Office of Scientific Research and Development, July 1945 (Washington, DC: United States Government Printing Office, 1945), https://www.nsf.gov/od/lpa/nsf50/vbush1945.htm
50 Milton Lomask, "A Minor Miracle: An Informal History of the National Science Foundation," (1976). National Science Foundation, Washington, DC., 1976. See: https://eric.ed.gov/?id=ED141130
51 Jeffrey Mervis, "Data Check: U.S. Government Share of Basic Research Funding Falls Below 50%," Science, March 9, 3017, https://www.science.org/content/article/data-check-us-government-share-basic-research-funding-falls-below-50
52 "Economic Security Is National Security," The White House, September 18, 2020, https://trumpwhitehouse.archives.gov/articles/economic-security-national-security/.
53 Jake Sullivan, "Remarks by National Security Advisor Jake Sullivan at the Special Competitive Studies Project Global Emerging Technologies Summit," The White House, September 16, 2022, https://www.whitehouse.gov/briefing-room/speeches-remarks/2022/09/16/remarks-by-national-security-advisor-jake-sullivan-at-the-special-competitive-studies-project-global-emerging-technologies-summit/.

54 National Quantum Coordination Office, "About the National Quantum Initiative," Quantum.gov, https://www.quantum.gov/about/#OVERVIEW.
55 See "National Artificial Intelligence Initiative Act of 2020," https://www.congress.gov/116/crpt/hrpt617/CRPT-116hrpt617.pdf#page=1210
56 Jeffrey Mervis, "NSF Hopes Big Data Will Finger Grantees Not Reporting Foreign Support," Science, September 30, 2022, https://www.science.org/content/article/nsf-hopes-big-data-will-finger-grantees-not-reporting-foreign-support
57 Gary Clyde Hufbauer and Megan Hogan, "CHIPS Act will Spur US Production But Not Foreclose China," Peterson Institute for International Economics Policy Brief 22–13 (2022).
58 Jonathan Lambert, "National Science Foundation Freezes Grant Review in Response to Trump Executive Orders," NPR, January 27, 2025, https://www.npr.org/sections/shots-health-news/2025/01/27/nx-s1-5276342/nsf-freezes-grant-review-trump-executive-orders-dei-science
59 Fishman, Edward, Chokepoints: American Power in the Age of Economic Warfare (Portfolio, 2025).
60 Nina V. Fedoroff, "Science Diplomacy in the 21st Century," Cell 136, no. 1 (2009): 9–11. Pierre-Bruno Ruffini, "What Is Science Diplomacy?," in Science and Diplomacy (Cham: Springer, 2017), 11–26.
61 See for general reference on the concept: Ben Murphy, "Chokepoints: China's Self-Identified Strategic Technology Import Dependencies," Center for Security and Emerging Technology, May 2022, https://cset.georgetown.edu/publication/chokepoints/.
62 Ben Murphy, "Chokepoints: China's Self-Identified Strategic Technology Import Dependencies," Center for Security and Emerging Technology, May 2022, https://cset.georgetown.edu/publication/chokepoints/.
63 "Xinhua Network Comment: Fight the Key (and) Core Technology Battle" [新华网评: 打好关键核心技术攻坚战], Xinhua, August 26, 2020, http://www.xinhuanet.com/2020–08/26/c_1126414617.htm.
64 See, for instance: "Commerce Adds Six to Entity List for Supporting PRC Military Modernization, Intelligence, and Reconnaissance Activities," Bureau of Industry and Security, February 10, 2023, https://www.bis.doc.gov/index.php/documents/about-bis/newsroom/press-releases/3220-2023-02-10-bis-press-release-six-prc-entities-final-3/file
65 "Commerce Implements New Export Controls on Advanced Computing and Semiconductor Manufacturing Items to the People's Republic of China (PRC)," Bureau of Industry and Security, October 7, 2022, https://www.bis.doc.gov/index.php/documents/about-bis/newsroom/press-releases/3158-2022-10-07-bis-press-release-advanced-computing-and-semiconductor-manufacturing-controls-final/file
66 Ben Jiang, "China's 2024 Chip Imports Surged 10.4% to US$385 Billion amid Tighter US Tech Sanctions," South China Morning Post, January 13, 2025, https://www.scmp.com/tech/tech-trends/article/3294570/chinas-2024-chip-imports-surged-104-us385-billion-amid-tighter-us-tech-sanctions
67 Ma Si, "Chipmakers Bank on 'White Knights' to Move Up the Ladder," China Daily, July 30, 2019, https://www.chinadaily.com.cn/a/201907/30/WS5d3fa161a310d83056401b4f.html.
68 Zhang Erchi and Mo Yelin, "Update: U.S. Puts Stranglehold on Chip Sales to Huawei," Caixin, August 18, 2020, https://www.caixinglobal.com/2020-08-18/us-announces-new-curbs-on-huawei-access-to-us-technology-101593815.html.
69 Ellen Nakashima and Jeanne Whalen, "U.S. Threatens Use of Novel Export Control to Damage Russia's Strategic Industries If Moscow Invades Ukraine," The Washington Post, January 23, 2022, https://www.washingtonpost.com/national-security/2022/01/23/russia-ukraine-sanctions-export-controls
70 Elaine Kurtenbach, "US Expands List of Chinese Technology Companies Under Export Restrictions," Associated Press, December 3, 2024, https://apnews.com/article/china-us-technology-chips-sanctions-bis-8f8ab1ab49b5bb57e5a290a3937fa939.

71 Liu Zhen, "Xi Jinping Calls On Science to Solve the Big Problems Choking China," South China Morning Post, September 13, 2020, https://www.scmp.com/news/china/science/article/3101376/xi-jinping-calls-science-solve-big-problems-choking-china.
72 William Alan Reinsch and Margot Putnam, "Addressing Gaps in U.S. Export Controls," CSIS, May 15, 2023, https://www.csis.org/analysis/addressing-gaps-us-export-controls
73 Eleanor Olcott, Qianer Liu and Demetri Sevastopulo, "Chinese AI Groups Use Cloud Services to Evade US Chip Export Controls," Financial Times, March 8, 2023, https://www.ft.com/content/9706c917-6440-4fa9-b588-b18fbc1503b9.
74 "TSMC Halts Chip Supply to Customer After Finding It in Huawei Product, Source Says," Reuters, October 23, 2024, https://www.reuters.com/technology/tsmc-halts-chip-supply-customer-after-finding-it-huawei-product-source-says-2024-10-23/.
75 Richard Vanderford, "Nvidia Chip Flow to China Should be Throttled Post-DeepSeek, U.S. Lawmakers Say," Wall Street Journal, January 3, 2025, https://www.wsj.com/articles/nvidia-chip-flow-to-china-should-be-throttled-post-deepseek-u-s-lawmakers-say-09f265df?utm_source=chatgpt.com
76 Jeanne Whalen," Computer chip Industry Begins Halting Deliveries to Russia in Response to U.S. Sanctions," The Washington Post, February 25, 2022, https://www.washingtonpost.com/technology/2022/02/25/ukraine-russia-chips-sanctions-tsmc/
77 Meaghan Tobin, "The Revolutionary Chipmaker: Your Guide to TSMC," Rest of World, July 7, 2020, https://restofworld.org/2020/taiwan-chipmaker-guide-to-tsmc/.
78 Becca Wasser, Martijn Rasser, and Hannah Kelley, "When the Chips Are Down: The Geopolitical Implications of Semiconductor Dependence," Center for a New American Security (CNAS), January 27, 2022, https://www.cnas.org/publications/reports/when-the-chips-are-down.
79 Jared M. McKinney & Peter Harris, "Broken Nest: Deterring China from Invading Taiwan," Parameters 51, no. 4 (2021): 23-36, doi:10.55540/0031-1723.3089.
80 Jeanne Whalen, "Russia's War in Ukraine Could Widen the Chip Shortage Crisis," The Washington Post, February 25, 2022, https://www.washingtonpost.com/technology/2022/02/25/ukraine-russia-chips-sanctions-tsmc/.
81 Dan Strumpf, Karen Hao, and Raffaele Huang, "Nvidia Offers Alternative Chip for China to Clear U.S. Export Hurdles," The Wall Street Journal, November 8, 2022, https://www.wsj.com/articles/nvidia-offers-alternative-chip-for-china-to-clear-u-s-export-hurdles-11667891718.
82 Brian Egan, "New U.S. Semiconductor Export Controls Signal Dramatic Shift in Tech Relations with China," Just Security, October 24, 2022, https://www.justsecurity.org/83744/new-us-semiconductor-export-controls-signify-dramatic-shift-in-tech-relations-with-china/.
83 "U.S. Tech Hegemony Is Unpopular" [美国的科技霸权不得人心], China Daily, November 27, 2021, https://cn.chinadaily.com.cn/a/202111/27/WS61a1b7eca3107be4979fa34b.html.
84 Alan Patterson, "TSMC Expansion in Arizona to Target 3-Nm Node," EE Times, November 14, 2022, https://www.eetimes.com/tsmc-expansion-in-arizona-to-target-3-nm-node/.
85 Zeyi Yang, "Chinese Chips will Keep Powering Your Everyday Life," MIT Technology Review, January 4, 2023, https://www.technologyreview.com/2023/01/04/1066136/chinese-legacy-chips-advantage/.
86 "Battered by COVID, China Hits Pause on Giant Chip Spending," Bloomberg, January 4, 2023, https://www.bloomberg.com/news/articles/2023-01-04/battered-by-covid-china-hits-pause-on-giant-chip-spending.
87 Shunsuke Tabeta, "China's 3rd Semiconductor 'Big Fund' Starts Spending $47bn War Chest,"Nikkei, January 7, 2025, https://asia.nikkei.com/Business/Tech/Semiconductors/China-s-3rd-semiconductor-Big-Fund-starts-spending-47bn-war-chest
88 "China's Big Fund Corruption Probe Casts Shadow Over Chip Sector," Financial Times, March 6, 2023, https://www.ft.com/content/8358e81b-f4e7-4bad-bc08-19a77035e1b4.

89 Keith Bradsher, "China Tightens Control Over Critical Minerals," The New York Times, December 9, 2024, https://www.nytimes.com/2024/12/09/business/china-critical-minerals.html.
90 Juliana Liu, "China Proposes Fresh Export Curbs on EV Technology," CNN, January 3, 2025, https://www.cnn.com/2025/01/03/tech/china-ev-tech-export-controls-intl-hnk/index.html.
91 Ashitha Shivaprasad, Amy Lv, and Lewis Jackson, "What Strategic Mineral Exports has China Restricted?" Reuters, April 4, 2025, https://www.reuters.com/world/china/chinas-curbs-exports-strategic-minerals-2025-02-04/.
92 Jeanne Whalen, "Sanctions Forcing Russia to Use Appliance Parts in Military Gear, U.S. says," The Washington Post, May 11, 2022, https://www.washingtonpost.com/technology/2022/05/11/russia-sanctions-effect-military/
93 "FACT SHEET: United States and G7 Partners Impose Severe Costs for Putin's War Against Ukraine," White House, https://www.whitehouse.gov/briefing-room/statements-releases/2022/05/08/fact-sheet-united-states-and-g7-partners-impose-severe-costs-for-putins-war-against-ukraine/
94 Jeanne Whalen, "Sanctions Forcing Russia to Use Appliance Parts in Military Gear, U.S. says," The Washington Post, May 11, 2022, https://www.washingtonpost.com/technology/2022/05/11/russia-sanctions-effect-military/
95 Jeanne Whalen, "Computer Chip Industry Begins Halting Deliveries to Russia in Response to U.S. Sanctions," The Washington Post, February 25, 2022, https://www.washingtonpost.com/technology/2022/02/25/ukraine-russia-chips-sanctions-tsmc/
96 Jeanne Whalen, "Russian Drones Shot Down Over Ukraine Were Full of Western Parts. Can the U.S. Cut Them Off?", The Washington Post, February 11, 2022, https://www.washingtonpost.com/technology/2022/02/11/russian-military-drones-ukraine/
97 Jeanne Whalen, "Sanctions Forcing Russia to Use Appliance Parts in Military Gear, U.S. says."
98 Conflict Armament Research, https://www.conflictarm.com/.
99 "The U.S. Technology Fueling Russia's War in Ukraine," Majority Staff Report, September 10, 2024, https://www.hsgac.senate.gov/wp-content/uploads/09.10.2024-Majority-Staff-Report-The-U.S.-Technology-Fueling-Russias-War-in-Ukraine.pdf
100 Jack Watling and Gary Somerville, "A Methodology for Degrading the Arms of the Russian Federation," Royal United Services Institute, June 26, 2024, https://rusi.org/explore-our-research/publications/occasional-papers/methodology-degrading-arms-russian-federation
101 During the Cold War, the KGB's Directorate T specialized in the illicit acquisition of Western technologies, especially concentrating on dual-use technologies, while the Coordinating Committee (COCOM), a system of export controls arranged among NATO members and Japan, worked to prevent the export of sensitive technologies to the Soviet Union. See for general reference: "Russian KGB Document: SU-0521," Federation of American Scientists, https://irp.fas.org/world/russia/kgb/su0521.htm.
102 "A Review of the President's Fiscal Year 2023 Funding Request for the Department of Commerce," May 11, 2022, https://www.appropriations.senate.gov/hearings/a-review-of-the-presidents-fiscal-year-2023-funding-request-for-the-department-of-commerce?source=email
103 Ellen Nakashima and Jeanne Whalen, "U.S. Prepares Export Controls on Russia to Stifle Economy," The Washington Post, January 23, 2022, https://www.washingtonpost.com/national-security/2022/01/23/russia-ukraine-sanctions-export-controls/.
104 TBC. https://www.bis.gov/press-release/commerce-tightens-export-controls-targets-illicit-procurement-networks-supplying
105 Ellen Nakashima and Jeanne Whalen, "U.S. Prepares Export Controls on Russia to Stifle Economy."
106 "Putin Says Russia Must Cut Dependence on Foreign Technology," AFP, May 20, 2022, https://www.themoscowtimes.com/2022/05/20/putin-says-russia-must-cut-dependence-on-foreign-technology-a77752

107 "Putin Says Russia Must Cut Dependence on Foreign Technology," AFP, May 20, 2022, https://www.themoscowtimes.com/2022/05/20/putin-says-russia-must-cut-dependenceon-foreign-technology-a77752
108 Laura Zhou, "China Will Boost Ties with Russia in Military Tech, Energy and Space, Top Envoy Says," The South China Morning Post, May 22, 2022, https://www.scmp.com/news/china/diplomacy/article/3177093/china-will-boost-ties-russia-military-tech-energy-and-space.
109 Dan Strumpf, "Chinese Tech Giants Quietly Retreat from Doing Business with Russia," The Wall Street Journal, May 6, 2022 https://www.wsj.com/articles/chinese-tech-giants-quietly-stop-doing-business-with-russia-11651845795
110 Dan Strumpf, "Chinese Tech Giants Quietly Retreat from Doing Business with Russia," The Wall Street Journal, May 6, 2022 https://www.wsj.com/articles/chinesetech-giants-quietly-stop-doing-business-with-russia-11651845795
111 Dan Strumpf, "Chinese Drone Giant DJI Suspends Business in Russia, Ukraine," The Wall Street Journal, April 27, 2022, https://www.wsj.com/articles/chinese-drone-giant-dji-suspends-business-in-russia-ukraine-11651058738?mod=article_inline
112 Joe McDonald, "China's Russia Dealings Irk US, But Don't Breach Sanctions," AP, June 1, 2022, https://apnews.com/article/russia-ukraine-putin-china-united-states-8c2466d549f6d082d0295cf0efb635d9
113 See, for instance: Lei Guang et al., "Chinese Talent, American Enterprise: Five Takeaways of How Chinese Talent Contributes to Biotech Innovation in the U.S," China Data Lab, University of California, San Diego, April 2022, https://chinadatalab.ucsd.edu/viz-blog/chinese-talent-american-enterprise-five-takeaways-of-how-chinese-talent-contributes-to-biotech-innovation-in-the-u-s/.
114 See this thoughtful commentary on the topic: Matt Sheehan, "Who Benefits from American AI Research in China?" MacroPolo, October 21, 2019, https://macropolo.org/china-ai-research-resnet/?rp=m.
115 Radu Stochita, "'The Divide between Russian and Global Sciences Is Happening,'" Al Jazeera, April 5, 2022, https://www.aljazeera.com/news/2022/4/5/the-divide-between-russian-and-global-sciences-is-happening
116 Mitch Ambrose, "Ukraine Invasion Triggers Push to Cut Science Ties with Russia," Physics Today, March 30, 2022, https://physicstoday.scitation.org/do/10.1063/PT.6.2.20220330a/full/?cookieSet=1
117 "G7 Science Ministers' Communiqué Frankfurt am Main, 12–14 June 2022," https://assets.publishing.service.gov.uk/government/uploads/system/uploads/attachment_data/file/1082498/g7-science-ministers-communiqu%C3%A9–2022.pdf
118 Philip Martin, "MIT Abandons Russian High-Tech Campus Partnership in Light of Ukraine Invasion," WGBH News, February 25, 2022, https://www.wgbh.org/news/local-news/2022/02/25/mit-abandons-russian-high-tech-campus-partnership-in-light-of-ukraine-invasion.
119 "ExoMars Suspended," European Space Agency (ESA), March 17, 2022, https://www.esa.int/Newsroom/Press_Releases/ExoMars_suspended.
120 Lana Koralova, "Alienated From West, Russia's Scientists Grapple with Working in Isolation," The Moscow Times, January 9, 2024, https://www.themoscowtimes.com/2024/01/09/alienated-from-west-russias-scientists-grapple-with-working-in-isolation-a83483
121 Lin Zhang, Zhe Cao, Gunnar Sivertsen and Dmitry Kochetkov, "Has War in Ukraine Isolated Russian science?", LSE Ideas, May 29, 2024, https://blogs.lse.ac.uk/impactofsocialsciences/2024/05/29/has-war-in-ukraine-isolated-russian-science/
122 Laura Zhou, "China Will Boost Ties with Russia in Military Tech, Energy and Space, Top Envoy Says," South China Morning Post, May 9, 2022, https://www.scmp.com/news/china/diplomacy/article/3177093/china-will-boost-ties-russia-military-tech-energy-and-space
123 He Qisong, "China-Russia Technology Cooperation in Space: Mutually Needed or Mutually Exclusive?" The Pacific Review (2022): 1–30.

124 "Mutual Interest Not Money Key to Russia-China Cooperation in AI, says Putin," TASS, November 20, 2019, https://tass.com/economy/1091195

125 "Russia Seeks Scientific Collaboration with India after Chinese Pause," The Economic Times, April 18, 2021, https://economictimes.indiatimes.com/news/india/russia-seeks-scientific-collaboration-with-india-after-chinese-pause/articleshow/90920784.cms?from=mdr

126 Bendett, S. and Kania E, "The Resilience of Sino-Russian High-Tech Cooperation," War on the Rocks, August 2020, https://warontherocks.com/2020/08/the-resilience-of-sino-russian-high-tech-cooperation/

127 "Russia, China discussing key projects for year of scientific cooperation," TASS, 25 December 2019, https://tass.com/science/1103515

128 "Mutual Interest Not Money Key to Russia-China Cooperation in AI, says Putin," TASS, November 20, 2019, https://tass.com/economy/1091195

129 "Chinese Academy of Sciences and Russian Academy of Sciences Signed Cooperation Agreement" [中国科学院与俄罗斯科学院签署合作协议], Chinese Academy of Sciences, June 10, 2018, http://www.cas.cn/tt/201806/t20180611_4652220.shtml

130 "Chinese Academy of Sciences President: S&T Innovation Cooperation Will Be the Main Line of the Future Two Years of China-Russia Bilateral S&T Cooperation" [中科院院长：科技创新合作将是未来两年中俄两国科技合作的主线], Sputnik News, November 8, 2019, http://sputniknews.cn/russia_china_relations/201911081030011508/

131 "After DeepSeek Leap Forward, Russia's Sberbank Plans Joint AI Research with China," Reuters, February 6, 2025, https://www.reuters.com/technology/artificial-intelligence/after-deepseek-leap-forward-russias-sberbank-plans-joint-ai-research-with-china-2025-02-06/

132 See, for instance: Ash Jain and Matthew Kroenig, "Toward a Democratic Technology Alliance: An Innovation Edge That Favors Freedom," Atlantic Council, June 13, 2022, https://www.atlanticcouncil.org/in-depth-research-reports/report/toward-a-democratic-technology-alliance-an-innovation-edge-that-favors-freedom/. Martijn Rasser, "The Case for an Alliance of Techno-Democracies," Center for a New American Security, October 19, 2021, https://www.cnas.org/publications/commentary/the-case-for-an-alliance-of-techno-democracies

133 "FACT SHEET: Quad Leaders' Tokyo Summit 2022," MAY 23, 2022, https://china.usembassy-china.org.cn/fact-sheet-quad-leaders-tokyo-summit-2022/

134 "FACT SHEET: Quad Leaders' Tokyo Summit 2022," May 23, 2022.

135 Melissa Price, "Australia's Defense Industry Minister: The Importance of AUKUS and a Strong Economy," December 6, 2021, https://www.defensenews.com/outlook/2021/12/06/australias-defense-industry-minister-the-importance-of-aukus-and-a-strong-economy/

136 Kirsty Needham, "AUKUS Needs to Pool Tech Resources as U.S. Edge Erodes, Australian Intelligence Chief Says," Reuters, April 4, 2023, https://www.reuters.com/world/aukus-needs-pool-tech-resources-us-edge-erodes-australian-intelligence-chief-2023-04-04/. Josh Luckenbaugh, "AUKUS Countries Team Up to Develop Key Quantum Capabilities," National Defense Magazine, February 17, 2023, https://www.nationaldefensemagazine.org/articles/2023/2/17/aukus-countries – team-up-to-develop – key-quantum-capabilities.

Bibliography

Barbieri, K. (2002). *The liberal illusion: Does trade promote peace?* University of Michigan Press.

Bendett, S. (2019, September 27). Russia's national AI center is taking shape. *Defense One*. https://www.defenseone.com/technology/2019/09/russias-national-ai-center-taking-shape/160219/

Bendett, S., & Kania E. (2020, August). The resilience of Sino-Russian High-Tech cooperation. *War on the Rocks*. https://warontherocks.com/2020/08/the-resilience-of-sino-russian-high-tech-cooperation/

Bloomberg. (2023, January 4). Battered by COVID, China hits pause on giant chip spending. *Bloomberg*. https://www.bloomberg.com/news/articles/2023-01-04/battered-by-covid-china-hits-pause-on-giant-chip-spending

Brainard, J., & Normile, D. (2021, September 16). China rises to first place for most-cited papers. *Science*. https://www.science.org/content/article/china-rises-first-place-most-cited-papers

Bush, V. (1945). *Science, the endless frontier: A report to the president by Vannevar Bush, Director of the Office of Scientific Research and Development, July 1945*. United States Government Printing Office. https://www.nsf.gov/od/lpa/nsf50/vbush1945.htm

Cheung, T. M. (2022). *Innovate to dominate: The rise of the Chinese techno-security state*. Cornell University Press.

China Daily. (2021, November 27). U.S. S&T hegemony is unpopular [美国的科技霸权不得人心]. https://cn.chinadaily.com.cn/a/202111/27/WS61a1b7eca3107be4979fa34b.html

Ding, J. (2024). The rise and fall of technological leadership: General-purpose technology diffusion and economic power transitions. *International Studies Quarterly*, *68*(2), sqae013.

Drezner, D. W., Farrell, H., & Newman, A. L. (Eds.). (2021). *The uses and abuses of weaponized interdependence*. Brookings Institution Press.

Egan, B. (2022, October 24). New U.S. semiconductor export controls signal dramatic shift in tech relations with China. *Just Security*. https://www.justsecurity.org/83744/new-us-semiconductor-export-controls-signify-dramatic-shift-in-tech-relations-with-china/

Fan, P., & Watanabe, C. (2006). Promoting industrial development through technology policy: Lessons from Japan and China. *Technology in Society*, *28*(3), 303–320.

Farrell, H., & Newman, A. L. (2019). Weaponized interdependence: How global economic networks shape state coercion. *International Security*, *44*(1), 42–79.

Fishman, Edward. (2025). *Chokepoints: American power in the age of economic warfare*. Portfolio.

Ganguli, I. (2014). Scientific brain drain and human capital formation after the end of the Soviet Union. *International Migration*, *52*(5), 95–110.

Hannas, W. C., & Tatlow, D. K. (Eds.). (2020). *China's quest for foreign technology: Beyond espionage*. Routledge.

He, Q. (2022). China-Russia technology cooperation in space: Mutually needed or mutually exclusive? *The Pacific Review*, 1–30.

Huang, Q. (2024, August 23). Strengthen the coordination of innovation resources and organization of forces [加强创新资源统筹和力量组织]. *People's Daily* [人民日报].

Hufbauer, G. C., & Hogan, M. (2022). *CHIPS act will spur US production but not foreclose China* (Peterson Institute for International Economics Policy Brief 22–13). Peterson Institute for International Economics, https://www.piie.com/publications/policy-briefs/2022/chips-act-will-spur-us-production-not-foreclose-china

Jain, A., & Kroenig, M. (2022, June 13). Toward a democratic technology alliance: An innovation edge that favors freedom. *Atlantic Council*. https://www.atlanticcouncil.org/in-depth/report/toward-a-democratic-technology-alliance-an-innovation-edge-that-favors-freedom/

Koralova, L. (2024, January 9). Alienated from West, Russia's scientists grapple with working in isolation. *The Moscow Times*. https://www.themoscowtimes.com/2024/01/09/alienated-from-west-russias-scientists-grapple-with-working-in-isolation-a83483

Lall, S. (1996). *Learning from the Asian tigers: Studies in technology and industrial policy*. Springer.

Majority Staff Report. (2024, September 10). The U.S. technology fueling Russia's war in Ukraine. *U.S. Senate Homeland Security and Governmental Affairs Committee*. https://www.hsgac.senate.gov/wp-content/uploads/09.10.2024-Majority-Staff-Report-The-U.S.-Technology-Fueling-Russias-War-in-Ukraine.pdf

Mervis, J. (2017, March 9). Data check: U.S. government share of basic research funding falls below 50%. *Science*. https://www.science.org/content/article/data-check-us-government-share-basic-research-funding-falls-below-50

Mervis, J. (2022, September 30). NSF hopes big data will finger grantees not reporting foreign support. *Science*. https://www.science.org/content/article/nsf-hopes-big-data-will-finger-grantees-not-reporting-foreign-support

Murphy, B. (2022, May). Chokepoints: China's self-identified strategic technology import dependencies. *Center for Security and Emerging Technology*. https://cset.georgetown.edu/publication/chokepoints/

Naughton, B. (2021). *The rise of China's industrial policy, 1978 to 2020*. Universidad Nacional Autónoma de México, Facultad de Economía.

Needham, K., *Reuters*. (2023, April 4). *AUKUS needs to pool tech resources as U.S. edge erodes, Australian intelligence chief says*. https://www.reuters.com/world/aukus-needs-pool-tech-resources-us-edge-erodes-australian-intelligence-chief-2023-04-04/

O'Mara, M. (2020). *The code: Silicon Valley and the remaking of America*. Penguin.

Porter, M. E. (1990, March/April). The competitive advantage of nations. *Harvard Business Review*. https://hbr.org/1990/03/the-competitive-advantage-of-nations

Rasser, M. (2020, September 9). The case for an alliance of techno-democracies. *Center for a New American Security*. https://www.cnas.org/publications/commentary/the-case-for-an-alliance-of-techno-democracies

Ruffini, P.-B. (2017). What is science diplomacy? In *Science and diplomacy* (pp. 11–26). Springer.

Simonite, T. (2019, March 13). China is catching up to the US in AI research – Fast. *Wired*. https://www.wired.com/story/china-catching-up-us-in-ai-research/

Sheehan, M. (2022, March). Who Benefits from American AI Research in China. *MacroPolo*. https://macropolo.org/china-ai-research-resnet/?rp=m

Strumpf, D., Hao, K., & Huang R. (2022, November 8). Nvidia offers alternative chip for China to clear U.S. export hurdles. *The Wall Street Journal*. https://www.wsj.com/articles/nvidia-offers-alternative-chip-for-china-to-clear-u-s-export-hurdles-11667891718

Stochita, R. (2022, April 5). The divide between Russian and global sciences is happening. *Al Jazeera*. https://www.aljazeera.com/news/2022/4/5/the-divide-between-russian-and-global-sciences-is-happening

Tabeta, S. (2025, January 7). China's 3rd semiconductor 'Big Fund' starts spending $47bn war chest. *Nikkei*. https://asia.nikkei.com/Business/Tech/Semiconductors/China-s-3rd-semiconductor-Big-Fund-starts-spending-47bn-war-chest

TASS. (2019, November 20). *Mutual Interest not money key to Russia-China cooperation in AI, says Putin*. https://tass.com/economy/1091195

Whalen, J. (2022, May 11). Sanctions forcing Russia to use appliance parts in military gear, U.S. says. *The Washington Post*. https://www.washingtonpost.com/technology/2022/05/11/russia-sanctions-effect-military/

Xinhua. (2016, March 13). Xi Jinping: Comprehensively implement an innovation-driven development strategy; promote realizing new leaps in national defense and military construction [习近平：全面实施创新驱动发展战略　推动国防和军队建设实现新跨越]. http://news.xinhuanet.com/politics/2016lh/2016–03/13/c_1118316426.htm

Xinhua. (2016b, May 30). Xi Jinping: Striving to build a global S&T power [习近平：为建设世界科技强国而奋斗]. http://news.xinhuanet.com/politics/2016–05/31/c_1118965169.htm

Xinhua Network Comment: Fight key core technology battles [新华网评：打好关键核心技术攻坚战]. (2020, August 26). *Xinhua*. http://www.xinhuanet.com/2020-08/26/c_1126414617.htm

Zhang, E., & Mo, Y. (2020, August 18). Update: U.S. puts stranglehold on chip sales to Huawei. *Caixin*. https://www.caixinglobal.com/2020-08-18/us-announces-new-curbs-on-huawei-access-to-us-technology-101593815.html

Zhang, L., Cao, Z., Sivertsen, G., & Kochetkov, D. (2024, May 29). Has war in Ukraine isolated Russian science? *Impact of Social Sciences*. https://blogs.lse.ac.uk/impactofsocialsciences/2024/05/29/has-war-in-ukraine-isolated-russian-science/

9 Diaspora Statecraft and Non-Military Warfare

*Audrye Wong**

In January 2022, UK intelligence services issued a security warning about a high-profile British-Chinese lawyer with close links to the Chinese Community Party who had made several political donations and was previously lauded by a former British prime minister. In 2018, the director of the Federal Bureau of Investigation (FBI) publicly described China as posing "not just a whole-of-government threat, but a whole-of-society threat" to the United States. By his account, ethnically Chinese students, professors, and scientists were infiltrating U.S. society and collecting intelligence on behalf of the Chinese government.

Amidst growing concern over authoritarian foreign influence operations, there has been renewed debate over how such governments are attempting to coopt certain groups and individuals to act on behalf of foreign interests. Unsurprisingly, diaspora communities of geopolitical rivals are often not only perceived by host countries as potential unfriendly agents but also viewed by home governments as a natural resource to pursue its political and security interests. Understanding the role of diaspora statecraft has important implications not just for the evolution of non-military warfare and foreign influence operations but also the healthy functioning of democratic systems and multicultural societies.

Diaspora management is also an increasing policy focus for many countries. Since the 1990s, the number of countries with diaspora institutions – including diaspora-focused ministries, departments, or other bureaucratic or political units – has increased substantially, to over half of all UN member states.[1] As will be discussed later, states often view – and attempt to control – diaspora behavior through the lenses of internal security as well as broader foreign policy goals.

In this chapter, I seek to bridge earlier scholarship that has discussed the role of diaspora politics in international relations with other interdisciplinary and sociological studies on diasporic migration and integration. I examine how the diaspora might serve as a tool of broader geopolitical influence, as well as the evolving role of diaspora statecraft in the emerging context of non-military warfare, with a particular eye to the contemporary rise of China as an authoritarian power with global geopolitical ambitions.

What a diaspora constitutes can be a contested subject; it is also a concept that is politically and socially constructed by home countries, host countries, and within diaspora communities themselves. In this chapter, I use diaspora to refer broadly to

DOI: 10.4324/9781003616993-9

emigrant communities, that is, people who have origins from a nation-state different from where they reside. This can include those who hold home state citizenship but live abroad, those who are citizens of the host country but born in the home state and have cultural and linguistic linkages there, or those who are descendants of emigrants from the home state but were born in the host country. There is almost certainly variation within the diaspora on their affinity to the homeland, assimilation into the host country, and their political and social identities. Given the focus of this volume on non-military warfare, I approach these topics through the lens of how a home state attempts to institutionalize and consolidate the diaspora as an instrument of foreign policy.[2]

While the main goal of the chapter is to develop some conceptual frameworks for thinking about diaspora statecraft and the implications for international security, I deeply recognize the important human agency of these communities and incorporate how such individual-level complexities may affect the ability of states to use diasporas as tools of foreign policy. Additionally, not all diaspora activity should be assumed to be "weaponized" by states; in fact, as will be discussed later, an overt approach to diaspora mobilization may well backfire for the home state.

After providing an overview of how diasporas have been discussed in existing international relations scholarship, I theorize the advantages of diaspora statecraft in non-military warfare. I then lay out the diasporic channels of foreign policy influence, as well as how states seek to control the diaspora for political purposes. In tandem, I examine the case of China as a rising authoritarian power and its approaches to diaspora statecraft. Finally, I discuss some challenges and constraints that states might face in using diaspora statecraft.

Diasporas in International Relations

While diaspora politics is not a new field of study in international relations, most scholarship has focused on the political and economic influence of diaspora communities back in their home states. For example, the diaspora – who tend to be foreign-educated or have overseas business experience – are often major sources of remittances, foreign direct investment, and skilled capital, particularly for developing countries.[3] Additionally, diaspora movements can help to consolidate state formation and nation-building processes.[4]

Diaspora communities also matter for home country politics. In fact, a powerful diasporic lobby can even alter homeland policies through their economic clout and overseas political voice, as in the case of the Armenian diaspora pushing Armenian foreign policy toward a more militant anti-Turkish stance. Political parties in democratic home countries also reach out to diaspora communities to gain electoral advantages, by targeting them with political propaganda and mobilizing them (or their in-country family networks) to vote.[5]

Non-democratic states may thus be wary of diaspora activity for these very reasons, seeking to control overseas populations so as to maintain regime stability and prevent dissension. Diaspora can transmit information back home about different political or social norms, including democratic values, that can threaten the home government's rule.[6] Exposure to foreign ideas, for example through educational or business interactions, can counter homeland propaganda and induce anti-regime activities among the diaspora. As a consequence, many authoritarian governments, such as in Morocco and Tunisia, have actively surveilled diaspora communities abroad and punished identified offenders.[7]

But diaspora engagement can also be driven by explicit foreign policy goals – seeking to use the diaspora to improve the home state's reputation, promote its geopolitical interests, or influence host country politics. Overseas citizens can serve as cultural or educational ambassadors, helping to inform and change public perceptions at the grassroots level. This often ties into broader public diplomacy and "soft power" efforts, but in authoritarian contexts can veer into what is sometimes called "sharp power," in which the diaspora is mobilized in more coercive and subversive ways.

In fact, diaspora populations are often instrumentalized for broader strategic objectives. Russia pursues an active "passportization" policy that fast tracks citizenship for Russian speakers as well as strategic non-Russian groups in neighboring states, such as Ukraine and Georgia, as a way to justify Moscow's political interference and even military intervention.[8] However, home governments may discourage diaspora repatriation from host countries where the home state is pursuing revisionist claims, in order to continue legitimizing its extraterritorial policies.[9] For example, Serbia promoted the return and integration of Serb refugees from Croatia and Bosnia and Herzegovina but not Serbs from Kosovo, because of Serbia territorial claims over the latter. India in its early days of independence refused to defend its diaspora's economic assets because it wanted to underscore the principle of national sovereignty over resources; subsequently, it embraced the Indian diaspora to legitimize needed economic reforms amidst globalization.[10]

The priorities and goals of diaspora management can change with a country's shifting objectives and global position. Whereas diaspora communities might have been predominantly seen as a source of capital and knowledge to drive homeland economic development, a rising power might now see the diaspora as a means to expand the home state's geopolitical influence and boost its overseas image.[11] Much scholarship has tended to examine diaspora politics in the context of a "weaker" home state,[12] but the case of China sheds new light on how the diaspora can be potentially marshaled by a powerful homeland for broader geopolitical influence, and as a tool of non-military warfare.

Diaspora Statecraft in Non-Military Warfare

As suggested in the introductory chapter by Käihkö and Jonsson (Chapter 1), warfare is changing. While military force certainly remains an important element of coercive diplomacy, competition for global influence and power has taken on new dimensions and is taking place in new arenas, aided by the spread of new technologies as well as globalized flows of people, information, and capital. This section discusses diaspora statecraft in the context of non-military warfare, including concepts of hybrid warfare, political warfare, and emerging technology. In addition, it examines some features and advantages of diaspora statecraft that lend itself to being used as a tool of geopolitical influence.

First, as in hybrid warfare, diaspora statecraft can be used in tandem with other military or non-military tools. Ethnic groups living across (sometimes tenuous or newly formed) political borders often serve as a source of interstate or intrastate conflict. Host states are suspicious of the political role of these ethnic groups, whereas home states may feel inclined to intervene to ensure the safety of their diasporas abroad. Transnational diaspora networks have provided funding, recruits, and resources that fuel violent civil conflicts.[13]

In that sense, diaspora politics plausibly facilitates potentially interventionist or expansionist designs in the guise of humanitarian goals – protecting the political, cultural, or linguistic rights of diaspora groups in other countries. Russia has often made such pronouncements toward its post-Soviet neighbors, although not frequently followed by actual policy responses. The Turkish government recently started to adopt an expansive definition of its diaspora that goes beyond Turkish citizens to the so-called kin communities encompassing broader Muslim populations and areas in the former Ottoman empire.[14]

Second, diaspora statecraft overlaps considerably with notions of political warfare, which entail the use of political subversion and covert action to support certain groups over others in another country's political system. By definition, diaspora statecraft involves the home state's attempts to shape the attitudes and behavior of diasporic individuals in ways that favor the homeland's strategic interests. The diaspora's position in host countries allows them to exert political voice, alter public discourse, or even change the domestic balance of power. Foreign influence and interference operations can thus be seen as part of political warfare.

Third, new technologies have had an interactive effect with the significance of diaspora statecraft. In fact, technology has provided an additional resource for the implementation and perhaps effectiveness of diaspora statecraft. The transnational nature of the internet and social media has radically altered the information landscape, enabling home states to communicate with diaspora populations more easily (and vice versa). Social media platforms provide new avenues for the home government to disseminate propaganda to a broader audience instead of relying on traditional print media. When needed, diaspora communities can be mobilized quickly by the government or among themselves, for example to gather at a pro-government rally or protest foreign criticism. Digital surveillance and internet monitoring technologies also allow home governments to keep better track of diasporic activity, including anti-regime activity. At the same time, easier access to multiple information sources could limit government monopoly over the flow of ideas and hence complicate efforts to control the diaspora.

Diaspora statecraft can certainly fit into the category of non-military warfare, broadly defined elsewhere in this volume as "non-kinetic tools of statecraft employed as means of strategic coercion and effort, employed with an intensity and purpose such that the target perceives them as significant escalations and/or commensurable to warfare" (see Chapter 1). Diaspora statecraft relies on apparently civilian activity rather than the use of force and is implemented in pursuit of the "home" government's strategic objectives overseas. The escalatory and coercive use of diaspora statecraft by authoritarian regimes such as China and Russia has also garnered the attention of many target countries concerned about malign infiltration and interference. At the same time, as this chapter will show, the modes and intensity of diaspora statecraft can vary considerably; some approaches may be more diffuse and less coercive. The feature of plausible deniability not only lends strategic advantages but also by nature seeks to avoid overt escalation to traditionally recognized levels of warfare, with sometimes intentional efforts to obfuscate recognition by the target. In that respect, diaspora statecraft ventures into broader notions of influence.

As will be discussed below, diaspora statecraft – similar to other non-military means of "warfare" or influence – can be challenging for governments to respond to effectively, because this requires some degree of securitization of elements of civilian activity. It can be hard to identify unambiguously the participants acting as meaningful and intentional proxies for state influence, which in turn makes legitimate and proportional policy responses more difficult.

The growing emphasis on shaping public and elite perceptions as part of geopolitical competition facilitates the use of diaspora statecraft. For a rising power such as China, non-military tools provide a way of consolidating global and regional influence short of escalating to war. Using what Beijing calls "discourse power"[15] to rebut criticisms and improve China's global reputation could help underscore China's growing military and economic clout. In that context, diaspora statecraft can serve as "soft" and "sharp" tools of influence.

On the soft power dimension, diaspora populations are uniquely poised to amplify China's voice in other countries, persuade the broader public of China's benign rise, and lobby elites to better reflect China's interests. By highlighting the human face of a rising power, diaspora statecraft could reassure other countries of the home state's intentions and emphasize the economic and cultural benefits of cooperation. This bolsters a legitimation strategy to achieve greater acceptance of the rising power's newfound geopolitical position.[16]

In the sharp power dimension, diaspora populations could be weaponized as coercive and subversive tools of influence. The diaspora of illiberal and authoritarian regimes are more likely to be vulnerable to such politicization and manipulation. In this reading, diasporic individuals seek to influence political discourse and decision-making processes through more illicit means or without declaring their links to the home state government. Tools of transnational authoritarianism, such as repression and cooptation, serve to keep diaspora populations in line with the home state's interests.

Beyond great power competition, diaspora statecraft also provides an opportunity for relatively weak states to play an outsized geopolitical role and shape a major power's foreign policy. Cross-border movement may occur for a number of reasons, and be either voluntary or forced, but the presence of human capital overseas provides an added resource – and greater international visibility – for a small or weak state who lacks the capacity to marshal extensive military and diplomatic resources. For example, Armenian diaspora in the United States have acted as a strong and well-organized lobbying network that secured favorable diplomatic and aid policies toward their homeland.[17] Similarly, the Cyprus government actively provides special privileges to and builds a sense of national identity among its diaspora in the United Kingdom in order to amplify the Cyprus state's political legitimacy and interests in its conflict with Turkey.[18]

One major advantage of diaspora statecraft as a non-military tool of influence is its plausible deniability. In many cases, home states prefer to portray diaspora activity as being driven by grassroots sentiment as opposed to government direction. During the 2008 war between Georgia and Russia, a U.K.-based Russian embassy official attended a diaspora event in an apparently unofficial capacity and emphasized that this was diaspora-organized, while simultaneously advocating for raising funds in support of South Ossetians.[19] For its part, the Chinese government often describes events such as overseas student protests as being driven by the will and anger of the Chinese people.

Plausible deniability is advantageous for a number of reasons. First, the apparently non-state nature of diasporic activity makes it less threatening and more subtle. This lowers the chance of immediate escalation. As individuals that live in host society and interact with locals, the diaspora are likely seen as more relatable and trustworthy than a foreign government official. This allows them to act as a bridge between home and host countries, helping to win hearts and minds in the targeted host. Longer-term grassroots engagement through cultural community events (such as food and festivals) also present a positive and non-political dimension that help to improve the public image of the home country.

Second, plausible deniability creates uncertainty and makes it harder for host countries to respond appropriately without over-escalating. This has parallels with the so-called military gray zone operations, in which the use of apparently civilian or paramilitary forces constrains the target's ability to respond with outright military force. As with Russia's deployment of "little green men" in Crimea or China's use of Coast Guard and maritime militia to assert its territorial claims, uncertainty over the government's role and the relative lack of equivalent response options allows diaspora statecraft to slip through the cracks more easily. Not all diasporic activity is necessarily driven by the home state, identifying links to foreign governments is inherently challenging given the often informal nature of diaspora-government interactions, and blunt tools to prevent diasporic influence may not be compatible with host country values, particularly in democratic contexts.

Relatedly, diaspora activities are frequently harder to detect because they are carried out in less conventional domains and communication also takes place more privately. This makes it more difficult to identify actors and government intent. For instance, to impede pro-Tibetan protests during the Chinese president's state visit to France in March 2019, Chinese diaspora leaders apparently tracked down a factory manufacturing Tibetan activist T-shirts and bought out all the apparel at a higher price.[20] This was in addition to more visible mobilization actions, such as organizing large crowds to wave Chinese flags on the roadside in support of the president's motorcade.

As such, plausible deniability and uncertainty may be advantageous for the success of diaspora statecraft. Promoting home government interests in the guise of grassroots mobilization could be more influential in host society politics while reducing effective resistance. This is different from the conventional display of military force where public signaling is seen to convey important information about the sender's capabilities and resolve.

In fact, diaspora statecraft may have the greatest geopolitical influence when it is invisible. Exposure of an overt government hand is likely to increase charges of foreign interference and subversion, and decrease receptivity to messaging. In that sense, public mobilization and demonstration of diaspora loyalty incurs political costs, while behind-the-scenes lobbying and networking, especially at the elite level, may be more effective. Publicity may also hinder the use of diasporic resources in the future. Thus, there is a tension between short-term diaspora mobilization that could more effectively send a strong signal (e.g., a street protest), and longer-term cultivation whose outcomes may be more diffuse and indirect (e.g., building broad-ranging relationships with political elites). As will be discussed later in this chapter, more overt diaspora mobilization is likely to lose the advantage of plausible deniability and lead to host country overreaction.

Diasporic Channels of Influence in Host Countries

What are the different ways in which the diaspora can exert political influence in host societies? Conceptualizing diaspora populations as actors in non-military warfare requires greater attention to their relative positions as interest groups in domestic politics, whether in the host or home countries. The influence on foreign and security policies, as well as on host-home relations, stems from their ability to exert political voice in both countries.

Broadly, diaspora influence can be broken down into three mechanisms: (i) agenda setting – influencing what policy issues and ideas get discussed, in particular those that are salient to the homeland; (ii) discourse framing – shaping public and elite discussions in line with the home state's interests and rhetoric; and (iii) political brokers – acting as intermediaries, facilitators, and organizers to link homeland interest groups with those in power in the host country. Diaspora populations can be public diplomacy ambassadors, participate politically (whether as voters, elected representatives, or donors), act as lobbying groups, engage in social movements, or sometimes spy on behalf of the home state.

Perhaps most innocuously, diaspora communities can increase the home state's cultural popularity. Making homeland cuisine more mainstream or organizing community festivals showcases the home country's cultural heritage, usually in a positive light. Less political activities can lay the groundwork for more positive public perceptions of the home state, while also further entrenching the diaspora as members of the host society and making them more trustworthy ambassadors. At the same time, as will be discussed later, these cultural events may also be coopted for the home state's political agenda, especially if diaspora organizations are dominated by pro-government agents.

Diaspora who are citizens in democratic host countries can exercise influence by voting, for instance for political candidates who support pro-homeland policies. Politicians running in districts with large concentrations of diaspora populations must court their votes and hence reflect their political preferences. In cases where these diaspora populations are relatively homogenous and aligned with the homeland, it becomes likely that politicians become more receptive to the home government's policy positions and interests.

Relatedly, diasporic individuals can also run for elected office, whether at the local or national levels. These political representatives have a larger platform and position of power with which to promote pro-homeland interests and exert more direct policy influence. Such influence can range from making public statements and introducing legislation that echo home government rhetoric to raising attention to specific issues and consolidating support among other politicians.

Additionally, diaspora can serve as political brokers and advisors to politicians, helping to organize campaign outreach events and providing talking points on relevant issues. Such brokers can play a particularly influential role if diaspora populations have recently emerged as new constituencies, and political candidates have relatively little knowledge of how to connect with these potential voters and donors, or have relatively limited background on diaspora-related issues.

Another important avenue of political participation is through lobbying. Diaspora groups can act as ethnic lobbies, seeking to persuade political elites and policymakers of the importance of homeland policy interests. Lobbying influence can also come through economic clout, whether as members of the business elite or as donors – groups that often have the ready ear of politicians. Major donors, lobbyists, and business leaders are granted access to top-level leaders through personal meetings, fundraisers, and other political events. This can give them (diasporic or not) disproportionate influence and voice in raising issues to the attention of political elites. Moreover, recent research suggests that "social lobbying" – lobbying outside of a formal office, such as in a bar or restaurant – makes elites more receptive to interest group messages.[21] The cultural context of diaspora statecraft, where lobbying easily takes place at community events or over dinners, could thus facilitate even greater potential influence.

As an example of diaspora political participation, there has been increased concern over the political influence of Chinese diaspora in Australia and New Zealand.[22] Chinese Communist Party (CCP)-linked Chinese businessmen have been significant campaign donors, meeting both national and state-level leaders, placing political advisors for Australian politicians, and shaping public elite statements on controversial issues such as Tibet and the South China Sea. In New Zealand, an ethnic Chinese MP was forced to resign after he was found to be a CCP member and had links to Chinese intelligence. Other evidence suggests that Chinese government lobbying makes U.S. legislators more likely to sponsor legislation favorable to Chinese interests and reduces U.S. media coverage of political tensions and threats from China.[23]

Diaspora mobilization can also take more publicly disruptive forms, such as rallies and protests. In the run-up to the 2008 Summer Olympics, Beijing successfully mobilized overseas Chinese to attend Olympic torch relays and wave the national flag, to counter protests around China's human rights violations. In the last few years, Chinese university students in the United States and Canada have protested against having the Dalai Lama as commencement speaker, protested against a campus talk given by a Uyghur activist, and repeatedly vandalized a "Lennon wall" of messages supporting Hong Kong pro-democracy protesters. For such social mobilization tactics, diaspora are more likely to have recently emigrated and maintain stronger personal and political ties with the home country.

Finally, diaspora statecraft can involve using the diaspora to acquire classified information and technology, that is espionage. Home government officials approach and cultivate specific members of the diaspora, capitalizing on their cultural or ideological affinities and offering economic benefits in exchange for the acquisition of internal government information, proprietary technologies, or technologies with military applications. While this is a common concern, it should also be noted that governments do not always have a good track record of identifying such incidents. The United States has seen a number of cases where Chinese Americans or ethnically Chinese individuals have been accused of spying for the Chinese government, despite a lack of evidence.

While these channels of influence have been discussed mostly in the context of diaspora acting on behalf of the home government, they can apply equally to diaspora seeking to shape the host country's foreign policy toward the homeland but not in ways that are aligned with the home regime. Home countries with substantial populations of dissident diasporas may find that diaspora activism in fact elevates international scrutiny and criticism of the home government.

Tools to Influence the Diaspora

Home countries adopt a range of material and ideational strategies, as well as a mix of sticks and carrots, to shape diaspora behavior. The most obvious form of diaspora control involves repression through a range of intimidation and coercive tactics, most often implemented by authoritarian regimes.[24] This can include surveillance and monitoring of activities, direct threats from government officials, coercion-by-proxy – targeting family members back home, forced return or disappearances, and assassination.

Diaspora engagement can also take more positive forms, seeking to coopt diaspora into acting on behalf of homeland interests. Patronage strategies include providing high-level political connections that can aid career or business opportunities, funding overseas study, or dangling direct financial benefits in exchange for activities such as espionage.[25]

On the ideational front, sending states can seek to legitimate their position and inculcate patriotic sentiments among diaspora communities. Cultural activities help to foster a sense of belonging with the home country, while government-sponsored trips to the homeland aim to strengthen political and cultural linkages and showcase successes of the home country – and at times push government narratives on politically controversial issues.[26] One of the most prominent examples is Israel's free birthright trips for American Jews.[27]

With the rise of global communications technology and social media, controlling the information environment of diaspora communities has also become a prominent tactic. Home governments can spread propaganda that attack host countries, praise the home government, and aim to drive a wedge between diaspora and host countries.[28] Sending states can also seek to limit what kinds of information and narratives diaspora populations are exposed to, by taking financial control of diaspora media outlets and harassing those outlets that are critical of the home regime. For example, the Chinese government and CCP-linked business actors own virtually all overseas Chinese media in Australia, by extension perpetuating its domestic propaganda and censorship apparatus and leaving little room for independent reporting.

Home governments have developed both formal and informal institutions for diaspora engagement. Many countries with large populations abroad have bureaucratic units, whether at the department or ministry level, dedicated to managing relations with the diaspora. China has traditionally managed diaspora affairs through the Overseas Chinese Affairs Office, although with shifting geopolitical aims (discussed below) more outward-facing bureaucracies have become more important.

Embassies and consulates in host countries – as an extension arm of the government – are also an important player in diaspora outreach and mobilization. As often the dominant representative and intermediary for diasporic access to citizen services, they are well-placed to coerce diaspora populations, monitor their activities, provide political backing, or mobilize action.

In many cases, diaspora organizations at the grassroots level are active in coordination and outreach, within the diaspora community and with more official government institutions. They may serve a variety of functions, from connecting diasporic members to their hometowns to organizing community events and facilitating business opportunities. These community organizations may vary in their degree of interactions with the home government, which can be seen as a source of financial and political support; some organizations, however, disavow official involvement to assert their independence as representatives of the diaspora.

Lines between official and grassroots are sometimes blurred. For instance, the Federation of Returned Overseas Chinese has been acting as a grassroots organization with the responsibility of communicating with the diaspora, in order to avoid host country suspicions of government interference and espionage.[29] Beijing has also set up ostensibly apolitical agencies to operate overseas, such as Friendship Associations and Reunification Promotion Associations, despite their links to the government.

Many Chinese diaspora organizations today are mostly dominated by CCP-affiliated individuals. This imbalance of power grew from a mix of the coercive tools – using threats and repression against dissenting voices – and positive incentives – coopting overseas Chinese eager for economic resources and political connections – described in this section. This trend has also complicated host countries' abilities to identify those who are acting on behalf of the Chinese government, as almost any diasporic individual will have inevitably had contact with CCP-linked representatives and organizations simply as a function of staying plugged into the community.

China's Policies of Diaspora Engagement and Mobilization

China presents an important case to understand the use of diaspora as non-military statecraft. First, as a rising power in an era where economic flows, information exchange, and human movement are perhaps more prominent than military force, diaspora mobilization presents a potential tool of expanding geopolitical influence at the intersection of these trends. Second, China has been a source of large-scale historical as well as contemporary overseas migration. Previous waves of emigrants moved for better economic opportunities or fled for political reasons, and may have mixed loyalties to the home regime (many are also from Taiwan or Hong Kong); more recent waves have been driven by economic growth and a new middle class, leading to more businesspeople and students with closer links to the Mainland.

Third, China's strong state capacity and propaganda apparatus provide a good indicator of what extensive diaspora mobilization can entail. Fourth, the authoritarian nature of China's political system sheds light on the export of such illiberal techniques, with implications for understanding the new terrain on which non-military statecraft might be conducted and by what rules. Taken together, these characteristics suggest that the Chinese government has ample motivation (domestic and foreign policy goals), opportunity (relatively receptive diaspora targets), and means (relatively well-developed institutional capacity, transnational authoritarian tools, and reduced dependence on diasporic resources) for diaspora engagement.[30]

Diaspora engagement, particularly in the present day, is deeply intertwined with a broader system of political control called the United Front. The United Front system consists of a coalition of government organizations, affiliated groups, and individuals that seeks to silence critics and mobilize allies of the CCP. Such activities take place not only within China but also well beyond China's geographic borders, from monitoring the activities of political dissidents abroad to courting foreign media and government elites.[31]

Unsurprisingly, Chinese diaspora communities are a major target of United Front work (along with other groups, such as entrepreneurs, ethnic minorities, and religious leaders).[32] From Beijing's perspective, their increased exposure to foreign ideas poses a threat to the CCP's domestic rule and calls for overseas propaganda and control – to rally patriotism and stamp out criticism. As with many other countries, diaspora engagement has been viewed in terms of consolidating government rule and internal stability.

This can be seen in the many ways the Chinese government has engaged with the diaspora over time. Overseas Chinese leaders and resources were key in the revolution leading to the fall of the Qing Dynasty in the early 20th century. In the aftermath of the Chinese civil war, the CCP and the KMT (Kuomintang) competed for diaspora loyalty to legitimize their claims to rule China, using ideological campaigns, economic incentives, and educational assistance. During the PRC's drive for economic modernization from the late 1970s, Chinese diaspora were courted as sources of investment and encouraged to return home. After the 1989 Tiananmen Square massacre, Beijing redoubled propaganda efforts to win over diaspora populations and promoted Chinese nationalism as a rallying force.[33] Instilling a sense of patriotism builds diaspora loyalty while constraining anti-CCP or pro-democracy movements that can endanger its grip on power.

Since the 2010s, however, China's diaspora policies have increasingly shifted from consolidating material support for internal matters – namely economic development and national unification – to managing the diaspora as a political means of expanding China's overseas influence.[34] This geopolitical stance is different from in the past, when Beijing did not actively protect overseas Chinese from discriminatory and nationalistic appropriation policies, and even renounced diaspora citizenship claims, in order to gain strategic allies in Southeast Asia during the Cold War.[35] In 2017, China's top diplomat Yang Jiechi called for new diaspora policies to serve China's overseas interests and consolidate China's growing global influence.[36] In the last several years, Chinese president Xi Jinping made several speeches calling for overseas Chinese students to serve as grassroots ambassadors, promote positive narratives about China, work more closely with embassies and consulates, and operate in line with China's diplomatic goals, such as the Belt and Road Initiative.[37] This strategic reorientation has also been reflected in China's diaspora institutions, with more overseas-facing bureaucrats and agencies assuming greater political power.[38] Such public rhetoric and policies have contributed to growing fears in host countries of a "weaponized" Chinese diaspora.

Whether for domestic political or foreign policy reasons, the Chinese government's messaging about the diaspora aims to blur the lines between Chinese nationals and those of ethnic Chinese descent.[39] By pushing a particular conception of an "overseas Chinese" as having an inevitable affinity and belonging to the homeland (read: the Party) regardless of their individual context, Beijing's policies intentionally homogenize and instrumentalize its diaspora communities.

The changing demographic of overseas Chinese populations – with recent migrants from the Mainland becoming more numerous – has also altered the dynamics of diaspora-homeland interactions. Chinese students have on occasion been vocal protestors and defenders of Beijing's policies, Chinese businesspeople have sometimes been prominent political donors, and diaspora organizations are increasingly dominated by CCP-affiliated individuals.

While the CCP has not hesitated to apply coercive tactics toward regime critics – in recent years, Beijing has kidnapped a Swedish-Chinese publisher and detained a Chinese-Australian journalist and Chinese-Australian writer on charges of espionage – government policy documents generally outline an approach of influencing diaspora populations through a subtle "guiding hand."[40] For example, the Chinese government uses a mix of patriotic propaganda, cultural outreach, state-sponsored programs (e.g., homeland tours in China), state-affiliated grassroots organizations, and the lure of political connections to engage with the diaspora. Anecdotal evidence suggests that Chinese consulates are in frequent contact with the many university-based Chinese Students and Scholars Associations in the United States.

Geopolitical Implications of Diaspora Statecraft

The mobilization of diaspora as tools of statecraft has potentially significant geopolitical consequences. Diaspora-driven activity can shift the balance of military and economic power. Successful espionage allows the transfer of technology that can boost the home state's military or economic capabilities, while transmitting politically sensitive information will likely undermine the security of host country personnel and strategic assets. On the other hand, the "soft power" approach to diaspora statecraft may promote the global diffusion of home country ideas and values. Improved public perceptions could boost the home country's reputation and contribute to broader legitimation and acceptance of its economic and military power.

Most fundamentally, diaspora statecraft can change the foreign policies of host countries. The conservative Cuban-American lobby has played a major role in limiting any U.S. opening to Cuba.[41] The pro-Israel lobby is seen as a powerful influence over U.S. policies in the Middle East (although there are also well-documented internal cleavages in the American Jewish diaspora).[42] Lobbying by the Indian diaspora helped to increase American support for Indian independence and, more recently, pushed Congressional support for the 2006 U.S.-India nuclear agreement.[43]

At the same time, diaspora statecraft as carried out by illiberal governments can have quite insidious effects. Mobilizing the diaspora for political purposes requires imposing a collective, homogenized narrative of diaspora-homeland relations, while silencing any dissent and criticism of home state policies. Diaspora statecraft may thus disproportionately amplify pro-government positions while drowning out alternative viewpoints. In the China case, this also feeds into broader overseas propaganda campaigns, where the Chinese government seeks to reclaim the upper hand in global narratives, defend its policies, and attack critics. As such, diaspora statecraft works in tandem with other informational tools and even as an extension of state-led propaganda, spreading and reinforcing Beijing's desired narratives.

As a result, host country actors may have the mistaken impression that the apparently dominant narrative is the homeland-propagated one. This distorts not just the representation of interests among the diaspora but also the perceived political incentives surrounding a particular issue, for example that a candidate cannot criticize the Uyghur genocide or support Taiwanese independence for fear of losing votes and donations. Universities may be more reluctant to host dissident speakers or politically sensitive events, having encountered public opposition and protests from student organizations. Diaspora mobilization pressures could also intensify the home government's perceived coercive clout. This may have a serious impact on public and elite discourse in the host country, leading to heightened self-censorship or more pro-homeland policies.

Beyond foreign policy impacts, such statecraft also adversely affects the healthy functioning of democratic political systems, including the liberties of diaspora as members of the host country. Those who do not agree with homeland policies are bullied into silence and criticized for their lack of loyalty, even while facing greater suspicion from the host country. Moreover, policies that seek to divide diaspora populations from their host countries exacerbate broader ethnic and social tensions. This extraterritorial authoritarian reach has implications for the transparency and integrity of democratic processes, as well as individual-level political and social freedoms.

Open democratic systems, being relatively permeable to a range of voices, tend to be more vulnerable to coercive or authoritarian uses of diaspora populations. Because it does not play by the same rules, authoritarian diaspora statecraft projects not only more influence but also undermines the democratic host country's own ability to respond and compete. Such consequences are even more worrying in the context of great power competition, where the loyalties of diaspora groups may be increasingly called (fairly or unfairly) into question. The currently perceived ideological contest between the United States and China, with each country attempting to demonstrate the superiority of its political model, has contributed to the weaponization of the Chinese diaspora. While Beijing sees diaspora statecraft (in tandem with informational campaigns) as key to ensuring internal loyalty and increasing geopolitical influence, Washington views the diaspora as a source of foreign influence and a tool of the Chinese state.

In the context of global trends in democratic backsliding, diaspora statecraft can also bolster illiberalism back in the home country. Successfully cultivating a sense of solidarity and patriotism among the diaspora – almost a prerequisite for mobilization – strengthens regime support and stability. The cooptation of diasporas through patronage or other special privileges turns them into status quo-preferring elites rather than political revolutionaries, further entrenching those in power.[44] Some have argued that remittances as unearned income can strengthen authoritarian rule, although others have found that remittances provide more resources for protests.[45]

Challenges and Constraints of Diaspora Statecraft

This is not to say that homeland governments are always easily able to manipulate diaspora communities as tools of influence. The heterogeneity of goals and actors within the home state as well as diaspora populations point to a complicated picture.[46] Effective diaspora mobilization is more likely with unified goals and close intergovernmental coordination. Even in China, lower-level diaspora officials may be more focused on capitalizing on economic resources from overseas Chinese rather than national-level geopolitical goals of expanding China's global influence.[47] Diaspora organizations also partake in local-level rivalries and political infighting, as in the case of Russian diaspora in the United Kingdom, making it harder for government officials to impose a standard line of action.[48]

There are also other constraints on home countries' ability to mobilize their diaspora. In terms of capacity, governments need economic and institutional resources to reach out to populations beyond their borders – funding and organizing community events or subsidizing educational and language materials may be beyond the ability of less-developed economies. In some cases, the wealthier diaspora of a weak state may be the tail wagging the dog, should they have different political agendas. In terms of motives, competing policy priorities affect states' interest in their diasporas, for instance concerns that diaspora engagement could destabilize ruling elites or encourage foreign intervention over internal ethnic conflicts.[49]

Importantly, diasporic resistance also matters. Diaspora communities themselves are not passive or monolithic agents. The notion of "diaspora" is often a political construct defined by the home government's interests and priorities.[50] For instance, Beijing chooses to treat all individuals who are ethnically Chinese, regardless of whether they have active ties to the Mainland, as members of the

Chinese diaspora. Naturally, this narrative encounters resistance from diaspora communities who may be from Taiwan or Hong Kong (both places with high levels of political contestation with Beijing), who fled China for political reasons, or who are emigrant descendants who were born and grew up in their host society.

Heterogeneity within diaspora communities means that mobilization efforts are likely to have varied impacts. Propaganda attempts to inculcate a sense of loyalty to the homeland may not always be effective. Individuals who are more integrated into host societies, bring their own political, social, and cultural experiences, or who are of later generations may be less persuaded by home government discourse and more inclined to challenge it.[51] In that regard, Chinese students or businesspeople who emigrated recently are likely to be more easily mobilized by the Chinese government, while long-standing overseas Chinese communities may feel much less attachment to Beijing. In places where they tend to live and do business within their ethnic communities, Chinese migrants in fact become more nationalistic and identify more with the Chinese state.[52]

Additionally, not all diaspora who appear to be agents of the home government are driven by loyalty. The need for economic or political resources from the home country, such as finding employment or maintaining political connections, can drive alignment with the home state and public display of state-driven narratives of homeland identity.[53] The psychology of status may also come into play: migrants who are courted by the home government now have elevated prominence in their host and home communities, making them feel more important and motivated to promote the homeland's interests. Additionally, diaspora groups may compete for financial and social resources from home governments in order to pursue their own projects.[54] Overseas Chinese students often participate in homeland tours for future career benefits or simply because it is a free social opportunity, and many remain largely indifferent to government propaganda efforts.[55] Finally, diaspora may be coerced, intimidated, or otherwise pressured into conformity by illiberal home governments.

The Potential Paradox of Diaspora Mobilization

Finally, there is the additional question of whether diaspora statecraft is effective at achieving geopolitical influence, conditional on successful diaspora mobilization. In other ongoing work, I argue that there is a double-edged sword: diaspora who are more easily mobilized are less likely to be integrated into host society and more likely to be marginalized; by extension, this makes them less politically influential for China's foreign policy purposes.

Relatedly, I suggest that paradoxically, active diaspora outreach and mobilization tends to raise the hackles of host countries, given the sensitive issue of foreign interference undermining national sovereignty. In particular, the Chinese government's attempt to homogenize its diaspora as an extension of China has sparked significant blowback. Diaspora statecraft can empower anti-cosmopolitan and hawkish elements in host country politics, including those skeptical of ethnic diversity. The heightened backlash to Beijing's heavy-handed and sometimes subversive approach bears some parallels with responses to China's economic statecraft.[56]

Difficulties in distinguishing between different elements within diaspora communities facilitate over-reactive policies. In the United States, the Justice Department's China Initiative along with previous FBI investigations have often targeted Chinese scientists or those of Chinese descent only to have cases fall apart on the lack of evidence, leading to charges of racial profiling. In Australia, China's perceived foreign influence activities have led to very strong elite and societal reactions, again casting the diaspora in a suspicious light and contributing to a much more hawkish turn in Australia's foreign policy toward China.

As a result, China's ability and desire to engage with its diaspora on a large-scale may have in fact undermined their position in host society and hence any potential influence. This threatens to marginalize diaspora communities economically, socially, and politically, making them victims rather than empowering them as agents of influence. This has happened across a range of host countries. During the Cold War, Indonesian elites tended to see ethnic Chinese as a monolithic group, despite major variations in ideology and socioeconomic status. Anti-Communist elites portrayed internal dissent as instigated by Beijing in order to justify domestic purges. Ethnic Chinese continue to be regarded with suspicion and have often been the target of communal violence in Indonesia.[57]

At the same time, host country overreaction, including the singling out of diaspora groups as untrustworthy or outsiders, could plausibly drive previously divided communities to consolidate their diasporic identity and advocate on behalf of their homeland.[58] Research has found that experiencing discrimination increases diaspora support for authoritarian rule back home, even among groups that tend to be more liberal-minded.[59] For liberal host countries to respond effectively to authoritarian diaspora statecraft, policies to prevent Beijing's targeting of the Chinese diaspora also need to avoid sowing further ethnic divisions and feeding into CCP narratives. Engaging and highlighting the diversity within diaspora communities complicates China's attempts to create a uniform diasporic narrative or utilize overseas Chinese as instruments of foreign policy. Working strategically with diaspora communities also makes host societies more resilient to continued efforts at foreign interference.

Conclusion

Diaspora statecraft is emerging as a non-military tool of geopolitical influence. Its plausible deniability as a government actor has made it less immediately threatening, harder to detect, and harder to respond to with existing tools, thus increasing the overall uncertainty in the realm of geopolitical competition. Governments seeking to bolster domestic legitimacy or promote foreign policy interests have a range of material and ideational tools to engage with diaspora communities. Having linkages to both the homeland and host country, diasporic individuals can participate in host country political processes, whether through voting, lobbying, or protesting. They can help to set the policy agenda, frame public and elite discourse, and influence policy choices. China's renewed efforts at mobilizing the diaspora demonstrate the potential significance of diaspora statecraft in geopolitical competition.

Moreover, the illiberal elements of diaspora statecraft can also undermine the integrity and diversity of democratic host countries, while constricting the freedoms of diasporic individuals. Diaspora communities can be both agents and victims.

This suggests several avenues for future work. How successful has diaspora statecraft actually been at achieving influence across a range of outcomes? The success of diaspora statecraft can also be hard to measure. Depending in part on the home government's articulated goals, success could entail a specific policy change in line with homeland interests, or a broader outcome of more favorable relations between the home and host countries. Successful diaspora statecraft also entails two stepwise components: an ability to mobilize the diaspora and an ability to influence political outcomes. Mobilizing diaspora populations as lobby groups points to the effectual use of diasporas as established constituencies in the host country's political system. One major challenge in evaluating the outcomes of diaspora statecraft is the latency of success – effective diaspora cultivation lays the foundation for future mobilization during a specific crisis or event, but such political payoffs (or lack thereof) may not be evident during "peacetime."

Disaggregating homeland-diaspora relations, host-diaspora interactions, and within-diaspora heterogeneity can help to explain when diaspora mobilization might be more or less effective as a foreign policy tool. At minimum, home governments need the capacity – coercive or persuasive – to control diaspora attitudes and behavior, and an alignment of diaspora and homeland political agendas. Extensive diaspora-facing institutions and targeted information channels (or even propaganda apparatuses) serve to monitor and interact with overseas communities in ways that align with homeland objectives. In that respect, diverse diaspora populations – who vary in the timing of migration, place of origin, language use, political orientation, or other characteristics – are more likely to have differing preferences and priorities. For example, political dissidents or refugees would tend to be more skeptical of the home government's efforts than recent migrants who come for business opportunities. The diasporas' relationship with the host country also matters – integration, assimilation, or interethnic mixing will tend to reduce homeland affinity and raise barriers to mobilization.

In addition, when might diaspora statecraft complement or undermine the effects of other tools of statecraft, and what determines whether a country is more likely to use diaspora statecraft relative to other military or non-military tools? Diaspora are not the only targets or agents of foreign influence operations. In the context of broader political warfare, host country political and business elites are also often coopted, sometimes as "useful idiots." Diaspora mobilization in tandem with informational and economic tools could together underscore the depth and scope of a country's influence for target countries.

More generally, what might the study of diaspora statecraft tell us about the utility of non-military means? Certainly, enlisting civilians to promote state goals opens up the range of tools of geopolitical influence, particularly in an era where military warfare has become less common and less useful. While diaspora communities may certainly be mobilized strategically to aid their home government and undermine the enemy during an active military conflict, diaspora statecraft is arguably gaining salience as a non-military tool in the context of peacetime as well as longer-term antagonistic competition. This is certainly the case for China's foreign policy strategy, as a rising power and major geopolitical rival of the United States. In that regard, diaspora statecraft can be seen as an alternative non-military means of shaping political discourse and policy attitudes to gain broader strategic influence in other countries. Compared to traditional military action, participants in diaspora statecraft are harder to detect and identify, making proportionate and legitimate policy responses more challenging.

Notes

* Assistant Professor of Political Science and International Relations, University of Southern California; Jeane Kirkpatrick Fellow, American Enterprise Institute. Contact: audryewong@gmail.com.

1 Alan Gamlen, Michael E. Cummings and Paul M. Vaaler, "Explaining the Rise of Diaspora Institutions," Journal of Ethnic and Migration Studies 45, no. 5 (2019): 492–516.

2 See also Latha Varadarajan, The Domestic Abroad: Diasporas in International Relations (Oxford University Press, 2010).

3 Benjamin A. T. Graham, Investing in the Homeland: Migration, Social Ties, and Foreign Firms (University of Michigan Press, 2019); David Zweig and Huiyao Wang, "Can China Bring Back the Best? The Communist Party Organizes China's Search for Talent," The China Quarterly 215 (September 2013): 590–615.

4 Gamlen et al., 2019; Fiona B. Adamson and Gerasimos Tsourapas, "The Migration State in the Global South: Nationalizing, Developmental, and Neoliberal Models of Migration Management," International Migration Review 54, no. 3 (2020): 853–82.

5 Michael Ahn Paarlberg, "Competing for the Diaspora's Influence at Home: Party Structure and Transnational Campaign Activity in El Salvador," Journal of Ethnic and Migration Studies 45, no. 4 (2019): 539–60; Michael Ahn Paarlberg, "Transnational Militancy: Diaspora Influence Over Electoral Activity in Latin America," Comparative Politics 49, no. 4 (2017): 541–59.

6 Clarisa Pérez-Armendáriz, "Cross-Border Discussions and Political Behavior in Migrant-Sending Countries," Studies in Comparative International Development 49, no. 1 (2014): 67–88.

7 Laurie A. Brand, Citizens Abroad: Emigration and the State in the Middle East and North Africa (Cambridge University Press, 2006).

8 See e.g. Neil Melvin, "Russia's Policy of Passport Proliferation," RUSI, May 1, 2020, https://rusi.org/explore-our-research/publications/commentary/russias-policy-passport-proliferation.

9 Harris Mylonas and Marko Zilovic, "Foreign Policy Priorities and Ethnic Return Migration Policies: Group-Level Variation in Greece and Serbia," Journal of Ethnic and Migration Studies 45, no. 4 (2019): 613–35.

10 Latha Varadarajan, The Domestic Abroad: Diasporas in International Relations (Oxford University Press, 2010).

11 Jiaqi M. Liu, "From 'Sea Turtles' to 'Grassroots Ambassadors': The Chinese Politics of Outbound Student Migration," International Migration Review 56, no. 3 (2022): 702–26.

12 Adamson & Tsourapas, 2020.

13 Fiona B. Adamson, "Crossing Borders: International Migration and National Security," International Security 31, no. 1 (2006): 165–99.

14 Bahar Baser and Ahmet Erdi Ozturk, "Positive and Negative Diaspora Governance in Context: From Public Diplomacy to Transnational Authoritarianism," Middle East Critique 29, no. 3 (2020): 319–34.

15 Nadège Rolland, China's Vision for a New World Order (National Bureau of Asian Research, 2020).
16 Stacie E. Goddard, When Right Makes Might: Rising Powers and World Order (Cornell University Press, 2018).
17 Yossi Shain and Aharon Barth, "Diasporas and International Relations Theory," International Organization 57, no. 3 (2003): 449–79.
18 Fiona B. Adamson and Madeleine Demetriou, "Reshaping the Boundaries of 'State' and 'National Identity': Incorporating Diasporas into IR Theorizing," European Journal of International Relations 13, no. 4 (2007): 489–526.
19 Andy Byford, "The Russian Diaspora in International Relations: 'Compatriots' in Britain," Europe-Asia Studies 64, no. 4 (2012): 715–35.
20 Jiaqi M. Liu, "When Diaspora Politics Meet Global Ambitions: Diaspora Institutions Amid China's Geopolitical Transformations," International Migration Review 56, no. 4 (2022): 1255–79.
21 Christian R. Grose, Pamela Lopez, Sara Sadhwani and Antoine Yoshinaka, "Social Lobbying," The Journal of Politics 84, no. 1 (2022).
22 See e.g. Anne-Marie Brady, Magic Weapons: China's Political Influence Activities Under Xi Jinping, Vol. 18 (Washington, DC: Wilson Center, 2017); Audrye Wong, "Peddling or Persuading: China's Economic Statecraft in Australia," Journal of East Asian Studies 21, no. 2 (2021): 283–304.
23 Erin Baggott Carter, Statement for the Record before the U.S.-China Economic and Security Review Commission," March 23, 2023, https://www.uscc.gov/sites/default/files/2023-03/Erin_Baggott_Carter_Statement_for_the_Record.pdf.
24 Gerasimos Tsourapas, "Global Autocracies: Strategies of Transnational Repression, Legitimation, and Co-optation in World Politics," International Studies Review 23, no. 3 (2021): 616–44.
25 Tsourapas, 2021.
26 See e.g. Liu, 2022a; Rilke Mahieu, "'We're Not Coming from Mars; We Know How Things Work in Morocco!' How Diasporic Moroccan Youth Resists Political Socialization in State-Led Homeland Tours," Journal of Ethnic and Migration Studies 45, no. 4 (2019): 674–91.
27 Shaul Kelner, Tours That Bind: Diaspora, Pilgrimage, and Israeli Birthright Tourism (NYU Press, 2010).
28 Patrick J. Chester and Audrye Wong, "Wedge Narratives and Diaspora Communities," Security Studies (2025): 1–52.
29 Liu, 2022b.
30 Shain & Barth, 2003.
31 Paul Charon and Jean-Baptiste Jeangène Vilmer, "Chinese Influence Operations: A Machiavellian Moment," IRSEM, October 2021; Alex Joske, "The Party Speaks for You," Australian Strategic Policy Institute 9 (2020).
32 To, James Jiann Hua. Qiaowu: Extra-Territorial Policies for the Overseas Chinese (Brill, 2014).
33 See e.g. Enze Han, "Bifurcated Homeland and Diaspora Politics in China and Taiwan Towards the Overseas Chinese in Southeast Asia," Journal of Ethnic and Migration Studies 45, no. 4 (2019): 577–94; To, 2014; Zhou, 2019.
34 Liu, 2022b.
35 Taomo Zhou, Migration in the Time of Revolution (Cornell University Press, 2019).
36 Yang Jiechi, "当好贴心人 成为实干家 凝聚侨心侨力同圆共享中国梦 ——深入学习贯彻习近平总书记关于侨务工作的重要指示 (Be a close friend, be a person of action, bring together diaspora hearts and strength to fully share in the China Dream – deeply learn and implement General Secretary Xi Jinping's important instructions on overseas Chinese affairs," Overseas Chinese Affairs Study No. 3 (2017), http://qwgzyj.gqb.gov.cn/tbbd/195/2914.shtml.
37 See Liu, 2022a.

38 Liu, 2022b.
39 Charon and Jeangène Vilmer, 2021.
40 To, 2014.
41 William M. LeoGrande, "Pushing on an Open Door? Ethnic Foreign Policy Lobbies and the Cuban American Case," Foreign Policy Analysis 16 (2020): 438–56.
42 John J. Mearsheimer and Stephen M. Walt, The Israel Lobby and U.S. Foreign Policy (Blackwell Publishing, 2007); Dov Waxman, Trouble in the Tribe: The American Jewish Conflict Over Israel (Princeton University Press, 2016).
43 Dinyar Patel, " 'One Man Lobby'? Propaganda, Nationalism in the Diaspora, and the India League of America During the Second World War," The Journal of Imperial and Commonwealth History 49, no. 6 (2021): 1110–40; Jason A. Kirk, "Indian-Americans and the U.S.-India Nuclear Agreement: Consolidation of an Ethnic Lobby?" Foreign Policy Analysis 4, no. 3 (July 2008): 275–300.
44 Brand, 2006; Laurie A. Brand, "Authoritarian States and Voting from Abroad: North African Experiences," Comparative Politics 43, no. 1 (2010): 81–99.
45 Faisal Z. Ahmed, "The Perils of Unearned Foreign Income: Aid, Remittances, and Government Survival," American Political Science Review 106, no. 1 (2012); Abel Escribà-Folch, Covadonga Meseguer and Joseph Wright, "Remittances and Protest in Dictatorships," American Journal of Political Science 62, no. 4 (2018): 889–904.
46 Alexandra Délano Alonso and Harris Mylonas, "The Microfoundations of Diaspora Politics: Unpacking the State and Disaggregating the Diaspora," Journal of Ethnic and Migration Studies 45, no. 4 (2019): 473–91.
47 Liu, 2022a.
48 Byford, 2012.
49 See e.g. Charles King and Neil J. Melvin, "Diaspora Politics: Ethnic Linkages, Foreign Policy, and Security in Eurasia," International Security 24, no. 3 (2000): 108–38.
50 Elaine Lynn-Ee Ho, " 'Claiming' the Diaspora: Elite Mobility, Sending State Strategies and the Spatialities of Citizenship," Progress in Human Geography 35, no. 6 (2011): 757–72.
51 Mahieu, 2019.
52 Jinpu Wang and Ning Zhan, "Nationalism, Overseas Chinese State and the Construction of 'Chineseness' among Chinese Migrant Entrepreneurs in Ghana," Asian Ethnicity 20, no. 1 (2019): 8–29.
53 Olga Zeveleva, "States and Standardization: Constructing the Co-ethnic Migrant Story in Germany," Journal of Ethnic and Migration Studies 45, no. 4 (2019): 636–55.
54 Byford, 2012.
55 Liu, 2022a.
56 Audrye Wong, "How Not to Win Allies and Influence Geopolitics: China's Self-Defeating Economic Statecraft," Foreign Affairs, May/June 2021.
57 Zhou, 2019.
58 Yehonatan Abramson, "Securing the Diasporic 'Self' by Travelling Abroad: Taglit-Birthright and Ontological Security," Journal of Ethnic and Migration Studies 45, no. 4 (2019): 656–73.
59 Yingjie Fan, Jennifer Pan, Zijie Shao and Yiqing Xu, "How Discrimination Increases Chinese Overseas Students' Support for Authoritarian Rule," 21st Century China Center Research Paper No. 5 (2020).

Bibliography

Abramson, Y. (2019). Securing the diasporic 'self' by travelling abroad: Taglit-Birthright and ontological security. *Journal of Ethnic and Migration Studies*, *45*(4), 656–673.

Adamson, F. B. (2006). Crossing borders: International migration and national security. *International Security*, *31*(1), 165–199.

Adamson, F. B., & Demetriou, M. (2007). Reshaping the boundaries of 'state' and 'national identity': Incorporating diasporas into IR theorizing. *European Journal of International Relations*, *13*(4), 489–526.

Adamson, F. B., & Tsourapas, A. (2020). The migration state in the global south: Nationalizing, developmental, and neoliberal models of migration management. *International Migration Review*, *54*(3), 853–882.

Ahmed, F. Z. (2012). The perils of unearned foreign income: Aid, remittances, and government survival. *American Political Science Review*, *106*(1).

Alonso, A. D., & Mylonas, H. (2019). The micro foundations of diaspora politics: unpacking the state and disaggregating the diaspora. *Journal of Ethnic and Migration Studies*, *45*(4), 473–491.

Baggott Carter, E. (2021, April 2). Statement for the Record before the U.S.-China Economic and Security Review Commission," March 23, 2023, https://www.uscc.gov/sites/default/files/2023-03/Erin_Baggott_Carter_Statement_for_the_Record.pdf.

Baser, B., & Ozturk, A. E. (2020). Positive and negative diaspora governance in context: From public diplomacy to transnational authoritarianism. *Middle East Critique*, *29*(3), 319–334.

Brady, A.-M. (2017). *Magic weapons: China's political influence activities under Xi Jinping* (Vol. 18). Washington, DC: Wilson Center.

Brand, L. A. (2006). *Citizens abroad: Emigration and the state in the Middle East and North Africa*. Cambridge University Press.

Brand, L. A. (2010). Authoritarian states and voting from abroad: North African experiences. *Comparative Politics*, *43*(1), 81–99.

Byford, A. (2012). The Russian diaspora in international relations: 'Compatriots' in Britain. *Europe-Asia Studies*, *64*(4), 715–735.

Charon, P., & Vilmer, J.-B. J. (2021, October). *Chinese influence operations: A machiavellian moment*. IRSEM.

Chester, P.J., & Wong, A. (2025). Wedge Narratives and Diaspora Communities. *Security Studies*, 1–52.

Fan, Y., Pan, J., Shao, Z., & Xu, Y. (2020), How discrimination increases Chinese overseas student's support for authoritarian rule. *SSRN Electronic Journal*, *2020*(5).

Gamlen, A., Cummings, M. E., & Vaaler, P. M. (2019). Explaining the rise of diaspora institutions. *Journal of Ethnic and Migration Studies*, *45*(5), 492–516.

Goddard, S. E. (2018). *When right makes might: Rising powers and world order*. Cornell University Press.

Graham, B. A. T. (2019). *Investing in the homeland: Migration, social ties, and foreign firms*. University of Michigan Press.

Grose, C. R., Lopez, P., Sadhwani, S., & Yoshinaka, A. (2022). Social lobbying. *The Journal of Politics*, *84*(1).

Han, E. (2019). Bifurcated homeland and diaspora politics in China and Taiwan towards the Overseas Chinese in Southeast Asia. *Journal of Ethnic and Migration Studies*, *45*(4), 577–594.

Ho, E. L.-E. (2011). 'Claiming' the diaspora: Elite mobility, sending state strategies and the spatialities of citizenship. *Progress in Human Geography*, *35*(6), 757–772.

Jiechi, Y. (2017). 当好贴心人 成为实干家 凝聚侨心侨力同圆共享中国梦 - 深入学习贯彻习近平总书记关于侨务工作的重要指示 (Be a close friend, be a person of action, bring together diaspora hearts and strength to fully share in the China Dream – deeply learn and implement General Secretary Xi Jinping's important instructions on overseas Chinese affairs. *Overseas Chinese Affairs Study*, (3).

Kelner, S. (2010). *Tours that bind: Diaspora, pilgrimage, and Israeli birthright tourism*. NYU Press.

King, C., & Melvin, N. J. (2000). Diaspora politics: Ethnic linkages, foreign policy, and security in Eurasia. *International Security*, *24*(3), 108–138.

Kirk, J. A. (2008, July). Indian-Americans and the U.S.-India nuclear agreement: Consolidation of an ethnic lobby? *Foreign Policy Analysis*, *4*(3), 275–300.
LeoGrande, W. M. (2020). Pushing on an open door? Ethnic foreign policy lobbies and the Cuban American case. *Foreign Policy Analysis*, *16*, 438–456.
Liu, J. M. (2022a). From "sea turtles" to "grassroots ambassadors": The Chinese politics of outbound student migration. *International Migration Review*, *56*(3), 702–726.
Liu, J. M. (2022b). When diaspora politics meet global ambition: Diaspora institutions amid China's geopolitical transformations. *International Migration Review*, *56*(4), 1255–1279.
Mahieu, R. (2019). "We're not coming from Mars; we know how things work in Morocco!": How diasporic Moroccan youth resists political socialization in state-led homeland tours. *Journal of Ethnic and Migration Studies*, *45*(4), 674–691.
Mearsheimer, J. J., & Walt, S. M. (2007). *The Israel lobby and U.S. foreign policy*. Blackwell Publishing.
Melvin, N. (2020, May 1). Russia's policy of passport proliferation. *RUSI*.
Mylonas, H., & Zilovic, M. (2019). Foreign policy priorities and ethnic return migration policies: Group-level variation in Greece and Serbia. *Journal of Ethnic and Migration Studies*, *45*(4), 613–635.
Paarlberg, M. A. (2017). Transnational militancy: Diaspora influence over electoral activity in Latin America. *Comparative Politics*, *49*(4), 541–559.
Paarlberg, M. A. (2019). Competing for the diaspora's influence at home: party structure and transnational campaign activity in El Salvador. *Journal of Ethnic and Migration Studies*, *45*(4), 539–560.
Patel, D. (2021). "One man lobby"? Propaganda, nationalism in the diaspora, and the India league of America during the second world war. *The Journal of Imperial and Commonwealth History*, *49*(6), 1110–1140.
Pérez-Armendáriz, C. (2014). Cross-border discussions and political behavior in migrant-sending countries. *Studies in Comparative International Development*, *49*(1), 67–88.
Rolland, N. (2020). *China's vision for a new world order*. National Bureau of Asian Research.
Shain, Y., & Barth, A. (2003). Diasporas and international relations theory. *International Organization*, *57*(3), 449–479.
To, J. J. H. (2014). *Qiaowu: Extra-territorial policies for the overseas Chinese*. Brill.
Tsourapas, G. (2021). Global autocracies: Strategies of transnational repression, legitimation, and co-optation in world politics. *International Studies Review*, *23*(3), 616–644.
Varadarajan, L. (2010). *The domestic abroad: Diasporas in international relations*. Oxford University Press.
Wang, J., & Zhan, N. (2019). Nationalism, overseas Chinese state and the construction of 'Chineseness' among Chinese migrant entrepreneurs in Ghana. *Asian Ethnicity*, *20*(1), 8–29.
Waxman, D. (2016). *Trouble in the tribe: The American Jewish conflict over Israel*. Princeton University Press.
Wong, A. (2021a). Peddling or persuading: China's economic statecraft in Australia. *Journal of East Asian Studies*, *21*(2), 283–304.
Wong, A. (2021b, May/June). How not to win allies and influence geopolitics: China's self-defeating economic statecraft. *Foreign Affairs*.
Zeveleva, O. (2019). States and standardization: Constructing the co-ethnic migrant story in Germany. *Journal of Ethnic and Migration Studies*, *45*(4), 636–655.
Zhou, T. (2019). *Migration in the time of revolution*. Cornell University Press.
Zweig, D., & Wang, H. (2013, September). Can China bring back the best? The communist party organizes China's search for talent. *The China Quarterly*, *215*, 590–615.

10 Conclusion

Oscar Jonsson and Ilmari Käihkö

War is the most extreme of human activities and one under constant change. War has traditionally been understood by the large-scale application of violence, but such definition has been challenged by an increasing utility of non-military means given the change in war. One of the biggest unknowns of modern times is how to understand and counter the use of non-military means. The intent of the concept of non-military warfare is not to directly challenge or replace adjacent ones as hybrid warfare or gray zone warfare but rather to liberate the discussion from excessive conceptual debates. This is done by distilling the analysis to simplest constituent parts, military and non-military warfare.

At the time of writing, most focus in war studies is centered around the war in Ukraine, which is the largest in Europe since World War II. While the invasion has prodded us to relearn the lessons of large-scale industrialized military warfare, it warrants as much thinking on the non-military parts of war as well. The war can either be seen as a culmination of the invasion in 2014 that was largely successful due to non-military means and driven by a plan to replicate the events on a larger scale throughout Ukraine. The other case that can be made is that the 2014 invasion showed the limits of what could be achieved with non-military means – which was a great deal – but that military means was required in order to achieve the complete submission of Ukraine desired by the Russian leadership. Regardless of which option one prefers, the Russian invasion in 2022 was premised primarily on a strategy of non-military means that would subvert, disorganize, and disable the Ukrainian will and capability to defend themselves. In the end, these means failed to deliver.[1] Only after that did the Russian regime transition to a strategy where military means became the foremost instrument.

On the other side, Western strategy to deter Russia from invading in the months leading up to February 24, 2022, was centered around economic means. The United States prepared "the mother of all sanctions against Russia."[2] After the start of the invasion, the Western response combined various forms of not only military assistance but also non-military means. The West's overriding strategy has been to punish Russia through political isolation, economic sanctions, and denying Russia access to critical components and technologies. Moreover, the United States completely overhauled their intelligence-sharing policy in a way that significantly empowered the Ukrainian Armed Forces.

DOI: 10.4324/9781003616993-10

In the case of Ukraine, in 2022, the use of military means needs to be investigated together with the use of non-military means. In Chapter 4, Sanshiro Hosaka provides an anatomy of Russian political warfare toward Ukraine leading up to the invasion. In it, he presents both the potential – which was vast – and the limitations in Russian political warfare toward Ukraine that then drove the decision to launch the full-scale invasion. Moreover, he also shows how political warfare was combined with the use of military forces.

Before Russian tanks crossed to the border to Ukraine, ones and zeroes did. On the afternoon of February 23, an aggressive form of malicious code never seen before spread among Ukrainian agencies, companies, and financial institutes, as well as to sub providers in Latvia and Lithuania. The virus, called "Hermetic Wiper," sought to erase all data on target computers.[3] As discussed in Benjamin Jensen's Chapter 5, cases exist where cyber warfare is used to substitute military force altogether. That is to say, cyberattacks have been used not only as a multiplier but also as a substitute of what could have been military attacks and as a tool of crisis management. An example was the successful Russian attack on February 23 against KA-SAT satellite, which the Ukrainian Armed Forces relied on for communication.[4]

In Chapter 6, written by Nina Kollars and Michael Poznansky, it is revealed how the increasing importance of the cyber domain forces us to rethink who is considered a participant in cyber warfare. In interstate warfare, participants are often (even if covertly) state agents, which could be responded to under international law. However, many of the largest cyber powers today are private companies. Their sensors are omnipresent, and they are often the first responders to cyber incidents. However, the dynamics within which they operate are also significantly different from state actors. They have different laws that govern them and different incentives. Private companies need to make revenue and showcase their capabilities. They are prone to publish their effective use of cyber defense and offense – whether real or not. Intelligence agencies does? mostly not show their success to protect their methods and sources. One particularly illustrating example is the U.S. company Palantir that claims their tools are behind "most of the targeting in Ukraine."[5] Both Microsoft and Google have made significant support to Ukraine to a degree that trumps many states. This includes providing free software (Project Shield) to not only safeguard Ukrainian infrastructure from DDoS attacks but also cyber threat intelligence and cyber response, and much more.

Another aspect of the privatization of war is illustrated in Audrye Wong's Chapter 9 on diaspora statecraft. The use of ethnic groups living beyond the borders of the one state has offered a plentitude of ways for the originator state to influence the host state. This includes not only the whole range of using intelligence officers and assets that are willingly cooperating but also ideological and information influence in favor of the originator state in the host state. Finally, it has also included the blackmailing of citizens to perform tasks against their host state, often through coercing them through their relatives. This said, diasporas can also provide opportunities for host countries for influencing originator states. Even here threats can be balanced with opportunities.

One reason for the advent of information warfare and cyber warfare is technological progress. Pioneering science and technology have become of such importance for the elements of national power, that technology in itself can become a domain of non-military warfare. Elsa B. Kania's Chapter 8 shows how technology and innovation have become both securitized and leveraged for international competition. A key part of the Western response to Russia's invasion of Ukraine has been the restriction of export of advanced technology such as electronics and semiconductors to Russia with the goal of impairing Russia's economy and ability to reconstitute its armed forces. If increasing the economic cost of Russia's invasion would not convince Russian leaders to give up, the remaining alternative was to try to ensure that they would not achieve their goals through use of force. Moreover, Kania further shows how limiting Huawei and other Chinese companies' access to critical technologies have been an important part of the U.S. strategy to contain the rise of China.

The expansion of conflict into new domains has necessitated the question of how law is able to mitigate both actors and means for antagonistic purposes. This rise of "lawfare" has given a stronger importance to the interpretation of what law is, and how it can be used as a means of strategic coercion. In Chapter 7, Justin Canfil provides the anatomy of how the process of legal negotiations in themselves have become an arena for great power competition: states seek to manipulate and shape legal processes to fit their interests. Canfil also provides a framework for categorizing legal actions to distinguish between the instrumentalization of law and everyday legal actions.

The increasing antagonistic use of information, technology, and other means also requires a fundamental rethink of the basics of international law. This includes, in particular, who can be deemed a participant in non-military warfare. In Chapter 3, Jack McDonald builds upon how international law regards participation, and pioneers a framework for deciding who can be deemed a participant in non-military warfare. For McDonald, participation in non-military warfare should be distinguished by four factors: direct and active participation, connection to an organized campaign of coercion, a state nexus, and finally, what is abnormal according to context. These four factors provide a framework for judging participation, itself a fundamental condition for states in their response to non-military warfare.

All the threads dealt with in this anthology boil down to the question addressed by Angstrom, in Chapter 2, about what can constitute victory in non-military warfare. A conclusive answer is almost impossible to validate empirically: we cannot causally account for how certain narratives lead to specific beliefs that over time yield desired outcomes. It is equally impossible to validate how restricting the export of semiconductors will impact Russian decision-making for launching, or not launching, a war in five years' time.

Nonetheless, there is a tangible value in non-military warfare. This is not least as it is perceived to be real among policymakers. We should regardless shift our expectations regarding what can be achieved in non-military warfare. Victory in non-military warfare can oftentimes be mere subversion and undermining adversaries in the long term,[6] rather than being as easily observable and conclusive as a military victory or defeat. In most cases the full impact of disinformation campaigns cannot be evaluated after a year, or necessarily even ten years.

One concrete example to manifest this point is a 2007 cable from the U.S. Embassy in Kyiv to the State Department that described Russian activities to create ethnic Russian NGOs in Crimea. At the time, they seemed low-level and unsuccessful. However, when the Russian leadership embarked on the 2014 invasion of the peninsula, the effective organization of movements on Crimea provided a critical cover for their claims of legitimacy. In retrospect, it can be seen as a very successful use of non-military means; but during the years between 2007 and 2014, it could have been seen as a wasted effort.

*

In the broad historical perspective, the advent of non-military warfare is real. While non-military means have always been a part of war, their utility has increased in an increasingly globalized, economically, informationally, and technologically interconnected world. The Russian Chief of General Staff famously suggested that the ratio was four to one in favor of non-military means against traditional military means. Non-military warfare is an integral part of great powers' arsenal ranging from the grand strategic level in economic statecraft, political warfare, and legitimacy. It is also increasingly integrated in military operations at the strategic-level with cyberattacks on critical communications capabilities or as a substitute of military means. Interestingly, we have also witnessed non-military means become a more prominent part of tactical military operations in Ukraine. Russian Armed Forces have targeted their Ukrainian opponents with text messages both for propaganda purposes, as well as in an effort to provoke them to communicate and emit electronic signals that can be targeted with military means. Ukraine's command platform DELTA is integrating target acquisition from cyber means (i.e., where does Russian targets connect from) and communicating using civilian infrastructure.

The increasing role of non-military means needs to be understood in the context of the stability-instability paradox. The paradox stipulates that nuclear weapons are so destructive that not only direct war between nuclear-armed blocs decreases but also increases the level of indirect conflict between those actors because the cost of a nuclear war would be so high.[7] That is to say, because of the costs inherent in nuclear war, the utility of non-military warfare increases between great powers.

This anthology suggests that we can think of these activities as warfare. We define non-military warfare as "non-kinetic tools of statecraft employed as means of strategic coercion and effect employed with an intensity and purpose such that the target perceives them as significant escalations and/or commensurable to warfare." Non-military warfare has been not only a critical complement to the use of military means but also at times a substitute for military means. Nonetheless, it is hard to draw clear boundaries of unequivocal boundaries for all non-military warfare, as the same means are also used in regular commerce and legal negotiations, for instance.

The criteria proposed in this volume emphasizes the perception that non-kinetic means are employed with such intensity and antagonistic purpose that the target perceives as comparable to warfare. This could be criticized for being too fluent a definition, but what constitutes war is ultimately (and always has been) a political decision rather than an independent or automatic judgment of objective factors. That underlines another rationale of applying the term "warfare" here. If states and practitioners coin the activity they are involved in warfare, as for instance cyber warfare or information warfare, it entails a description of hostile intent and calls for specific resources and mandates to respond. The implication is that threats of non-military nature are open to securitization, but it is also an inescapable condition for states to respond to such a threat they perceive of such consequence and intensity that they consider it warfare.

It is our hope that this volume contributes to the understanding the phenomenon of non-military warfare. We have sought to do so by not only investigating its domain but also addressing it in a common fashion. The broader questions on how we should think about utility, participation, victory, and defeat as well as its relation to military warfare can only be answered in such way. Our hope is that this is the beginning of a fruitful debate and a contribution for practitioners handling these questions.

Notes

1 Jack Watling, Oleksandr V. Danylyuk and Nick Reynolds, " Preliminary Lessons from Russia's Unconventional Operations During the Russo-Ukrainian War, February 2022–February 2023," Royal United Services Institute, Special Report, March 29, 2023, https://rusi.org/explore-our-research/publications/special-resources/preliminary-lessons-russias-unconventional-operations-during-russo-ukrainian-war-february-2022

2 Economist, "America Prepares the 'Mother of All Sanctions' against Russia", February 4, 2022, https://www.economist.com/finance-and-economics/america-prepares-the-mother-of-all-sanctions-against-russia/21807487

3 CISA, "Destructive Malware Targeting Organizations in Ukraine," April 28, 2022, https://www.cisa.gov/news-events/cybersecurity-advisories/aa22-057a

4 C. Vasquez and E. Groll, "Satellite Hack on the Eve of Ukraine War Was a Coordinated Multi-pronged Assault," CyberScoop, August 10, 2023, https://cyberscoop.com/viasat-ka-sat-hack-black-hat/

5 J. Dastin, "Ukraine Is Using Palantir's Software for 'Targeting,' CEO says", February 2, 2023, reuters.com/technology/ukraine-is-using-palantirs-software-targeting-ceo-says-2023-02-02/

6 As argued by Harknett, Fischerkeller, and Goldman's, "Cyber Persistence Theory", whose analysis is primarily concerned with the cumulative strategic gains over time; see also Thomas Rid, "Cyber War Will Not Take Place", where he mainly sees cyberspace about subversion (e.g. through espionage and sabotage).

7 Glenn Herald Snyder, *The Balance of Power and the Balance of Terror* (San Francisco: Chandler, 1965).

Bibliography

CISA. (2022, April 28). Destructive malware targeting organizations in Ukraine. *Cybersecurity and Infrastructure Security Agency*. https://www.cisa.gov/news-events/cybersecurity-advisories/aa22-057a

Dastin, J. (2023, February 2). Ukraine using Palantir's software for targeting, CEO says. *Reuters*. https://www.reuters.com/technology/ukraine-is-using-palantirs-software-targeting-ceo-says-2023-02-02/

Economist. (2022, February 4). America prepares the 'mother of all sanctions' against Russia. *The Economist*. https://www.economist.com/finance-and-economics/america-prepares-the-mother-of-all-sanctions-against-russia/21807487

Harknett, R., Fischerkeller, B., & Goldman, E. (2022). *Cyber persistence theory*. Oxford University Press.

Rid, T. (2013). *Cyber war will not take place*. Oxford University Press.

Snyder, G. H. (1965). *Balance of power and balance of terror*. Chandler Publishing Company.

Vasquez, C., & Groll, E. (2023, August 10). Satellite hack on the eve of Ukraine war was a coordinated multi-pronged assault. *CyberScoop*. https://www.cyberscoop.com/viasat-ka-sat-hack-black-hat-2022/

Watling, J., Danylyuk, O. V., & Reynolds, N. (2023, March 29). Preliminary lessons from Russia's unconventional operations during the Russo-Ukrainian war, February 2022-February 2023. *Royal United Services Institute*. https://rusi.org/explore-our-research/publications/special-resources/preliminary-lessons-russias-unconventional-operations-during-russo-ukrainian-war-february-2022

Index

Note: *Italic* page numbers refer to *figures*